Deep Learning Applications in Remote Sensing for Climate Change Monitoring

Dekai Li
https://orcid.org/0000-0003-3126-9339
Hainan University, China

Uzair Aslam Bhatti
Hainan University, China

Sibghat Ullah
https://orcid.org/0000-0003-3042-5977
Balochistan University of Information Technology, Engineering, and Management Sciences, Takatu, Pakistan

Mir Muhammad Nizamani
https://orcid.org/0000-0002-8709-9212
Hainan University, China

IGI Global
Scientific Publishing
Publishing Tomorrow's Research Today

Vice President of Editorial	Melissa Wagner
Director of Acquisitions	Mikaela Felty
Director of Book Development	Jocelynn Hessler
Production Manager	Mike Brehm
Cover Design	Jose Rosado

Published in the United States of America by
IGI Global Scientific Publishing
701 East Chocolate Avenue
Hershey, PA, 17033, USA
Tel: 717-533-8845 | Fax: 717-533-7115
Website: https://www.igi-global.com E-mail: cust@igi-global.com

Library of Congress Cataloging-in-Publication Data

LCCN: 2025022716 (CIP Data Pending)
ISBN13: 9798337335018
Isbn13Softcover: 9798337335025
EISBN13: 9798337335032

British Cataloguing in Publication Data
A Cataloguing in Publication record for this book is available from the British Library.

Table of Contents

Detailed Table of Contents

COVID-19 pandemic prove the value of AI, IoT, and telemedicine for continuity of care. Some key obstacles like data privacy, infrastructure constraints, interoperability are also reviewed. Ethical frameworks, capacity building and technology integration that build on disaster-resilient healthcare systems are highlighted by the chapter. Eventually, it calls for stakeholder cooperation in order to develop equitable, data-based solutions for vulnerable populations in the event of an emergency.

Studying land cover through remote sensing technologies has become essential for coastal zone management since these approaches let users make important environmental observations regarding habitat degeneration while monitoring ecosystem changes. The paper investigates current developments together with implementation areas and remote sensing obstacles in coastal zone land cover cartography. The evaluation reviews diverse remote sensing systems which begin with satellite systems then advance to flying sensors and aerial LiDAR and uncrewed aerial vehicles showing their tracking functionality for shoreline identification and their uses in monitoring vegetation and analyzing wetland conditions and urban coastal development. This paper examines classification algorithms with detection methods for changes alongside a discussion of data fusion techniques as well as evaluation of machine learning technologies and deep learning approaches to enhance map classification qualities.

To enhance the air quality monitoring and decision-making, this research investigates the combination of Big Data technologies, Machine learning, Deep learning models, and Artificial Intelligence. Using IOT sensor networks, satellite imaging, and meteorological data will be beneficial to meet the volume, velocity, and variety of big data characteristics. Advanced models like CNNs, LSTMs, and CNN-LSTM hybrid models are employed for accurate and precise spatiotemporal forecasting. Such models are evaluated using metrics such as RMSE, MAE, and R^2. Domain adaptation and transfer learning are used in regions where there is limited data. AI-powered public health analysis, wearable sensors for real-time exposure and pollution detection in various countries like China and India, are some real-world applications. Despite the limitations in the model's performance and data dependability, this research outlines a future path integrating ethical AI, citizen science, and integration with smart cities for sustainable pollution management.

Improved climate change surveillance arises from rapid remote sensing via satellites, UAVs, and sensor networks. However, real-time data processing faces storage and format challenges due to growing data volumes. Deep learning enhances remote sensing for climate monitoring when integrated with big data tools in cloud environments. Scientists use multispectral, hyperspectral, and radar systems for analyzing surface temperatures, deforestation, sea levels, and glaciers. Research on Google Earth Engine's structure explores tool performance between AWS and Azure for geolocation-based deep learning. Image analysis benefits from cloud clusters and statistical models to detect climate anomalies. Cloud-powered systems monitor air pollution and floods. To ensure effectiveness, large-scale systems must follow protocols for data processing, optimization, and quality control. The section concludes with data privacy, then shifts to hybrid edge-cloud models and affordable real-time climate analytics. These methods inform policy with critical adaptation data.

Disaster management refers to the preparation, response, and recovery activities to mitigate or minimize the losses caused by a disaster. Remote sensing plays a crucial role in disaster management by gathering information about the affected areas through satellite and other aerial imagery platforms. This chapter explores the role of deep learning techniques in processing remote sensing data and supporting the decision-making process in disaster management. We explain how various deep learning architectures, such as convolutional neural networks, sequential models, transformers, and autoencoders, can be used in real-life disaster situations. The chapter details how multi-modal and multi-temporal data fusion integrates optical, synthetic-aperture radar, and light detection and ranging imagery into a single analytic pipeline. A discussion on datasets and benchmarking protocols is followed by model deployment and real-time considerations. The chapter concludes with limitations of the discussed techniques, open challenges, and future research directions.

AI is rapidly becoming used for predictive modelling in all sectors of life. It is slowly replacing physical model-based climate projections in weather prediction. This poll gathers our knowledge on ML's weather and climate prediction capabilities. This intriguing sector is becoming increasingly significant in agriculture and catastrophe management. Eight strategies were identified as promising for improving short-term environmental circumstances and long-term climate forecasts. It evaluated over 20 academic methodologies. The survey indicated that ML is good at short-term climate prediction but not mid- to permanent climate prediction because to numerous environmental elements and data limits. Short-term or moderate-to-long-term weather estimation is usually the focus of the literature. It rarely considers the relationship

Preface

INTRODUCTION

In an era defined by unprecedented environmental challenges, from accelerating coastal erosion and frequent natural disasters to worsening air pollution and climate variability—and the growing urgency of safeguarding ecological and public health systems, technology has emerged as a critical pillar of solutions. Remote sensing, big data analytics, artificial intelligence (AI), and cloud computing have transcended their siloed applications to form an interdisciplinary toolkit, enabling humans to monitor, predict, and manage complex environmental and societal systems with unprecedented precision. This book, titled Advances in Remote Sensing, AI, and Big Data for Environmental Monitoring and Disaster Resilience, brings together 11 cutting-edge chapters from leading researchers worldwide, each contributing unique insights into how these technologies intersect with real-world challenges. By bridging technical innovation with practical application, the book aims to serve as a comprehensive reference for anyone seeking to understand, adapt, or advance the use of data-driven tools in environmental stewardship, disaster management, and public health protection.

OVERVIEW OF THE SUBJECT MATTER

The core subject matter of this book revolves around the integration of advanced technologies—remote sensing, AI (including machine learning and deep learning), big data analytics, and cloud computing—into environmental monitoring, disaster resilience, and public health management. Unlike texts that focus on a single technology or application area, this book adopts a cross-disciplinary lens, exploring how these tools work in tandem to address pressing global issues.

Key thematic threads unify the chapters:

Remote Sensing as a Foundation: From satellite imagery and UAVs to LiDAR and hyperspectral sensors, remote sensing provides the high-volume, high-spatiotemporal-resolution data that powers downstream analytics.

AI and Machine Learning as Enablers: Algorithms (e.g., Random Forest, CNNs, LSTMs, U-Net) transform raw data into actionable insights—whether for classifying coastal land cover, forecasting air pollution, or mapping forest mycorrhizal networks.

Big Data and Cloud Computing as Infrastructure: These technologies solve the logistical challenges of storing, processing, and sharing massive datasets (e.g., real-time climate data, multi-source disaster imagery) at scale.

Application-Centric Solutions: All chapters tie technical advancements to real-world problems, such as coastal erosion, flood risk, chronic disease management during disasters, and forest conservation—ensuring theory is grounded in practice.

By weaving these threads together, the book creates a holistic narrative: how technology can turn "data" into "decision-making," and how this transformation can build more resilient, sustainable systems.

WHERE THE TOPIC FITS IN THE WORLD TODAY

Today's global landscape is marked by interconnected crises that demand data-driven, agile solutions—making the book's topic not just relevant, but essential. Here's how the subject aligns with contemporary priorities:

Climate Change Mitigation and Adaptation

The Intergovernmental Panel on Climate Change (IPCC) emphasizes the need for "high-quality, timely data" to track climate indicators (e.g., sea-level rise, deforestation, glacier melt). This book's focus on remote sensing and AI-powered climate monitoring directly addresses this need, providing tools to quantify changes and forecast impacts.

Disaster Risk Reduction

The United Nations Office for Disaster Risk Reduction (UNDRR) reports that natural disasters have increased by 35% in the past two decades, with climate-driven events (floods, hurricanes) leading to billions in losses. Chapters on flood risk assessment, disaster management, and chronic disease care during catastrophes offer actionable technologies to reduce vulnerability—such as real-time flood mapping via Sentinel satellites or AI-powered resource deployment.

Environmental Degradation and Public Health

Coastal zones (home to 40% of the global population) face rapid degradation from human activity and climate change; air pollution causes 7 million premature deaths annually (WHO). The book's chapters on coastal land cover monitoring and air pollution trend analysis provide critical tools to protect these vulnerable ecosystems and populations.

Digital Transformation of Public Sectors

Governments, NGOs, and private entities are increasingly adopting "smart" solutions—e.g., Google Earth Engine for climate analysis, IoT sensors for air quality monitoring. This book bridges the gap between technical innovation and operational adoption, showing how to implement these tools in real-world contexts (e.g., hybrid edge-cloud models for low-resource regions).

In short, the book's topic sits at the intersection of three global imperatives: technological advancement, environmental sustainability, and human resilience. It addresses a urgent need for practical, cross-disciplinary knowledge to navigate an increasingly unpredictable world.

DESCRIPTION OF THE TARGET AUDIENCE

This book is designed for a diverse, interdisciplinary audience, each group gaining unique value from its content. The target audience is divided into four key categories:

Academic Researchers and Graduate Students

Focus Areas

Environmental science, remote sensing, AI/ML, big data, public health, and disaster management.

Value

The book offers comprehensive reviews of state-of-the-art technologies (e.g., unsupervised learning for climate applications, deep learning for flood mapping) and identifies critical research gaps (e.g., data privacy in cloud-based remote sensing, model interpretability in disaster AI). It serves as a launchpad for new research projects and a reference for literature reviews.

Practitioners and Engineers

Focus Areas

Environmental monitoring, disaster response, public health informatics, and geospatial technology.

Value

Chapters provide practical, application-focused insights—such as how to integrate UAVs and LiDAR for coastal mapping, or how to use CNN-LSTM models for air pollution forecasting. Case studies (e.g., Hurricane Katrina, COVID-19) offer real-world examples of technology deployment, helping practitioners adapt tools to their own work.

Policy Makers and NGOs

Focus Areas

Climate policy, disaster risk reduction, environmental conservation, and public health planning.

Value

The book translates technical jargon into actionable recommendations—e.g., using cloud-powered systems for real-time climate analytics to inform adaptation policies, or stakeholder collaboration frameworks for equitable disaster healthcare. It highlights how technology can support evidence-based decision-making, ensuring policies are both effective and inclusive.

Industry Professionals

Focus Areas

Aerospace (satellite technology), IoT (sensor networks), cloud computing (e.g., AWS, Azure), and AI/ML solution development.

Value

Chapters identify emerging market needs—such as demand for hybrid edge-cloud models for low-bandwidth regions, or AI-powered tools for sustainable forestry. They also address technical challenges (e.g., data fusion for multi-sensor systems) that can guide product development and innovation.

IMPORTANCE OF EACH CHAPTER SUBMISSION

Each chapter contributes a unique piece to the book's overarching narrative, covering distinct technologies, applications, and challenges. Below is a detailed description of their importance:

Chapter 1: A Review of Machine Learning Approaches for Coastal Land Cover Segmentation and Change Detection in Remote Sensing

This chapter lays the foundational framework for understanding how machine learning powers coastal monitoring—the most frequently addressed application area in the book. It systematically reviews supervised (Random Forest, SVMs), unsupervised (K-means), and deep learning (CNNs, U-Net) approaches, comparing their strengths and limitations for land cover segmentation and change detection. By focusing on coastal regions—among the most dynamic and vulnerable ecosystems, this chapter fills a critical gap in existing literature, which often treats machine learning as a generic tool rather than tailoring it to coastal-specific challenges (e.g., tidal fluctuations, mixed land cover types). It is essential for researchers and practitioners new to coastal remote sensing, as well as those seeking to compare algorithm performance.

Chapter 2: Advancements in Remote Sensing for Coastal Zone Land Cover Segmentation and Change Detection; A Comprehensive Review

While Chapter 1 focuses on machine learning, Chapter 2 shifts to remote sensing technology itself, exploring how advancements in sensors (high-resolution, multi-spectral, hyperspectral) and platforms (satellites, UAVs) have transformed coastal monitoring. It addresses key practical challenges—cloud interference, low observation resolution, and data fusion needs—that often hinder real-world application. By integrating these technical advancements with machine learning (e.g., CNNs for land cover modification detection), the chapter provides a holistic view of the "data-to-insight" pipeline. Its emphasis on multi-source data integration is particularly valuable, as modern coastal monitoring increasingly relies on combining data from multiple sensors. This chapter serves as a bridge between hardware (sensors) and software (algorithms), making it indispensable for practitioners designing end-to-end coastal monitoring systems.

Chapter 3: Analytics for Health Resilience: Managing Chronic Diseases During Natural Disasters

This chapter represents the book's first cross-sectoral pivot, moving beyond environmental monitoring to public health—specifically, how big data analytics supports chronic disease management during disasters. It addresses a critical, understudied gap: disasters disrupt healthcare access for the 1.9 billion people living with chronic diseases (WHO), yet few resources exist to guide data-driven response. By showcasing case studies (Hurricane Katrina, Haiti earthquake, COVID-19), the chapter demonstrates how wearable data, EHRs, and social media can enable early warning, remote monitoring, and smart resource deployment. It also tackles pressing ethical and operational issues (data privacy, infrastructure constraints), making it a key resource for public health practitioners, disaster responders, and policy makers. Its focus on equity—ensuring vulnerable populations benefit from these tools—aligns with global health priorities, adding unique value to the book's interdisciplinary focus.

Chapter 4: Applications of Remote Sensing in Coastal Zone Land Cover Mapping; Challenges, Advances, and Future Directions

Chapter 4 builds on Chapters 1 and 2 by focusing on practical applications and future trends in coastal land cover mapping. It expands the technology scope to include aerial LiDAR and uncrewed aerial vehicles (UAVs), which are increasingly used for high-resolution shoreline mapping and wetland analysis. Unlike earlier chapters, it emphasizes "end-use"—how land cover maps support coastal management (e.g., habitat restoration, urban planning) rather than just technical methods. It also identifies future directions (e.g., improving classification accuracy via deep learning, addressing data gaps in developing regions) that guide researchers toward impactful work. This chapter is critical for practitioners implementing coastal mapping projects and policy makers using these maps to inform land-use decisions.

Chapter 5: Big Data Analytics and AI Driven Approaches for Air Pollution Trend Analysis

Air pollution is a global public health crisis, and this chapter provides a state-of-the-art guide to using big data and AI for its monitoring and forecasting. It covers the full data lifecycle: from data sources (IoT sensors, satellites, meteorological data) to advanced models (CNNs, LSTMs, hybrid CNN-LSTM) and evaluation metrics (RMSE, MAE, R^2). It addresses a key practical challenge: limited data in low-resource regions, proposing domain adaptation and transfer learning as solutions. Real-world applications (e.g., wearable sensors in China and India) make the content actionable, while discussions of ethical AI and citizen science ensure the work is inclusive. This chapter is unique in its integration of public health impact (e.g., real-time exposure detection) with technical modeling, making it valuable for both environmental engineers and public health researchers.

Chapter 6: Big Data and Cloud Computing for Climate-Oriented Remote Sensing

This chapter fills a critical infrastructure gap in the book, focusing on how big data and cloud computing enable scalable climate monitoring. It addresses a major bottleneck: remote sensing generates petabytes of data, which traditional computing systems cannot process in real time. By reviewing cloud platforms (Google Earth Engine, AWS, Azure) and their application to climate analytics (e.g., surface temperature tracking, deforestation mapping), the chapter shows how cloud environments enable global, real-time climate surveillance. It also tackles operational issues (data format standardization, quality control) and future trends (hybrid edge-cloud models), making it essential for researchers and industry professionals building large-scale climate monitoring systems. Its focus on policy relevance—how cloud-powered data informs climate adaptation—also adds value for policy makers.

Chapter 7: Deep Learning in Remote Sensing for Disaster Management and Humanitarian Aid

This chapter expands the book's disaster focus beyond public health (Chapter 3) to end-to-end disaster management (preparation, response, recovery). It explores how deep learning architectures (CNNs, transformers, autoencoders) process remote sensing data (satellite, aerial imagery) to support decision-making, e.g., mapping disaster damage, identifying displaced populations, or forecasting secondary hazards. A key strength is its focus on multi-modal data fusion (optical, SAR, LiDAR), which is essential for disaster scenarios where weather (e.g., clouds) limits single-sensor utility. It also addresses practical deployment challenges (dataset availability, real-time processing) and benchmarking protocols, making it a valuable reference for disaster response organizations and AI researchers. Its emphasis on humanitarian aid ensures technology is framed as a tool for equity, aligning with the book's resilience focus.

Chapter 8: Unsupervised, Semi-Supervised, and Self-Supervised Learning for Climate Applications

Most AI chapters in the book focus on supervised learning, which requires large, labeled datasets—often scarce in climate science. Chapter 9 fills this gap by exploring low-data AI approaches (unsupervised, semi-supervised, self-supervised learning) for climate applications. It reviews 20+ methodologies and identifies 8 promising strategies for short-term weather prediction and long-term climate forecasting. It also addresses a critical limitation of existing research: the divide between short-term and long-term climate modeling. By examining frameworks that span multiple time scales, the chapter provides a roadmap for more integrated climate prediction. This is essential for researchers working in data-scarce regions (e.g., parts of Africa, Southeast Asia) and policy makers seeking reliable climate forecasts to guide long-term adaptation.

Chapter 9: Wood Wide Web: Integrating Artificial Intelligence in Forest Communication

This chapter represents the book's most innovative, niche focus: the intersection of AI and forest mycorrhizal networks (the "Wood Wide Web"). It explores how machine learning, sensor networks, and predictive models can uncover hidden patterns in fungal-root communication—patterns that traditional ecology cannot detect. By proposing a new framework for mapping underground networks using neural models, the chapter opens a new research frontier: using AI to understand and protect forest ecosystems. It is unique in its integration of computational science with ecological theory, making it valuable for forest ecologists, conservationists, and AI researchers interested in biodiversity. Its focus on sustainable forestry also aligns with global climate goals (forests absorb 30% of global CO_2 emissions), adding practical relevance.

CONCLUSION: HOW THE BOOK IMPACTS THE FIELD AND CONTRIBUTES TO THE SUBJECT MATTER

This book makes three transformative contributions to the fields of remote sensing, AI, big data, and environmental resilience:

First, it integrates fragmented knowledge into a cohesive framework. Existing literature often focuses on a single technology (e.g., CNNs for remote sensing) or a single application (e.g., coastal monitoring). This book unifies these silos, showing how remote sensing, AI, big data, and cloud computing work together across multiple domains—coastal ecosystems, disaster management, public health, forest conservation, and climate science. By doing so, it helps readers understand the "big picture" of data-driven environmental stewardship, rather than isolated technical details.

Second, it bridges theory and practice with a focus on equity. Many technical texts prioritize algorithms or sensors without addressing real-world challenges (e.g., data scarcity, infrastructure constraints) or equitable deployment (e.g., ensuring vulnerable populations benefit from AI tools). This book centers case studies (Hurricane Katrina, COVID-19, air pollution in India) and ethical frameworks (data privacy, stakeholder collaboration), making it a practical guide for implementation. It also addresses understudied topics—such as chronic disease management during disasters, low-data AI for climate prediction, and forest mycorrhizal networks—filling critical gaps in the literature.

Third, it guides future research and action. Each chapter identifies clear research gaps (e.g., model interpretability in flood AI, data fusion for coastal sensors) and practical recommendations (e.g., hybrid edge-cloud models, citizen science for air pollution monitoring). For researchers, this means a roadmap for impactful work; for practitioners, it means actionable tools to address real-world challenges; for policy makers, it means evidence to support equitable, sustainable policies.

In the long term, this book will serve as a catalyst for collaboration across disciplines—bringing together environmental scientists, AI researchers, public health workers, and policy makers to develop data-driven solutions for a more resilient world. It is not just a reference text, but a call to action: to use technology not just for innovation, but for impact.

Chapter 1

A Review of Machine Learning Approaches for Coastal Land Cover Segmentation and Change Detection in Remote Sensing

Mughair Aslam Bhatti
https://orcid.org/0009-0000-8039-0173
Department of Robotics and Aritificial Intelligence, SZABIST University, Karachi, China

Muhammad Zaffar Hashmi
Health Service Academy, Islamabad, Pakistan

Tang Hao
Trinity College, Ireland

Ahmad Hanain
Fudan University, China

Mengyao Yang
School of Information and Communication Engineering, Hainan University, China

Songpeng Gong
School of Information and Communication Engineering, Hainan University, China

ABSTRACT

The coastal regions operate as active environmental areas and they face substantial risks of land surface transformations because of the actions of nature and human interventions. The identification of such changes proves vital to protect both coastal areas and environmental resources as well as develop suitable policies. Remote sensing applications using machine learning techniques enable highly effective and

DOI: 10.4018/979-8-3373-3501-8.ch001

automated processes to detect coastal land cover transitions. An examination takes place which evaluates how machine learning techniques operate on remote sensing information for recognizing coastal land cover distributions while detecting their changes. The paper covers application of supervised learning algorithms such as Random Forest and Support Vector Machines together with unsupervised learning through K-means Clustering and deep learning implementations using Convolutional Neural Networks and U-Net architectures. .

INTRODUCTION

Environmental conservation alongside management practices heavily depend on continuous coastal land cover monitoring together with change detection capabilities. Areas with coastal zones maintain ecological, economic and social importance so they face multiple natural and human-influenced activities that include coastal erosion together with urban development and deforestation alongside vegetation changes. Tracking continuous land cover modifications in these zones matters heavily to scientists because they need to determine ecosystem health and execute urban development planning and manage natural resources and detect climate change effects. The discovery of coastal land cover modifications allows us to understand the environmental consequences from human influence and climate change and other structural factors modifying the landscape which leads to more sustainable management methods.

Coastal land cover mapping traditions require extensive field surveys together with manual interpretation of aerial photos as well as classification technologies that analyze remote sensing data. These techniques encounter multiple constraints including high expenses along with time restrictions and restricted area and time monitoring capabilities. Field-based surveys require substantial resources to survey coastal regions properly since large areas or distant territories might remain uncovered. The interpretation of remotely sensed images through manual methods depends on expert knowledge because it remains both subject to human judgment and also prone to errors. The process of traditional classification presents obstacles when dealing with the variable coastal environments which reduces detection capabilities for subtle transformations.

The emerging technology of Machine Learning (ML) solves the existing issues in coastal land cover segmentation and change detection processes while transforming their operational approach. The capabilities of ML algorithms to handle big volumes of remotely sensed data provide an excellent solution for sorting complicated coastal landscapes. Through trained data supervision these algorithms become better at classifying data while detecting important patterns which enable the detection of minimal land cover changes that standard methods may omit. The three types of ML approaches such as supervised learning and unsupervised learning and deep learning have been implemented for diverse coastal land cover classification problems. The monitoring techniques enhance data handling capabilities to manage large information quantities and provide improved spatial detail so they outperform conventional strategies in terms of efficiency scalability and accuracy.

This paper evaluates the rising application of machine learning techniques in improving remote sensing segmentation combined with change detection for coastal areas. This paper investigates dominant machine learning approaches in the field along with their operational success metrics and it analyzes implementation barriers and possibilities. The study examines advancing trends which unite multi-source information with hybrid ML models in addition to time-sensitive coastal environment surveillance mechanisms. A thorough analysis of machine learning together with remote sensing technologies shows how they help enhance coastal land cover administration and change detection practices in this research.

OVERVIEW OF COASTAL LAND COVER SEGMENTATION AND CHANGE DETECTION

Successful coastal area monitoring depends on two fundamental processes which include segmentation of land cover types along with detecting alterations within them. The coastal regions encompass different earth features that include beaches together with wetlands and forests and cities and aquatic systems, (Camilleri et al., 2017). The process of effective segmentation requires mapping different land cover types from remotely sensed images to create precise land cover information. Change detection systems aim to reveal temporal differences in land cover because this reveals both storm-induced and sea-level rise impacts together with urbanization and industrial development outcomes, (Nicholls, 2011). The analytical methods combine to generate vital data that allows coastal planning and zone administration while helping to detect environmental situations, (Brown et al., 2013).

Coastal land cover mapping traditionally implements manual satellite image interpretation using aerial pictures together with minimal ground measurements, (Malthus & Mumby, 2003). The ways which address coastal environments typically encounter challenges when managing their diverse nature as well as their complicated characteristics. The accuracy levels achieved using these methods depend heavily on three factors: analyst proficiency, image resolution quality as well as ground data availability, (Green et al., 1996). All these factors tend to be limited in coastal regions. The traditional approach to classification fails to obtain precise coastal land cover detection because it encounters limitations from mixed pixels and spectral confusion combined with image noise, (Latella et al., 2021).

Machine learning techniques (ML) enable advanced coastal land cover segmentation and change detection operations according to recent research, (Civco et al., 2002). The adoption of machine learning models has grown due to their effective solution of classical methodology limitations since they incorporate supervised learning, unsupervised learning and deep learning techniques. Support Vector Machines (SVM) together with random forests serve as supervised classification models that consume labeled data to develop precise segmentation of various land cover features, (Schindler, 2012). The unsupervised method called k-means clustering operates without labeled data yet it succeeds at finding patterns in data making it appropriate for initial data research and limited training input situations. The segmentation of complex coastal regions achieves higher accuracy and efficiency with deep learning models specifically convolutional neural networks (CNNs) because they automatically extract features from raw data, (Zhang, Tang, & Zhao, 2021).

Applications of machine learning have demonstrated successful outcomes yet researchers face ongoing obstacles when working with coastal land covers and identifying alterations to these areas. The main task stands in handling diverse data characteristics because images captured by different sensors or at different observation periods frequently display various properties. Multiple sources of data complexity affect optical images by introducing cloud cover and seasonal variations and haze disturbances or data complications from radar and LiDAR systems, (Whitcraft et al., 2015). Different types of data as well as coastal environments present obstacles that impede the successful application of machine learning models across various scenarios. Models become limited in their development of robust systems because accurate and up-to-date labeled datasets remain scarce, (Li et al., 2022). The solution requires additional investigation into data methods for multiple data sources alongside consistency techniques as well as adaptable learning techniques for different coastal settings, (Abouhalima et al., 2024; Gambin et al., 2021).

Figure 1. Workflow of Land Use/Land Cover (LULC) mapping using satellite data and machine learning

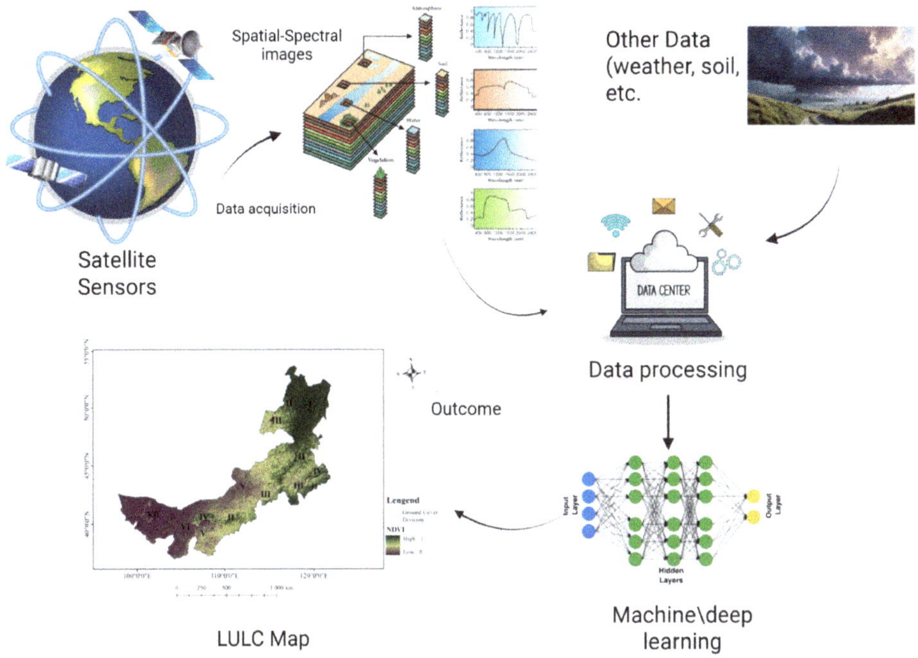

MACHINE LEARNING TECHNIQUES IN REMOTE SENSING

New techniques in remote sensing use Machine learning (ML) to revolutionize the way satellite imagery analysis is conducted for detecting changes and segmenting land cover features, (Zhu et al., 2017). The processing of massive datasets produced by remote sensing platforms becomes effective through ML algorithms which extract features along with performing classification and analysis, (Cracknell & Reading, 2014). The algorithms learn patterns from datasets to conduct automated land cover classification as well as time-based detection without human programmers or manual intervention, (Lary et al., 2016). Large datasets alongside complex input-output relationships can be managed by ML systems thus making them necessary components for contemporary remote sensing applications, (Kvzoglu, Bilucan, & Teke, 2020).

Remote sensing utilizes supervised learning approaches which demand an algorithm receives training by labeled data points that connect to specific labels. The remote sensing field applies three primary supervised ML algorithms: decision trees as well as support vector machines (SVM) together with random forests according to Kvzoglu, Bilucan, and Teke (2020). The algorithms achieve exceptional results in classifying data elements into separate categories through their ability to learn spectral feature-land cover type associations. Supervised learning algorithms achieve accurate predictions thanks to high-quality and sufficient labeled data so data labeling and collection acts as a fundamental process requirement, (Sesnie et al., 2010). Supervised learning produces accurate outcomes in labeled datasets yet its performance decreases when data labeling coverage is limited and the data distribution ranges extensively, (Rodriguez-Galiano et al., 2015).

Unsupervised learning methods operate without needing labeled database information for their operation. These methods utilize various algorithms to discover natural patterns inside the data collection without any need for human input, (Usama et al., 2019). Programs of this type provide valuable insights in data collection scenarios where obtaining ground truth information proves to be complicated or impractical. NDI utilizes k-means clustering and hierarchical clustering as two widely adopted unsupervised algorithms according to Chander and Gopalakrishnan (2023). The clustering algorithm combines data points according to spectral similarity between them for automated land cover type segmentation even when no pre-existing labels exist. The flexibility and reduced need for manual data labeling of unsupervised methods leads to lower accuracy in their results when the land cover types display significant heterogeneity, (Khatami, Mountrakis, & Stehman, 2016).

Deep learning which belongs to machine learning has emerged as a prominent subfield because its systems automatically obtain significant features from basic sensory information and create advanced data distributions, (Najafabadi et al., 2015; Denker et al., 1987). The image analysis functions of land cover segmentation and change detection work best with Convolutional neural networks (CNNs) which are deep learning models made for these operations, (Bengio, 2012). The processing method of CNNs includes successive filtering of input images to extract spatial hierarchies and complex patterns from different levels of analysis. Advanced segmentation capabilities emerge from this technology because of its ability to deliver better results in complex coastal environments which conventional methods tend to fail at, (Inman & Brush, 1973). Deep learning models face two major obstacles in remote sensing applications since they need multiple large datasets for training and powerful computational resources that might not exist due to infrastructure or data limitations, (Meliadou et al., 2012).

Figure 2. Machine learning techniques for remote sensing applications

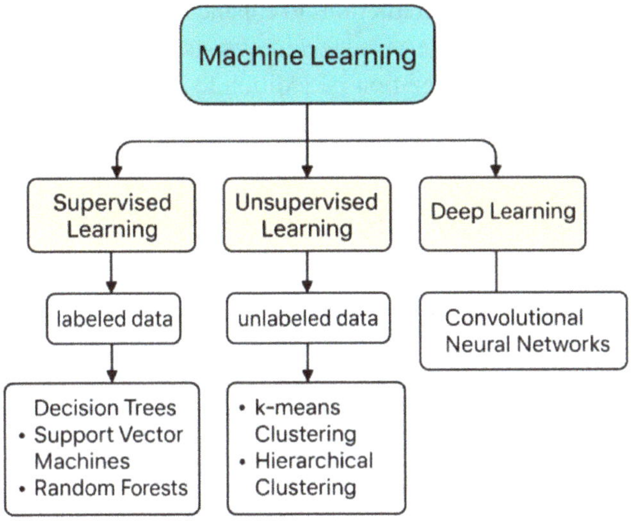

APPLICATIONS OF MACHINE LEARNING FOR COASTAL LAND COVER SEGMENTATION

Coastal land cover segmentation through machine learning has proven as a superior technique than conventional methods due to its accurate and efficient performance with remote sensing data. Shared coastal zone mapping proves difficult because the land cover collection comprises varied types including water bodies in addition to vegetation elements and both urban areas and natural landscape features. Operating systems that detect coastal zones face challenges differentiating between these minimal distinctions when observing coastal areas which show dynamic and heterogeneous characteristics, (Li & Lam, 2018). Machine learning models operate effectively on complex situations because they study extensive datasets to establish advanced linkages between different features which leads to precise classifying of land covers, (Barry & Dayton, 1991).

The coastal land cover segmentation process utilizes support vector machines (SVM) together with random forests and decision trees as supervised machine learning algorithms. The algorithms demonstrate high capability to distinguish different land cover types through the process of training with labeled information. The SVM algorithm functions as a classifier of pixels by identifying the perfect hyperplane that divides specified land cover classes in the characteristic feature domain, (Gambin et al., 2021). Random forests provide ensemble learning through which several decision trees produce combined predictions to enhance classification accuracy while eliminating overfitting. The supervised techniques successfully classify mangroves and coral reefs and coastal vegetation in addition to urban development scenarios as per research in Alhatami et al. (2023).

Figure 3. Workflow for preprocessing and machine learning in remote sensing image analysis

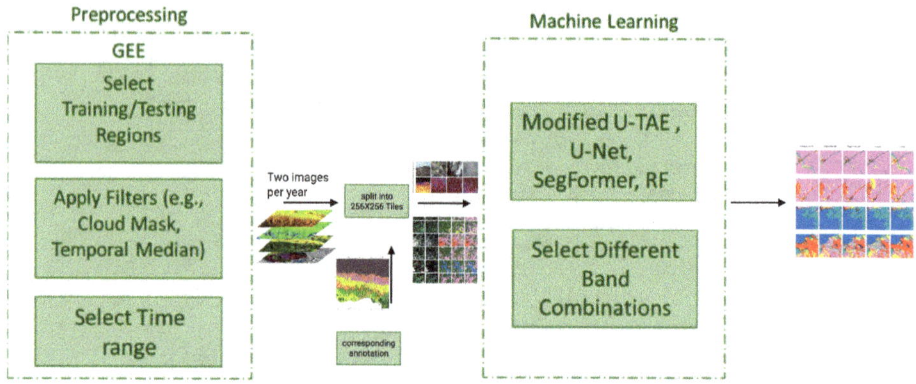

The classification methods of k-means clustering and hierarchical clustering serve as strong alternatives for coastal land cover segmentation because they work well when labeled data exists in limited amounts. The algorithms detect natural data clusters according to spectral similarity which makes them suitable for exploratory studies. Through data clustering procedures remote sensing information becomes organized into separate sections which detect coastal elements including water bodies and urban structures and vegetation types regardless of label availability. Terms in the field perform better with less supervision although they give less precise outcomes when the ground truths lack clear edges or contain considerable noise, (Aamir et al., 2023; Ahasnain et al., 2024; Bhatti et al., 2016).

The segmentation of coastal land cover benefits from deep learning methods especially convolutional neural networks (CNNs) which demonstrate substantial potential according to research in Bhatti, Song, and Hasnain (2023). CNNs constitute an ideal solution for remote image analysis through their built-in capacity to extract hierarchical data features from unprocessed information. The ability of CNNs enables them to detect challenging patterns and structures like shoreline boundaries combined with various coastal vegetation types, (Cui et al., 2021). Extremely detailed satellite imagery benefits from CNN applications which generate outstanding segmentation results to precisely identify different coastal land cover components at high spatial requirements. The ability of deep learning systems to process abundant data with automated feature learning capabilities makes them a strong instrument to improve coastal land cover segmentation, (Dong et al., 2023; Joysoyal et al., 2024).

APPLICATIONS OF MACHINE LEARNING FOR COASTAL CHANGE DETECTION

Coastal change detection uses machine learning as a vital tool because it enables the process of detecting significant land cover variations across time periods, (Yang et al., 2024). These time-based land cover alterations prove essential for environmental surveillance and resource management. Various factors such as climate change together with urbanization and natural disasters and human intervention contribute to the high dynamics found in coastal regions. The detection techniques presented in Yang et al. (2024), experience difficulties when mapping subtle changes in coastal environments because either post-classification comparison or simple image differencing cannot adequately recognize these changes. Through machine learning techniques operators gain a stronger change detection solution by utilizing temporal data and substantial datasets to achieve precise and accurate change identification, (Wu et al., 2018; Zhang et al., 2024).

Supervised classification represents a main machine learning approach for coastal change detection since it trains models with time-period labeled datasets containing different coastal land cover types, (Zhang et al., 2024). Positive detections of different land types and water conditions in satellite images depend on trained technical methods that include random forests and SVM and neural networks to break down images into pixel categories. The detection of urban expansion alongside beach and wetland regressing occurs through algorithmic comparison of image classifications from different periods. The detection of large and easily observed changes in the coastal environment remains highly successful through these methods although they become less accurate when dealing with small and gradual changes, (Zhang et al., 2023).

Coastal changes can effectively be uncovered through unsupervised learning analysis especially when labeled data is unattainable, (Zhang et al., 2023). The K-means clustering along with Self-Organizing Maps (SOM) helps detect intrinsic data patterns to determine both land cover types and their alterations absent human basis, (Cao, Li, & Bhatti, 2024). Unsupervised learning algorithms provide important capabilities to identify previously unforeseen or spontaneous changes in coastal environments which include altered vegetation regions and modified patterns of erosion as well as newly appearing water features. The results produced from unsupervised techniques need additional post-processing stages to separate genuine data alterations from random noise patterns, (Hasnain et al., 2023; Li, Bhatti, & Zhang, 2024; Zeng et al., 2024).

The recent success of deep learning through convolutional neural networks (CNNs) changed coastal change detection methods because these networks automatically extract image features without human-manipulated features, (Bhatti, Li, & Nawaz, 2023). CNNs have established themselves as a primary tool for analyzing high-resolution remote sensing data while performing coastal land cover change detection applications. The training of CNNs on satellite images covering different times reveals the significant changes located in areas like expanding coastal settlements and shifting mangrove areas along with sea level changes, (Bhatti et al., 2024). CNNs demonstrate excellent performance in spotting coastal zone changes because their capability to understand complicated spatial arrangements and their resistance against noise-based distortions make them perfect for detecting subtle environmental influences in the area, (Cao et al., 2023).

CHALLENGES IN MACHINE LEARNING FOR COASTAL LAND COVER SEGMENTATION AND CHANGE DETECTION

The substantial progress made from machine learning (ML) methods in coastal land cover segmentation together with change detection has multiple obstacles which restricts adoption of its complete potential. The primary obstacle stands in the way of adequate training data availability and data quality, (Li et al., 2022). Machine learning algorithms require numerous and diverse data collections that have precise labels for delivering their best performance. Entropy levels in coastal environments make data collection for extensive ground truth measurements consistently challenging because of the complex environmental conditions. The intricate changes observed in coastal land areas between vegetation seasons and shore movement create extra obstacles for collecting useful data. Limited datasets that encompass various coastal ecosystems together with diverse cover types present challenges for training generalized models that work across different coastal regions, (Li et al., 2020).

One major obstacle exists in the various forms coastal land covers present, (Li et al., 2021). The coast exhibits considerable land cover diversity because it includes ecosystems such as water bodies, wetlands, forests, beaches together with urban regions which frequently undergo temporal changes. The substantial variations within classes and similarities between classes create accuracy issues for machine learning models when they try to differentiate between different categories, (Yang et al., 2024). Variations in land and urban area characteristics are hard to distinguish by satellites in conditions such as drought or seasonal change periods. Penalized decisions regarding land and coastal observations occur when the environment demonstrates minor alterations such as coastal deterioration and water alterations due to limitations in both optical observation methods and atmospheric influence and cloud interference, (Li et al., 2020; Li et al., 2020).

Figure 4. Framework and Challenges in Deep Learning-Based LULC (Land Use/ Land Cover) classification

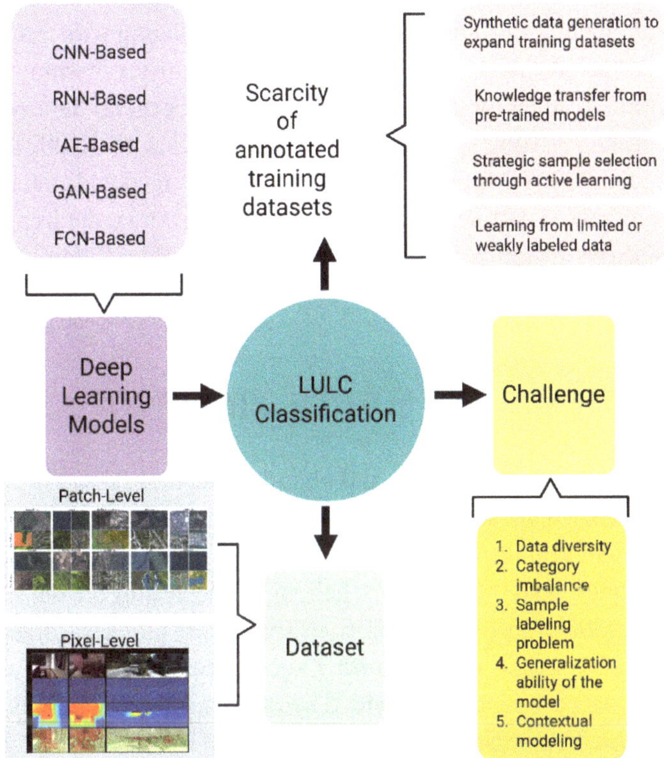

The intricate nature of remote sensing data creates obstacles for machines when detecting changes in coastal area land cover characteristics. Remote sensing images contain multiple data quality issues because of sensor noise combined with atmospheric interferences and cloud cover interference that creates degraded data conditions, (Zhang et al., 2023). The problems mentioned lead to accuracy issues with ML models when operating in areas that often experience haze or cloud coverage. Light variations and tidal fluctuations together with sea wave activities make image analysis in coastal regions significantly more complicated. The noisy data from environmental factors needs advanced preprocessing approaches that produce resource-heavy and time-consuming procedures, (Yu et al., 2024).

The problem of overfitting and generalization affects deep learning models because of their excessive data needs, (Li et al., 2020). The segmentation successes of deep learning models especially convolutional neural networks (CNNs) fail to extend to new coastal regions because of excessive training data dependence, (Bhatti, Ali, & Aslam, 2018). The training data performance drives such models to excel but they become ineffective when facing new data or alternative coastal regions. The development of robust techniques to enhance model performance remains essential because data augmentation and regularization methods along with transfer learning should be used. Experimental results indicate that transfer learning provides an effective solution to adapt models trained on extensive general datasets for coastal region applications utilizing sparse labeled data, (Shoukat et al., 2021; Shao et al., 2024; Zeng et al., 2024). Model training and accuracy face additional difficulties when processing mixed data sources that include optical and radar imagery because these sources differ in terms of data format, resolution and temporal alignment according to research in Gao et al. (2024).

EMERGING TRENDS AND FUTURE DIRECTIONS

Coastal land cover segmentation and change detection research benefits from new trends regarding multi-source remote sensing data integration, (Bhatti & Zhang, 2024). Multiple environmental conditions found in coastal areas make obtaining precise data through a solitary sensor challenging according to Bai et al. (2024). Clouds together with atmospheric conditions selectively hide coastal features while optical images become less reliable during such occurrences. Weather conditions do not affect the data consistency provided by radar imagery especially when using Synthetic Aperture Radar (SAR), (Li & Bhatti, 2021; Bhatti & Liu, 2019; Xiong et al., 2020). The use of LiDAR data gives researchers accurate elevation measurements and land surface data to study coastal features effectively. The unified machine learning framework that includes multiple data source types will deliver complete coastal environment observation which leads to better time-dependent detection and segmentation results. Internationally-sourced data that undergoes multi-source fusion strengthens the accuracy and resilience of machine learning models helping to monitor coastal regions.

The application of convolutional neural networks (CNNs) under deep learning principles has gained dominance as a tool for coastal land cover segmentation during the recent years. The processing capabilities of images benefit greatly from

CNNs because these networks allow them to identify hierarchical spatial features automatically, (Rahman et al., 2023). Through deep learning models gain the ability to appropriately distinguish complex coastal land cover types between urban areas and water bodies and vegetation. The detection of changes relies on recurrent neural networks (RNNs) and their improved version known as long short-term memory (LSTM) networks which have become increasingly sought after. These models demonstrate exceptional strength when handling time-series data because temporal changes in coastal environments require such processing. Deep learning will achieve new advancements through attention mechanisms and transformers to enhance coastal change detection models while better processing relationships in big remote sensing datasets, (Wang, Li, & Bhatti, 2024).

Transfer learning presents promising capabilities for coastal land cover segmentation and change detection because it tackles data restriction problems in coastal settings, (Zhang, Li, & Bhatti, 2024). The acquisition of appropriately labeled data remains a primary challenge when training remote sensing models because such datasets prove hard to collect and maintain in remote coastal areas, (Xu, Liu, & Bhatti, 2023). Using transfer learning models can extract information from domains with plenty of labeled data to successfully address the issues of domains with scarce data. Pre-trained models achieve better performance after receiving fine-tuning through smaller region-specific datasets which helps reduce the requirement of big data collection efforts, (Zha et al., 2023). Transfer learning provides maximum benefits in ecosystems that have varied coastal land cover patterns because it helps models quickly adapt to different coastal zones using small additional datasets, (Wu, Gao, & Bhatti, 2024).

Figure 5. Workflow of a machine learning pipeline for data-driven analysis

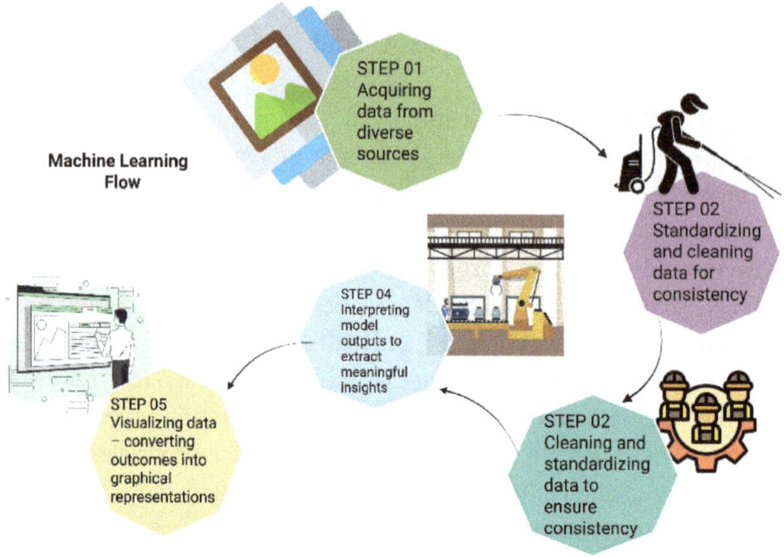

The promotion of explainable artificial intelligence (XAI) represents a key developmental aspect in coastal zone monitoring, (Hasnain & Bhatti, 2022). Deep learning models together with machine learning algorithms excel at detecting and segmenting coastal land change detection but these methods operate as opaque systems which make it difficult to identify the decision-making processes. The predictive outputs from model predictions need complete transparency because they directly influence strategic policy making regarding coastland regulation and ecosystem protection. The field of research now gives priority to creating approaches which enable comprehension and explanation of machine learning model outcomes, (Li, Yu, & Bhatti, 2024). Researchers utilize various techniques including attention-based models for identifying crucial image elements in predictions and post-hoc explanation approaches including SHAP and LIME to enhance the transparent analysis of coastal monitoring models. User trust in AI systems will increase as explainable AI technology continues development thus leading to broader acceptance of machine learning tools in coastal zone management, (Zhang & Bhatti, 2023).

Modern research indicates that coastal land cover segmentation and change detection technology through machine learning techniques shows excellent prospects for development. The expansion of coastal land cover monitoring depends on multi-source remote sensing data fusion alongside deep learning model progress and transfer learning techniques for limited data challenges and explainable AI significance. Machine learning will gain stronger importance in coastal environment monitoring and management due to these emerging trends that continue developing.

Better remote sensing technologies will emerge through advancing combinations and improved solutions to data issues and heterogeneity which will establish remote sensing as essential management tools for sustainably protecting coastal environments.

CONCLUSION

Machine learning technologies enhance the accuracy levels together with operational efficiency of remote sensing technologies that classify coastal land cover. Several machine learning approaches optimized for coastal monitoring extend from supervised learning algorithms to deep learning models along with transfer learning techniques. The detection power of remote sensing technology gets improved through machine learning technologies which surpass traditional detection systems by monitoring small environmental changes in coastal regions effectively and providing precise coastal observation. The conservation and management of coastal zones requires recent advancement technologies to deliver detailed and timely environmental data which allows instant decision-making for critical environmental issues.

Implementation obstacles remain plentiful which prevent the use of machine learning methods in coastal land cover assessment. Data heterogeneity represents the primary challenge in this operation because the available analysis resources possess varying levels of detail and measure precision. Unstable and unreliable data collection occurs in coastal areas because their environmental conditions demonstrate significant changes. Machine learning models encounter problems during development because remote and under-investigated coastal areas limit data accessibility of tagged information. Multiple land cover and land cover changes in complex coastal ecosystems make it difficult to detect land uses authentically while segmenting geographic areas.

New trends like multi-source data fusion and explainable AI are predicted to resolve these present issues. Different remote sensing platforms that combine optical imagery with SAR and LiDAR data will create a thorough understanding of coastline movement. Enhanced transparent and interpretable model development will create better understanding between complex machine learning algorithms and end-users so they can confidently implement these technologies in actual practice. The advancement of these fields will create improved access for machine learning tools in coastal zone monitoring operations.

The future of machine learning for coastal land cover segmentation and change detection analysis appears highly promising. Machine learning models benefit from increased data availability together with expanded computational power which allows them to perform accurate and real-time coastal area monitoring of large diverse datasets. Deep learning architecture advancement combined with expanded multi-source data software integration as well as explainable artificial intelligence method improvement will be essential to develop advanced remote sensing-based coastal observation systems. The developed technologies will support critical choices and lead to better sustainable coastal ecosystem management.

REFERENCES

Aamir, M., Rahman, Z., Abro, W. A., Bhatti, U. A., Dayo, Z. A., & Ishfaq, M. (2023). A progressive approach to generic object detection: A two-stage framework for image recognition. *Computers, Materials & Continua*, *75*(3).

Abouhalima, M., das Neves, L., Taveira-Pinto, F., & Rosa-Santos, P. (2024). Machine learning in Coastal Engineering: Applications, challenges, and perspectives. *Journal of Marine Science and Engineering*, *12*(4), 638.

Ahasnain, A., Hashmi, M. Z., Khan, S., Bhatti, U. A., Min, X., Yue, Y., & He, Y. (2024). Predicting ambient PM2.5 concentrations via time series models in Anhui Province, China. *Environmental Monitoring and Assessment*, *196*(5), 487. PMID: 38687422

Alhatami, E., Bhatti, U. A., Huang, M. X., & Feng, S. L. (2023). Review and enhancement of discrete cosine transform (DCT) for medical image fusion. *Proceedings of International Conference on Information Technology and ....*

Bai, Z., Zhang, Y., Bhatti, U. A., & Liu, J. (2024). Feature Enhanced fusion of U-Net and VIT Transformer for Brain Tumor MRI Image Segmentation. *IEEE Journal of Biomedical and Health Informatics*.

Barry, J. P., & Dayton, P. K. (1991). Physical heterogeneity and the organization of marine communities. In *Ecological heterogeneity* (pp. 270–320). Springer New York.

Bengio, Y. (2012, June). Deep learning of representations for unsupervised and transfer learning. In Proceedings of ICML workshop on unsupervised and transfer learning (pp. 17-36). JMLR Workshop and Conference Proceedings.

Bhatti, U. A., Ali, S., & Aslam, U. (2018). Data-driven switched-gain observation and control. *Nonlinear Dynamics*, •••, 1–19.

Bhatti, U. A., Li, J., & Nawaz, S. A. (2023). *Deep Learning for Multimedia Processing Applications: Volume One: Image Security and Intelligent Systems for Multimedia Processing*. CRC Press.

Bhatti, U. A., & Liu, J. (2019). *Medical image watermarking algorithm based on DCT-DFRFT*. Innovation in Medicine and Healthcare Systems, and Multimedia.

Bhatti, U. A., Song, Z., & Hasnain, M. A. (2023). Predicting the impact of change in air quality patterns due to COVID-19 lockdown policies in multiple urban cities of Henan: A deep learning approach. *Atmosphere*, *14*(5), 902.

Bhatti, U. A., Yu, K., Mamyrbayev, O. Z., Aitkazina, A., Hao, T., & Zhumazhan, N. O. (2024). Recommendations for Healthcare: An Interpretable Approach Using Deep Learning. *7th International Conference on Pattern Recognition and Artificial Intelligence.*

Bhatti, U. A., Zhang, Y., Feng, W., & Huang, M. (2016). Research on the smartphone based eHealth systems for strengthening healthcare organization. *International Conference on Smart Health*, 91-101.

Bhatti, U. A., & Zhang, Z. (2024). *Deep Learning for Multimedia Processing Applications: Volume One: Image Security and Intelligent Systems for Multimedia Processing*. CRC Press.

. Brown, S., Nicholls, R. J., Woodroffe, C. D., Hanson, S., Hinkel, J., Kebede, A. S., ... & Vafeidis, A. T. (2013). Sea-level rise impacts and responses: a global perspective. Coastal hazards, 117-149.

Camilleri, S., De Giglio, M., Stecchi, F., & Pérez-Hurtado, A. (2017). Land use and land cover change analysis in predominantly man-made coastal wetlands: Towards a methodological framework. *Wetlands Ecology and Management*, *25*, 23–43.

. Cao, L., Li, J., & Bhatti, U. A. (2024). Robust Zero Watermarking Algorithm for Medical Images Based on PJFM and 3D-DCT. Deep Learning for Multimedia Processing Applications, 31-44.

Cao, P., Li, J., Liu, J., Li, Y., & Bhatti, U. A. (2023). Robust Zero Watermarking Algorithm for Medical Images Based on STAR-TEBLID-DCT. *International KES Conference on Innovation in Medicine and Healthcare*, 231-245.

Chander, B., & Gopalakrishnan, K. (2023). Data clustering using unsupervised machine learning. In *Statistical Modeling in Machine Learning* (pp. 179–204). Academic Press.

Civco, D. L., Hurd, J. D., Wilson, E. H., Song, M., & Zhang, Z. (2002, April). A comparison of land use and land cover change detection methods. In *ASPRS-ACSM Annual Conference* (Vol. 21, pp. 18-33).

Cracknell, M. J., & Reading, A. M. (2014). Geological mapping using remote sensing data: A comparison of five machine learning algorithms, their response to variations in the spatial distribution of training data and the use of explicit spatial information. *Computers & Geosciences*, *63*, 22–33.

Cui, W., Li, J., Liu, J., Fang, Y., Xiao, X., Bhatti, U. A., & Nawaz, S. A. (2021). A zero-watermarking algorithm for medical images based on Gabor-DCT. Cyberspace Safety and Security: 12th International Symposium, CSS 2020.

Denker, J., Schwartz, D., Wittner, B., Solla, S., Howard, R., Jackel, L., & Hopfield, J. (1987). Large automatic learning, rule extraction, and generalization. *Complex Systems*, *1*(5), 877–922.

Dong, F., Li, J., Bhatti, U. A., & Li, Y. (2023). Robust Zero-Watermarking Algorithm for Medical Images Based on Direction Gradient Histogram and DCT. Artificial Intelligence and Security: 6th International Conference, ICAIS.

Gambín, Á. F., Angelats, E., González, J. S., Miozzo, M., & Dini, P. (2021). Sustainable marine ecosystems: Deep learning for water quality assessment and forecasting. *IEEE Access : Practical Innovations, Open Solutions*, *9*, 121344–121365.

Gao, J., Li, D., Zhang, Y., & Bhatti, U. A. (2024). EF-VPT-Net: Enhanced Feature-Based Vision Patch Transformer Network for Accurate Brain Tumor Segmentation in Magnetic Resonance Imaging. *IEEE Journal of Biomedical and Health Informatics*.

Green, E. P., Mumby, P. J., Edwards, A. J., & Clark, C. D. (1996). A review of remote sensing for the assessment and management of tropical coastal resources. *Coastal Management*, *24*(1), 1–40.

. Hasnain, A., Hashmi, M. Z., Yue, Y., Wei, G., & Bhatti, U. A. (2023). Investigating the spatio-temporal pattern of PM2.5 concentrations in Jiangsu Province, China. E3S Web of Conferences, 379, 01001.

Hasnain, M. Z., & Bhatti, U. A. (2022). Improved multi-step dissolved oxygen prediction in aquaculture using adaptive temporal convolution. *Computers and Electronics in Agriculture*.

Inman, D. L., & Brush, B. M. (1973). The Coastal Challenge: Fragile ribbons which border our land require more understanding, new technology, and resolute planning. *Science*, *181*(4094), 20–32. PMID: 17769816

Joysoyal, R., Uddin, S. S., Islam, T., Sarker, S. K., Li, L., Ahsan, F., & Bhatti, U. A. (2024). Blockchain for sustainable city transformation: A review on Bangladesh. *Engineering Reports*, •••, 12948.

Khatami, R., Mountrakis, G., & Stehman, S. V. (2016). A meta-analysis of remote sensing research on supervised pixel-based land-cover image classification processes: General guidelines for practitioners and future research. *Remote Sensing of Environment*, *177*, 89–100.

Lary, D. J., Alavi, A. H., Gandomi, A. H., & Walker, A. L. (2016). Machine learning in geosciences and remote sensing. *Geoscience Frontiers*, *7*(1), 3–10.

Latella, M., Luijendijk, A., Moreno-Rodenas, A. M., & Camporeale, C. (2021). Satellite image processing for the coarse-scale investigation of sandy coastal areas. *Remote Sensing*, *13*(22), 4613.

Li, D., Bhatti, U. A., & Zhang, Y. (2024). EF-VPT-Net: Enhanced Feature-Based Vision Patch Transformer Network for Accurate Brain Tumor Segmentation in Magnetic Resonance Imaging. *IEEE Journal of Biomedical and Health Informatics*.

Li, D., Yu, M., & Bhatti, U. A. (2024). Enhanced feature-based network for accurate segmentation of MRI images. *IEEE Journal of Biomedical and Health Informatics*.

Li, J., & Bhatti, U. A. (2021). Robust zero watermarking algorithm for medical images based on Tetrolet-DCT. Cyberspace Safety and Security: 12th International Symposium, CSS 2020.

Li, J., Bhatti, U. A., Shoukat, M. U., & Nawaz, S. A. (2020). Image fusion based on discrete cosine transform with high compression. *International Conference on Signal and Image Processing*.

Li, J., Bhatti, U. A., Shoukat, M. U., & Nawaz, S. A. (2020). Medical image hybrid watermark algorithm based on frequency domain processing and inception v3. *Advanced Intelligent Systems*, •••, 2400654.

Li, J., Nawaz, S. A., Liu, J., & Bhatti, U. A. (2021). A robust digital image watermarking algorithm based on IWT-SVD. *Indian Journal of Engineering*, *17*(47), 110–116.

Li, J., Nawaz, S. A., Shoukat, M. U., & Bhatti, U. A. (2020). Improved watermarking algorithm based on SURF and SVD with wavelet transformation against geometric attacks. *Proceedings of the 2020 3rd International Conference on Image and Graphics*.

Li, J., Yu, Z., Yuan, L., Nawaz, S. A., & Bhatti, U. A. (2022). SURF based security of remote sensing images by encrypted watermark. *Proceedings of the 7th International Conference on Signal and Image Processing (ICSIP)*.

Li, K., & Lam, N. S. (2018). A spatial dynamic model of population changes in a vulnerable coastal environment. *International Journal of Geographical Information Science*, *32*(4), 685–710.

Li, Z., Shen, H., Weng, Q., Zhang, Y., Dou, P., & Zhang, L. (2022). Cloud and cloud shadow detection for optical satellite imagery: Features, algorithms, validation, and prospects. *ISPRS Journal of Photogrammetry and Remote Sensing*, *188*, 89–108.

. Malthus, T. J., & Mumby, P. J. (2003). Remote sensing of the coastal zone: an overview and priorities for future research.

Meliadou, A., Santoro, F., Nader, M. R., Abou Dagher, M., Al Indary, S., & Abi Salloum, B. (2012). Prioritising coastal zone management issues through fuzzy cognitive mapping approach. *Journal of Environmental Management*, *97*, 56–68. PMID: 22325583

Najafabadi, M. M., Villanustre, F., Khoshgoftaar, T. M., Seliya, N., Wald, R., & Muharemagic, E. (2015). Deep learning applications and challenges in big data analytics. *Journal of Big Data*, *2*, 1–21.

Nicholls, R. J. (2011). Planning for the impacts of sea level rise. *Oceanography (Washington, D.C.)*, *24*(2), 144–157.

Rahman, Z., Abro, W. A., Aamir, M., Bhatti, U. A., Dayo, Z. A., & Ishfaq, M. (2023). Object recognition using machine learning techniques. *Computers, Materials & Continua.*

Rodriguez-Galiano, V., Sanchez-Castillo, M., Chica-Olmo, M., & Chica-Rivas, M. J. O. G. R. (2015). Machine learning predictive models for mineral prospectivity: An evaluation of neural networks, random forest, regression trees and support vector machines. *Ore Geology Reviews*, *71*, 804–818.

Schindler, K. (2012). An overview and comparison of smooth labeling methods for land-cover classification. *IEEE Transactions on Geoscience and Remote Sensing*, *50*(11), 4534–4545.

Sesnie, S. E., Finegan, B., Gessler, P. E., Thessler, S., Ramos Bendana, Z., & Smith, A. M. (2010). The multispectral separability of Costa Rican rainforest types with support vector machines and Random Forest decision trees. *International Journal of Remote Sensing*, *31*(11), 2885–2909.

Shao, K., Li, D., Tang, H., Zhang, Y., Xu, B., & Bhatti, U. A. (2024). Adaptive Temporal Convolutional Network (Atcn) and Internally-Optimized Transformer for Dissolved Oxygen Prediction in Aquaculture. Available at *SSRN* 5039613.

Shoukat, M. U., Nawaz, S. A., Raza, M., Bhatti, U. A., & Mehmood, A. (2021). A robust watermarking algorithm for encrypted face recognition based on DWT-DCT and tent map. Cloud Computing and Security: 4th International Conference, ICCCS 2018.

Usama, M., Qadir, J., Raza, A., Arif, H., Yau, K. L. A., Elkhatib, Y., & Al-Fuqaha, A. (2019). Unsupervised machine learning for networking: Techniques, applications and research challenges. *IEEE Access : Practical Innovations, Open Solutions*, *7*, 65579–65615.

Wang, B., Li, J., & Bhatti, U. A. (2024). Robust Zero-Watermarking Algorithm for Medical Images Based on Direction Gradient Histogram and DCT. *Computers, Materials & Continua.*

Whitcraft, A. K., Vermote, E. F., Becker-Reshef, I., & Justice, C. O. (2015). Cloud cover throughout the agricultural growing season: Impacts on passive optical earth observations. *Remote Sensing of Environment, 156*, 438–447.

Wu, D., Huang, M., Zhang, Y., Bhatti, U. A., & Chen, Q. (2018). Strategy for assessment of disaster risk using typhoon hazards modeling based on chlorophyll-a content of seawater. *EURASIP Journal on Wireless Communications and Networking, 2018*, 1–12.

Wu, H., Gao, J., & Bhatti, U. A. (2024). An advanced system approach for disaster risk assessment. *EURASIP Journal on Wireless Communications and Networking.*

Xiong, J., Li, J., Bhatti, U. A., & Shoukat, M. U. (2020). Medical Image Watermarking Algorithm based on DCT. *Journal of Medical Imaging and Health Informatics.*

Xu, B., Liu, Y., & Bhatti, U. A. (2023). Robust watermarking method in wavelet domain based on sift features. *Proceedings of the 2020 3rd International Conference on Artificial Intelligence.*

Yang, M., Li, J., Bhatti, U. A., Shao, C., & Chen, Y. W. (2024). Robust Zero-Watermarking Algorithm for Medical Images Based on Non-Subsampled Shearlet Transform and Schur Decomposition. *Computers, Materials & Continua, 75*(3).

Yang, M., Li, J., Bhatti, U. A., & Yuan, Y. (2024). Robust Zero Watermarking Algorithm for Medical Images Based on Non-Subsampled Shearlet Transform and Schur Decomposition. *Computers, Materials & Continua, 75*(3).

Yang, M., Li, J., Bhatti, U. A., Yuan, Y., & Zhang, Q. (2024). A Novel Robust Watermarking Algorithm for Encrypted Medical Images Based on Non-Subsampled Shearlet Transform and Schur Decomposition. Deep Learning for Multimedia Processing Applications: Volume One. *Image.*

Yu, M., Zeng, Z., Tang, H., Bhatti, U. A., & Fahim, A. (2024). Research on smartphone-based eHealth systems for strengthening healthcare organization. *International Conference on Smart Health*, 91-101.

Zeng, Y., Tang, H., Gao, J., & Bhatti, U. A. (2024). A Domain Knowledge Graph Construction Method Based on Joint Extraction of GlobalPointer. *7th International Conference on Pattern Recognition and Artificial Intelligence.*

Zeng, Z., Tang, H., Gao, J., Bhatti, U. A., & Lu, Y. (2024). A Domain Knowledge Graph Construction Method Based on Joint Extraction of GlobalPointer. *7th International Conference on Pattern Recognition and Artificial Intelligence*.

Zha, Y., Yue, Y., Zhang, Y., & Bhatti, U. A. (2023). Simulation optimization of AOA estimation algorithm based on MIMO UWB communication system. *Proceedings of the 2nd International Conference on Information Technologies*.

Zhang, L., Wang, H., & Zhang, B., Xu, Li, J., & Bhatti, U. A. (2024). A New ANN Technique for Short-Term Wind Speed Prediction Based on SCADA System Data in Turkey. *Atmosphere*, *14*(10).

. Zhang, P., Li, J., & Bhatti, U. A. (2024). Robust Multi-Watermarking Algorithm Based on DarkNet53. Deep Learning for Multimedia Processing Applications, 74-91.

. Zhang, P., Li, J., Liu, J., Bhatti, U. A., Chen, Y. W. (2023). A robust watermarking algorithm for encrypted medical volume data based on PJFM and 3D-DCT. Deep Learning for Multimedia Processing Applications, 31-44.

Zhang, Q., Li, J., Liu, J., Bhatti, U. A., & Chen, Y. W. (2023). Robust Zero-Watermarking Algorithm for Medical Images Based on GFTT-KAZE and DCT. 2023 26th ACIS International Winter Conferenceon. *Software Engineering*.

Zhang, Q., Li, J., Liu, J., Bhatti, U. A., & Chen, Y. W. (2023). Robust Zero-Watermarking Algorithm for Medical Images Based on SUSAN-DCT. 2023 26th ACIS International Winter Conference on Software Engineering.

Zhang, W., Tang, P., & Zhao, L. (2021). Fast and accurate land-cover classification on medium-resolution remote-sensing images using segmentation models. *International Journal of Remote Sensing*, *42*(9), 3277–3301.

Zhang, X., & Bhatti, U. A. (2023). Image fusion based on discrete cosine transform for medical data. *Journal of Medical Imaging and Health Informatics*.

Zhang, X., Ma, Z., Liu, Y., & Bhatti, U. A. (2024). Robust Zero-Watermarking Algorithm for Medical Images Based on Maximally Stable Extremal Regions and DCT. IEEE 6th International Conference on Pattern Recognition and Artificial Intelligence.

Zhu, X. X., Tuia, D., Mou, L., Xia, G. S., Zhang, L., Xu, F., & Fraundorfer, F. (2017). Deep learning in remote sensing: A comprehensive review and list of resources. *IEEE Geoscience and Remote Sensing Magazine*, *5*(4), 8–36.

Chapter 2

Advancements in Remote Sensing for Coastal Zone Land Cover Segmentation and Change Detection: A Comprehensive Review

Mughair Aslam Bhatti
https://orcid.org/0009-0000-8039-0173
Department of Robotics and Aritificial Intelligence, SZABIST University, Karachi, China

Mohammad Asif
Hunan University, China

Huang Mengxing
Hainan University, China

Muhammad Aamir
https://orcid.org/0000-0002-7679-0980
Huanggang Normal University, China

Mengyao Yang
School of Information and Communication Engineering, Hainan University, China

Songpeng Gong
School of Information and Communication Engineering, Hainan University, China

ABSTRACT

Remote sensing technologies have experienced substantial advancements which improve management and monitoring of coastal zone changes in land cover segmentation. Sustainable management of coastal ecosystems depends on effective monitoring because these ecosystems show high dynamics from natural sources and human activities. Modern satellite systems together with aerial vehicles and unmanned aerial vehicles combined with advanced machine learning algorithms have brought a total transformation to how coastal lands and their time-evolved

DOI: 10.4018/979-8-3373-3501-8.ch002

changes get detected and mapped. The research investigates modern advancements of coastal zone land cover segmentation through high-resolution imagery and multi-spectral and hyperspectral sensors together with multi-source data integration. The introduction details how machine learning along with deep learning approaches especially CNNs serves the detection of land cover modifications. The review examines coastal change detection obstacles caused by cloud visibility and observation resolution together with data fusion requirements.

INTRODUCTION

Background and Importance of Coastal Zones

Remote sensing technologies have experienced substantial advancements which improve management and monitoring of coastal zone changes in land cover segmentation. Sustainable management of coastal ecosystems depends on effective monitoring because these ecosystems show high dynamics from natural sources and human activities. Modern satellite systems together with aerial vehicles and unmanned aerial vehicles combined with advanced machine learning algorithms have brought a total transformation to how coastal lands and their time-evolved changes get detected and mapped. This article evaluates contemporary remote sensing technology improvements in coastal zone land cover segmentation through analysis of high-resolution images alongside multispectral and hyperspectral sensing alongside multi-source information amalgamation. The introduction details how machine learning along with deep learning approaches especially CNNs serves the detection of land cover modifications. The research evaluates the difficulties of coastal change detection by studying cloud cover as well as spatial resolution and data fusion and shows how advanced technologies produce enhanced accurate and rapid monitoring capabilities for coastal areas. End of the review exists a presentation of developing trends in real-time monitoring and the future prospects of next-generation remote sensing technologies for coastal zone management improvement.

Remote sensing technology employs the combination of land cover segmentation, change detection, machine learning, and deep learning concepts to interpret Coastal zone applications.

The Role of Remote Sensing in Coastal Zone Management

The effective management of coastal zones benefits from remote sensing methods because they give stakeholders an economical and unobtrusive method to monitor areas that are both active and difficult to access. Requirements for continuous monitoring arise from coastal zones' sensitivity to natural along with human-made alterations since these regions need sustained observations of land cover variations and biodiversity patterns and environmental changes. Remotely sensed data provides diverse acquisition methods to produce complete spatial and temporal coverage of extensive regions essential for following coastal zonal processes involving erosional patterns and ecological condition declines in addition to water quality modifications. Remote sensing reaches its goal through satellite and drone platforms alongside other aerial systems to obtain data that proves difficult or costly to acquire with traditional field methods.

Remote sensing yields extensive mapping data with detailed imagery because it enables thorough monitoring of coastal process evolution and land cover transformations. Through satellite imagery it becomes possible to track large geographical regions to study changes in coastal shorelines as well as identify conditions of vegetation and aquatic system health. The method proves valuable when monitoring secluded regions or service areas with restricted ground-based assessment capabilities. Remote sensing data analysis occurs at multiple spatial scales so researchers study extensive coastal patterns together with very detailed neighborhood trends throughout the coastal region.

Remote sensing technology allows scientists to track the moving dynamics of coastal events such as tide cycles as well as storms and growing sea levels. Knowledge of such natural processes' patterns generates vital information to enable effective coastal zone management by affecting both landforms and ecosystems of coastal regions. Remote sensing data tracks temporal processes that gives useful knowledge about climate change effects and extreme weather events to create predictive models for forecasting and developing adaptive mitigation plans. To develop minimally reactive management plans that respond to shifting environmental risks scientists need data modeling and change prediction abilities.

Remote sensing obtains additional value for coastal zone management by uniting with Geographic Information Systems (GIS) analytics alongside additional analytical tools. Decision-makers use GIS technology to evaluate spatial trends together with remote sensing data for assessing how different management approaches would impact their environments. Through GIS integration combined with remote sensing technology organizations make better data-based choices about resource distribution and habitat protection while promoting sustainable coastal development. Coastal zone monitoring and protection and management have found essential tools in remote sensing technology as environmental strain and human-linked development increase in our modern world.

Objective of the Review

The main purpose behind a review is to conduct an objective evaluation of a particular object or service. The review aims to analyze different characteristics including performance alongside usability and total worth. The evaluation includes complete information on beneficial aspects as well as unfavorable ones to assist consumers in making knowledgeable choices. The evaluation focuses on both informative description along with effective analysis which requires market comparison between alternatives whenever appropriate.

During the evaluation phase the reviewer focuses on thorough assessment of important characteristics and operational abilities. The main assessment objective evaluates the delivery quality of product performance alongside the promises stated by manufacturers or service providers. The reviewer examines both user accessibility features alongside system usability and real-world product practicability to provide consumers with an accurate view of their expected experiences.

The review effectively explains the practical benefits which the product offers to users. The reviewer needs to evaluate the price of the product alongside its delivered benefits. The investment offers worth but consumers need to know if other options deliver equivalent value for their money. Products acquire their value through multiple elements like how good they perform and how well they function and also how the customer service treats clients after purchase. Reviews executed successfully enable users to evaluate those factors which match their requirements while staying within their financial means.

A recommendation based on the analysis performed should be included at the end of the review. At the conclusion the reviewer gives a clear recommendation regarding whether to buy the product or not or which section of the audience would find it appropriate. Readers obtain effective takeaways and understanding regarding the recommendation from this conclusion section. The review should transmit useful details about the product's performance which support buyers' purchasing decisions.

OVERVIEW OF REMOTE SENSING FOR COASTAL ZONE MONITORING

Remote Sensing Technologies and Sensors

Remote sensing technologies transformed the process of Earth and environmental data acquisition. The sensors on specialized remote sensing technology platforms include satellite-based along with aerial and ground-based systems which enable distant information capture. Remote sensing technology focuses on surface observation and monitoring which lets scientists and researchers and decision-makers gain complex process understanding for better decision-making. Remote sensing systems help professionals in environmental monitoring together with agriculture and urban planning as well as disaster management.

The main components of remote sensing consist of sensors used to measure electromagnetic radiation that reflects or emits from Earth surface features. The sensors function within a wide electromagnetic spectrum because different kinds need distinct types of data. Passive sensors obtain information through sunlight yet active sensors generate their own signals corresponding to radar processes. The various types of sensors were designed to detect different ground surface characteristics including temperature, vegetation health and surface roughness which makes them essential for multiple applications.

Remote sensing technologies organize themselves according to spatial distribution as well as spectral capabilities and temporal measurement abilities. The resolution of details in remote images depends on spatial resolution because it determines what minimum features the system detects. Sensor capabilities regarding spectral bands determine how well the instrument can detect unique Earth surface features and materials to generate specific spectral signatures. A sensor's ability to record data at specific locations is reflected in its temporal resolution because this measure determines how often sensors can capture data to detect change occurrences for instance deforestation and natural disaster development.

Remote sensing technology development resulted in creating improved sensors coupled with sophisticated platforms through high-resolution imaging satellites together with unmanned aerial vehicles (UAVs). The developed innovations facilitate collected data with higher levels of precision and detail never achieved before. Through continuous advancement remote sensing technology becomes more essential for global problem assessment including climate change mitigation and natural resource protection and disaster relief efforts. Remote sensing will deliver exponentially greater value in resource management because of continual sensor technological innovation and data processing advancements.

Spatial and Temporal Resolution of Remote Sensing Data

Remote sensing data achieves effective outcomes in applications based on two fundamental factors which are resolution in space and time. The ability of sensors to reflect spatial resolution depends on the size of image pixels which defines the smallest detectable object or feature. The sensor achieves higher accuracy in detecting surface details when it possesses increased spatial resolution while urban planning and small plot monitoring depend most heavily on this feature. Higher spatial resolution datasets become heavier to process while the methods for reaching this resolution have both financial and technological complexity.

A sensor measures data from a single Earth surface location at intervals that determine its temporal resolution ability. Multiple fields need this capability to track long-term modifications such as forest fire development along with crop development and flood impact assessment. The ability of sensors to produce updated information at regular intervals becomes vital for instant decision-making processes throughout disaster preparedness as well as climate observation activities. The need for rapid image collection through frequent measurements can reduce the amount of available detailed data information.

Achieving the correct match between spatial detail and time-based data collection forms the basis for determining appropriate remote sensing data selection. Large-scale land-cover mapping tasks typically succeed with moderate spatial resolution sensors which have high temporal resolution capabilities to track broad changes between time periods. For visual tasks that must identify small objects such as examining trees in forests or detecting physical structures in cities operators should use sensors with enhanced spatial detail. When multiple sensors with various resolutions exist it can be advantageous to integrate their data in order to achieve a broader understanding of the situation under observation.

Figure 1. Types of resolution in remote sensing data: Spatial and temporal resolutions

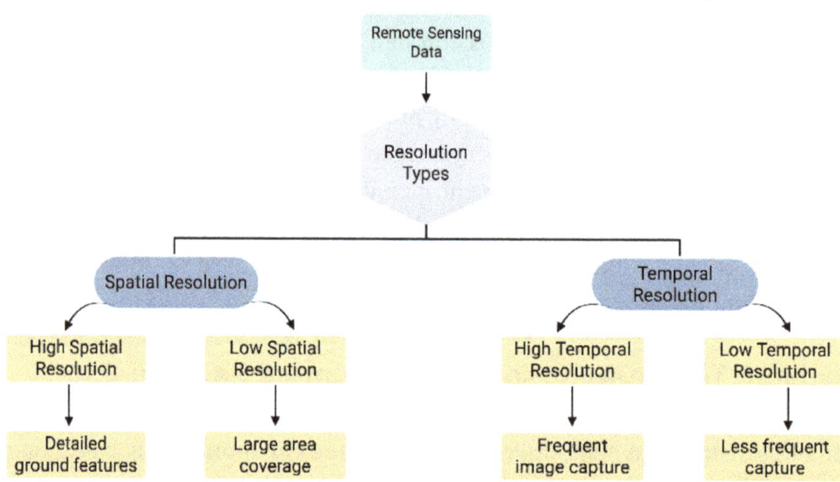

Modern advancements in remote sensing technology continue to improve both remote sensing resolution elements through space and time domains. The advanced satellites alongside their sensor systems provide high-resolution data which needs fewer intervals for delivering it so researchers can monitor events in more dynamic ways. The ability to monitor using high-resolution data in real-time became possible through the combination of various data sources which include UAVs and ground-based sensors. The advancement of remote sensing technologies enables users to alter spatial and temporal resolutions which enhances the operational value and enhanced accuracy of remote sensing applications.

COASTAL ZONE LAND COVER SEGMENTATION TECHNIQUES

Traditional Segmentation Methods

Remote sensing professionals use traditional division methods to break down images into segmented areas that represent uniform areas by using characteristics such as texture or color and intensity. The early image processing solutions utilized these methods to obtain useful information from complicated datasets. Traditional segmentation depends on fixed threshold standards as well as straightforward procedures to assign image pixels similar categories based on identification criteria. Traditional segmentation produces effective results in situations which show distinct separation between different land types during the identification of water elements and urban structures from vegetation.

Thresholding stands as one of the standard traditional segmentation approaches which assigns pixel values exceeding or below particular intensity thresholds to separate classifications. When features of interest have distinct features from the background this technique provides effective results. Thresholding technique faces challenges during segmentation because image features such as lighting changes and shadow effects alongside background noise can cause wrong results. Better segmentation methods emerged to track object boundaries because region-growing and edge-detection algorithms processed pixel values and gradients in images.

Region-growing methods utilize seed pixels to expand neighboring regions that match intensity or additional properties thus generating regions that become easy to analyze. Region-growing shows high effectiveness when objects have either uniform textures or color patterns. The selection process for seed points creates a potentially large sensitivity to the method while its performance degenerates in images featuring diverse areas. Edge detection performs boundary detection through pixel value variations examination which employs Sobel and Canny operators. The detection methods work best for identifying objects that have obvious distinct boundaries yet become ineffective when used on regions with inadequate contrast and diffused transition zones, (Neal, 2013).

The scrap nature of typical segmentation methods interferes with successful image segmentation mainly due to imaging complexity alongside noise patterns. The methods have limited success when applied to environments which contain unclear features along with substantial lighting and texture differences. Because traditional approaches fail to integrate environmental or contextual signal they cannot adapt to modifications in their operating environment. Traditional segmentation techniques serve as important tools but new learning-based segmentation approaches rise to meet modern remote sensing demands through better handling of complex data.

Figure 2. Traditional image segmentation methods: Techniques for dividing remote sensing data

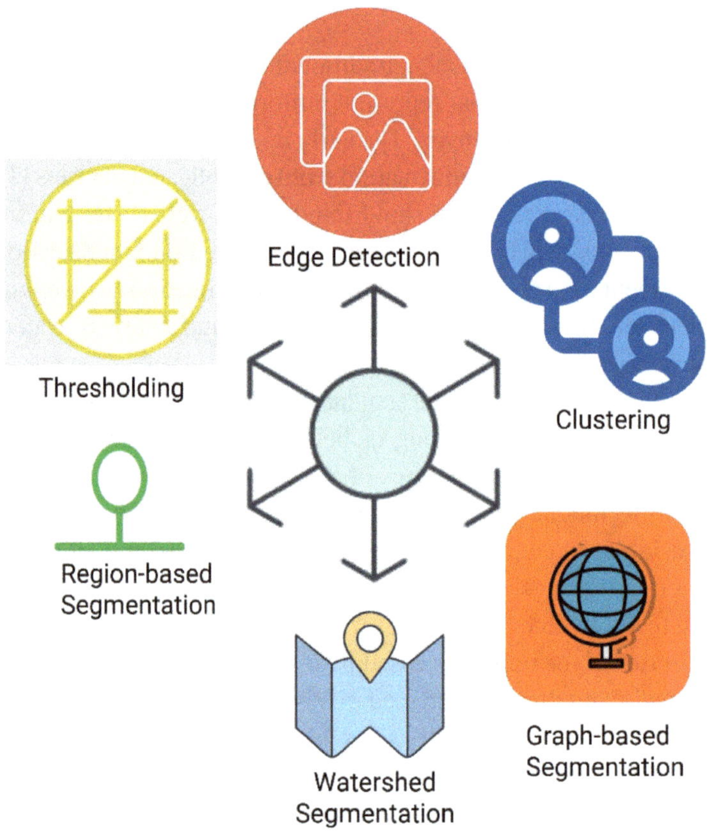

Machine Learning-Based Segmentation

The extensive execution of machine learning-based segmentation methods surpasses conventional techniques through their ability to effectively deal with complex remote sensing datasets. The algorithms base their decisions on data learning capabilities and recognize patterns through pixel intensities together with texture analysis and spatial relationship investigation. Supplied training sets allow machine learning models to distinguish distinct categories by classifying pixel or region areas such as land cover types and objects of interest. The segmentation technique brings superior results when you use high-dimensional images like multispectral or hyperspectral which traditional techniques find challenging to differentiate between features, (Thota et al., 2023).

The major benefit of employing machine learning-based segmentation lies in its superior capability to deal with large heterogeneous datasets better than conventional techniques. The automatic classification of regions using complex data patterns depends on three important algorithms which include decision trees, support vector machines (SVM), and k-nearest neighbors (KNN). The capability of SVM to locate its optimal separating hyperplane between different classes enables it to identify features that look alike spectrally. Such methods enable better segmentation because they include features which analyze spatial relationships between pixels as well as surface textures thus delivering more precise results in complex settings such as urban areas and mixed vegetation areas.

The application of deep learning methods especially convolutional neural networks (CNNs) represents the highest level of performance for machine learning-based segmentation systems. CNNs effectively master the process of detecting image spatial relationships which enables them to excel at image segmentation tasks. Technical models learn pertinent image characteristics from plain pixel information while eliminating the necessity to conduct mechanical feature detection processes. Deep learning models demonstrate exceptional accuracy with their ability to process extensive datasets so they can deliver outstanding performance across images that contain diverse lighting conditions and shadows as well as complex scenes. These systems efficiently detect detailed aspects while maintaining an understanding of different scale patterns which makes them ideal for applications requiring land cover mapping solutions and building detection services and monitoring the changes in the environment.

The impressive features of machine learning-based segmentation methods lead to specific obstacles that developers need to tackle. The major challenge stems from requiring large datasets of labeled data which effectively trains learning models. Obtaining properly annotated datasets with high quality data demands significant allocation of time and financial resources. Large computational systems are required for deep learning models during their training period as well as during inference time thus making these models inaccessible to many users or applications. The main drawback of model implementation is overfitting because it leads to excellent training metrics but limited capacity to adapt to unknown observation data. Machine learning-based segmentation continues to establish itself as a valuable tool for remote sensing data analysis through the combination of transfer learning along with data augmentation advancements and rising computational power capabilities.

Deep Learning Approaches

Remote sensing has experienced a revolution because of deep learning tools that create superior methods to analyze complex multidimensional information. Artificial neural networks based methods together with other techniques have the power to find automatic features while building hierarchical data structures from raw datasets. Deep learning models particularly convolutional neural networks (CNNs) identify both patterns and image spatial connections automatically since they do not need direct human interaction. Deep learning possesses special strength in image classification and object detection and segmentation of remote sensing data because of its automatic feature extraction capabilities.

CNNs achieve exceptional results in remote sensing through their powerful ability to handle image data efficiently, (Zhang et al., 2022). Deep convolutional neural networks use sequential convolutional layers that examine specific image patterns starting with basic features such as edges until they build compound elements. The tiered model structure enables CNNs to detect details from images together with comprehensive spatial patterns that makes them valuable for spaceborne satellite photography processing. When used with multispectral or hyperspectral image types CNNs create more precise detection methods for different environmental situations.

Remote sensing has experienced a revolution because of deep learning tools that create superior methods to analyze complex multidimensional information. Artificial neural networks based methods together with other techniques have the power to find automatic features while building hierarchical data structures from raw datasets. Deep learning models particularly convolutional neural networks (CNNs) identify both patterns and image spatial connections automatically since they do not need direct human interaction. Because of its automatic feature learning capability deep learning is highly effective for remote sensing applications such as image classification and detecting objects and segmenting imagery.

CNNs achieve exceptional results in remote sensing through their powerful ability to handle image data efficiently, (Zhang et al., 2022). A network of convolutional layers in CNNs successively detects input image patterns starting from local features until they produce elaborate combinations of features. The tiered model structure enables CNNs to detect details from images together with comprehensive spatial patterns that makes them valuable for space borne satellite photography processing. By combining CNN with other image data types including multispectral and hyperspectral images users achieve superior classifications and detections results within diverse environmental settings.

Hybrid Methods and Multi-Source Data Integration

The combination of hybrid methods and multi-source data integration presents itself as popular approaches in remote sensing which enhances analysis accuracy and robustness according to Meng et al. (2019). Hybrid methods unite separate algorithms and techniques thus making their combined capabilities extract better solutions than independent methods can achieve alone. Multiple remote sensing methods work together through common hybrid strategies where approaches like thresholding and region-growing algorithms team up with machine learning methods that utilize support vector machines (SVMs) or deep learning models. Complex image scenarios with low contrast and multiple land cover types become manageable through the combination of classical and modern approach strengths in hybrid methodologies.

The concept of Multi-source data integration exceeds hybrid methods by sharing information between various sensing instruments that include radar detection and optical and LiDAR systems. Multiple sensors provide distinct environmental measurements about surface texture elevation and spectral reflectance which enable combined assessment of Earth's surface characteristics. Multiple data platforms supply unique information levels to users where optical images show colors in detail while radar imagery reveals surface characteristics and LiDAR systems create 3D extraction models. The combination of various data types creates a complete realization of complex terrain which proves beneficial in disaster response and urban development planning as well as vegetation monitoring activities.

The combination of multi-source data demands sensor data alignment because sensors often operate with variable spatial resolution together with temporal resolution and spectral resolution parameters. Multiple sophisticated processing methods including image registration and data fusion and machine learning model applications allow users to properly connect various data sets. Pixel-level fusion operates to combine sensors with different resolution capabilities and machine learning algorithms work to find significant patterns in merged datasets. When analysts merge data that enhances each other they can produce better classifications alongside more accurate detections as well as more precise predictions, (Al-Mhiqani et al., 2020).

Hybrid methods and multi-source data integration excel at managing both the rising data complexity and large data amount that comes from contemporary remote sensing platforms. Modern advancements in satellite and aerial platforms have resulted in a booming increase of both data quantity and diversity available for analysis. Researchers and practitioners benefit from adopting hybrid and multi-source integration techniques to maximize their use of large amounts of information through solutions for data redundancy and spatial-spectral inconsistency alongside varying data quality levels. These combination methods demonstrate potential to enhance vital decisions when assessing environmental data and climate change impacts as well as reacting to natural hazards.

CHANGE DETECTION TECHNIQUES IN COASTAL ZONES

Classical Change Detection Methods

Change detection using classical methods serves as technology that uses remote sensing data to locate and measure surface changes across different timescales. The detection methods use different time point data which get compared against each other to discover transformations of land use or urban growth or deforestation patterns, (Guild, Cohen, & Kauffman, 2004). The difference between two temporally separated images is created through pixel value subtraction between them. Significant changes within the studied area become visible when comparing remote sensing data because big differences between time periods show greater modification. The image differencing method is easy to execute but becomes unreliable when users consider alterations in lighting along with changes in atmospheric climate or sensor variations leading to erroneous results.

Post-classification comparison stands as a common technique which involves independent classification of images taken from different time periods followed by the comparison of results to spot changes, (Srivastava et al., 2012). Land cover type comparisons utilizing this method establish definite separations between pre-event and post-event classifications of different land surface features. This technique proves effective yet its outcomes depend greatly on classification accuracy and the initial classification precision. The methodology struggles to identify minor and gradual alterations because its analytical method centers around specific defined land cover categories.

Figure 3. Spatial interaction of ecological and human spaces

Change vector analysis (CVA) functions as another established method which detects modifications in multi-band remote sensing information acquired from multispectral or hyperspectral images. The process of CVA starts with measuring vector differences between equivalent pixel positions in multiple images followed by assessing these vector change metrics. CVA provides ideal results during vegetation health monitoring and other environmental tracking since this method utilizes multiple spectral band variations. The performance of CVA might decrease when images contain small contrast between them or when seasonal variations modify spectral signatures of the land cover, (Fomalont, 1999).

The standard change detection procedures show restricted capability to detect complicated alterations that result from urban growth or natural disasters. The detection of slow environmental changes along with heterogeneous terrains remains problematic for these techniques while also yielding incorrect positive results and undocumented variations in the environment. The use of accurate pixel comparison demands precise image registration as an essential requirement for classical methods since working with different sensors and acquisition conditions can be challenging. Even so classical change detection methods now integrate with more advanced techniques of machine learning and deep learning-based approaches because these novel methods deliver improved detection capacity across multiple change processes over time.

Multi-Temporal Analysis with Remote Sensing

The analysis of changes within Earth's surface over time takes place through remote sensing by comparing images obtained across different points in time. Using remote sensing data along the temporal dimension facilitates researchers to detect active changes and monitor both natural and anthropogenic transformations, (Hirschmugl et al., 2017). The detection of short-term together with long-term changes occurs through the use of imagery acquired from satellites and drones or other remote sensing platforms within Multi-temporal analysis. Multiple series of images analyzed together generates important understanding of seasonal patterns along with trends and abnormal events in environmental factors like land cover and vegetation.

The main benefit of applying multi-temporal analysis is its capacity to monitor and document extended-time-based societal and environmental transformations. Visual data sets spanning multiple years enable researchers to detect environmental changes like urbanization together with deforestation and climate-based impacts on ecosystems. The evaluation technique detects steady alterations in addition to unexpected changes to deliver extensive environmental system comprehension. The long-term observation of seasonal events enables the efficient monitoring of agricultural patterns and seasonal flooding so organizations can plan their resource use proficiently.

The identification of differences between images collected during different time periods requires change detection methods as a common analysis technique for multi-temporal research. The analysis employs basic differencing techniques between images as well as sophisticated PCA methods to produce substantial modifications of reduced data dimensions. Multi-temporal analysis employs time-series analysis to examine data from various time points for the purpose of detecting patterns together with trends and anomalies. Multi-temporal analysis proves most beneficial when investigating gradual processes including land deterioration and forest changes since historical data enables predictions of upcoming patterns.

Multiple time analysis provides many advantages yet contains specific obstacles that researchers need to overcome. The main challenge in this approach stems from achieving proper image alignment because differences between sensors or space conditions alongside orbital alterations lead to degraded data quality levels. The temporal resolution of remote sensing data exhibits different collection patterns because some sensors operate with set intervals yet others produce images with irregular schedules, (Ganesan et al., 2004). The different variations present analysis challenges that become particularly difficult during rapid or short-term change detection. Remote sensing benefits significantly from multi-temporal analysis although it deals with its respective obstacles because this method delivers essential information about environmental changes and urban development and tracking natural disasters.

Machine Learning and Deep Learning for Change Detection

The analysis of changes within Earth's surface over time takes place through remote sensing by comparing images obtained across different points in time. The temporal element of remote sensing data enables researchers to observe and monitor dynamic processes including land-use change urban growth together with environmental shifts. The detection of short-term together with long-term changes occurs through the use of imagery acquired from satellites and drones or other remote sensing platforms within Multi-temporal analysis. Multiple series of images analyzed together generates important understanding of seasonal patterns along with trends and abnormal events in environmental factors like land cover and vegetation.

By following changes across multiple time intervals multi-temporal analysis provides substantial information about environmental and societal transformations of past times. Time series images examined across multiple years or decades allow experts to study the growth patterns of urban areas and track the loss of forests and track changes caused by climate effects on natural systems. The evaluation technique detects steady alterations in addition to unexpected changes to deliver extensive environmental system comprehension. The technique allows continuous tracking of periodic occurrences like agricultural patterns as well as seasonal flooding patterns essential for planning resource distribution.

The identification of differences between images collected during different time periods requires change detection methods as a common analysis technique for multi-temporal research. The analysis employs basic differencing techniques between images as well as sophisticated PCA methods to produce substantial modifications of reduced data dimensions. Time-series analysis enables pattern detection of temporal trends as well as anomalies using information collected from various time points in multi-temporal analysis. Multi-temporal analysis proves most beneficial when investigating gradual processes including land deterioration and forest changes since historical data enables predictions of upcoming patterns, (Dale, 1997).

Multiple time analysis provides many advantages yet contains specific obstacles that researchers need to overcome. The most crucial barrier in multi-temporal analysis involves achieving precise image alignment from different time points since variations in sensor properties and atmospheric factors and satellite orbit dynamics influence the data quality. The temporal resolution of remote sensing data exhibits different collection patterns because some sensors operate with set intervals yet others produce images with irregular schedules, (Li & Hsu, 2020). The different variations present analysis challenges that become particularly difficult during rapid or short-term change detection. The ability of multi-temporal analysis to function as a vital remote sensing instrument enables crucial monitoring of environmental changes together with urban development and natural disasters despite facing various data collection challenges.

Integrating Remote Sensing with Ground Truth and Auxiliary Data

The analysis of changes within Earth's surface over time takes place through remote sensing by comparing images obtained across different points in time. Through this method researcher can assess temporal remote sensing data to track and identify time-sensitive processes like environmental transformations and urbanization patterns. The detection of short-term together with long-term changes occurs through the use of imagery acquired from satellites and drones or other remote sensing platforms within Multi-temporal analysis. Multiple series of images analyzed together generates important understanding of seasonal patterns along with trends and abnormal events in environmental factors like land cover and vegetation.

Multi-temporal analysis enables extensive time-based assessment which produces exclusive knowledge about long-term environmental transformations and societal evolutions. Repeated image analyses extending over yearly to decades-long periods enable scientists to detect patterns including urban growth as well as deforestation and ecological effects because of climate change. The evaluation technique detects steady alterations in addition to unexpected changes to deliver extensive environmental system comprehension. Monitoring recurrent natural cycles including agricultural cycles and seasonal flooding has become possible through this approach which helps resource managers in their planning initiatives.

The identification of differences between images collected during different time periods requires change detection methods as a common analysis technique for multi-temporal research. The analysis employs basic differencing techniques between images as well as sophisticated PCA methods to produce substantial modifications of reduced data dimensions. Time-series analysis is part of multi-temporal analysis because it examines data collection points across different moments to identify patterns and anomalies and emerging trends. Multi-temporal analysis proves most beneficial when investigating gradual processes including land deterioration and forest changes since historical data enables predictions of upcoming patterns, (Wallace, Caccetta, & Kiiveri, 2004).

Multiple time analysis provides many advantages yet contains specific obstacles that researchers need to overcome. The main obstacle for maintaining exact image alignments during different times consists of sensor characteristic modifications and atmospheric conditions and inconsistent satellite orbits which degrade data quality and precision. The temporal resolution of remote sensing data exhibits different collection patterns because some sensors operate with set intervals yet others produce images with irregular schedules, (Simboura et al., 2016). The different variations present analysis challenges that become particularly difficult during rapid or short-term change detection. The analytical abilities of multi-temporal analysis continue to deliver vital information needed for observing environmental changes in addition to urban growth and both environmental disasters.

CASE STUDIES IN COASTAL ZONE LAND COVER SEGMENTATION AND CHANGE DETECTION

Coastal Zone Mapping and Monitoring Using Remote Sensing

The analysis of changes within Earth's surface over time takes place through remote sensing by comparing images obtained across different points in time. Remote sensing data analysis based on time helps scientists detect and track essential dynamic environmental processes such as land-use evolution as well as urban expansion and natural ecological modifications, (Klemas, 2001). The detection of short-term together with long-term changes occurs through the use of imagery acquired from satellites and drones or other remote sensing platforms within Multi-temporal analysis. Multiple series of images analyzed together generates important understanding of seasonal patterns along with trends and abnormal events in environmental factors like land cover and vegetation.

Multi-temporal analysis enables tracking of prolonged changes through time because it provides historians with an environmental and societal development retrospective view. Researchers studying patterns through time-based image collections can detect essential changes such as urban development and forest depletion together with environmental ecosystem modifications due to climate change. The evaluation technique detects steady alterations in addition to unexpected changes to deliver extensive environmental system comprehension. The analysis enables continuous monitoring of regular events including agricultural patterns along with seasonal flood occurrences for effective resource management and development planning.

The identification of differences between images collected during different time periods requires change detection methods as a common analysis technique for multi-temporal research. The analysis employs basic differencing techniques between images as well as sophisticated PCA methods to produce substantial modifications of reduced data dimensions. Time-series analysis represents another part of multi-temporal analysis because it uses dataset information from different periods to detect trends and anomalies over time. Multi-temporal analysis proves most beneficial when investigating gradual processes including land deterioration and forest changes since historical data enables predictions of upcoming patterns, (Krause et al, 2020).

Multiple time analysis provides many advantages yet contains specific obstacles that researchers need to overcome. The quality and accuracy of data suffer due to sensor characteristic variations and atmospheric conditions and satellite orbit alterations during the process of aligning images taken at different times. The temporal resolution of remote sensing data exhibits different collection patterns because some sensors operate with set intervals yet others produce images with irregular schedules, (Beechie et al., 2008. The different variations present analysis challenges that become particularly difficult during rapid or short-term change detection. The technical complexities do not diminish the effectiveness of multi-temporal analysis as a vital remote sensing method that delivers essential environmental tracking records about ecological shifts and urban growth and natural disaster situations.

Application of Remote Sensing for Coastal Ecosystem Protection

The analysis of changes within Earth's surface over time takes place through remote sensing by comparing images obtained across different points in time. The use of this approach in remote sensing takes advantage of time data to help researchers track developing environmental changes including land-use alterations and urban expansion and ecosystem modifications, (Adjoyu et al., 2023). The detection of short-term together with long-term changes occurs through the use of imagery acquired from satellites and drones or other remote sensing platforms within Multi-temporal analysis. Multiple series of images analyzed together generates important understanding of seasonal patterns along with trends and abnormal events in environmental factors like land cover and vegetation.

Multi-temporal analysis gives researchers an ability to detect long-term environmental changes by allowing them to study historical events across multiple time dimensions. Analysts can monitor long-term trends through the study of pictures collected across various periods from years up to decades which reveals patterns regarding urbanization and deforestation and climate change effects on ecological systems. The evaluation technique detects steady alterations in addition to unexpected changes to deliver extensive environmental system comprehension. This methodology makes it possible to track recurring environmental occurrences like agricultural cycles together with seasonal flooding since these events need monitoring for resource planning objectives.

The identification of differences between images collected during different time periods requires change detection methods as a common analysis technique for multi-temporal research. The analysis employs basic differencing techniques between images as well as sophisticated PCA methods to produce substantial modifications of reduced data dimensions. Multiple time points of analyzed data through time-series analysis reveal essential trends and patterns as well as abnormal behaviors. Multi-temporal analysis proves most beneficial when investigating gradual processes including land deterioration and forest changes since historical data enables predictions of upcoming patterns, (Zielinski et al., 2009).

Multiple time analysis provides many advantages yet contains specific obstacles that researchers need to overcome. Must-on data alignment represents the primary analytic limitation since sensor system variations and atmospheric factors as well as orbital perturbations affect the precision and clarity of data measurements collected at different points in time. The temporal resolution of remote sensing data exhibits different collection patterns because some sensors operate with set intervals yet others produce images with irregular schedules, (Estes & Loveland, 1999). The different variations present analysis challenges that become particularly difficult during rapid or short-term change detection. The powerful data analysis method of multi-temporal science captures essential insights about environmental changes and urban development along with natural disaster development in spite of various data challenges.

Figure 4. Remote sensing data processing workflow

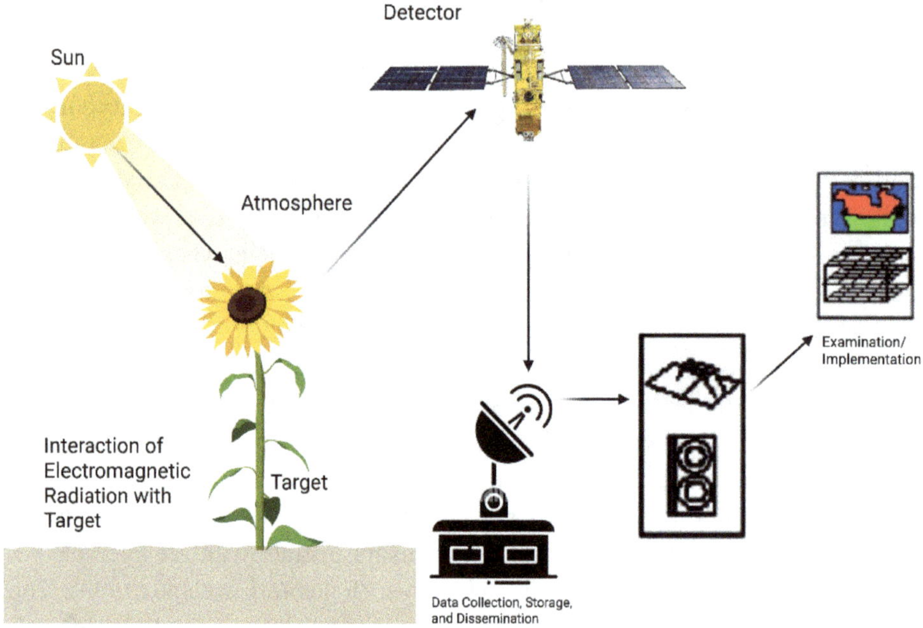

Coastal Disaster Monitoring and Response

Response capabilities for monitoring coastal disasters help decrease the effects of hurricanes alongside storm surges and tsunamis and coastal flooding incidents. Remote sensing technologies serve as critical instruments for delivering instant data collection and disaster site awareness purposes during these events. The assessment of disaster damage and the identification of critical disaster zones rely on satellites and drones and other aerial systems which gather high-quality information and sensor inputs. The immediate delivery of accurate data represents a vital resource that helps emergency responders guide their actions and deploy available resources towards disaster-hit areas as well as offer essential assistance to local communities following the disaster.

Remote sensing solutions used for coastal disaster monitoring prove essential for identifying storm systems through their tracking processes which include hurricanes and cyclones. Devices installed on satellites enable the collection of measurement data regarding storm power together with wind speed reports and precipitation distribution patterns, (Estes & Loveland, 1999). Remote sensing technology enables meteorological experts to forecast storm trajectories and possible consequences thus enabling them to warn residents in advance of protecting their safety through evacuation operations. Remote sensing systems maintain the ability to evaluate storm surge and coastal flooding impacts through comparison of event-pre and event-post images. Satellite images help identify areas affected by flooding as well as analyze structural damage together with changes in coastal landscapes for immediate disaster evaluations and reconstruction tasks.

Remote sensing systems track coast after disasters along with their management needs. Remote sensing satellites show in their images how severe flooding caused by tsunamis and major floods affected coastal areas by revealing what happened to buildings and land features. The recovery planning process depends on accurate assessment of damage and prompt intervention needs because of this data. Remote sensing detects environmental damages through its ability to detect oil spills and coastal ecosystem destruction which enables proper coordination of long-term restoration efforts, (Williams & Hoffman, 2020).

The widespread use of remote sensing remains vital for disaster monitoring yet specific difficulties need to be managed while employing this technology, (Hussain et al., 2009). The rapid conditions in coastal regions including shifting tides and cloud cover impacts the way remote sensing data reports its quality levels and accuracy metrics. Different disaster monitoring sources need advanced data fusion methods to combine aerial imagery with ground-based observations and sensor network information for complete disaster impact understanding. Remote sensing remains a vital element of disaster response tactics that offers vital information for coastal hazard regions to analyze risks and restore from damage and assess losses through essential data collection.

CHALLENGES AND LIMITATIONS IN COASTAL ZONE LAND COVER SEGMENTATION AND CHANGE DETECTION

Data Quality and Availability

The effectiveness of remote sensing applications throughout separate fields strongly depends on both the quality and availability of acquired data. The precision of analysis together with decision-making and environmental interpretation demands only the use of high-quality data for environmental and agricultural and urban and disaster-related phenomena, (Gupta & Nair, 2011). A remote sensing data's suitability for analysis depends on its accuracy level and data consistency together with resolution capabilities as well as reliability features. Reliable conclusions along with sound decisions become impossible when data quality becomes inconsistent because of sensor calibration errors or atmospheric conditions and cloud cover presence. Remote sensing systems generate effective actionable insights through precise findings only when data quality receives proper attention.

Remote sensing data availability strongly influences the effectiveness with which this technology can serve different applications. A growing number of satellites combined with UAVs and additional data collection platforms produces increasing availability of remote sensing data worldwide. The cost of commercial satellite data remains high and certain sensitive areas prevent public access to address privacy and security needs. Superior availability of open-source and accessible remote sensing data emerged during recent years because of NASA and ESA's and other organizations increased production of freely accessible datasets, (Rhind, 1992).

The temporal as well as spatial resolution of remote sensing data constitutes a part of data availability. Remote sensing applications that need time-based observation rely on high temporal resolution monitoring abilities to track alterations throughout periods. These uses include disaster relief operations and plant habitat management along with metropolitan growth observations. Spatial resolution needs to be high when analyzing small areas because detailed analysis depends on it therefore mapping infrastructure and land cover classification both require this level of resolution. The analysis requires an appropriate combination between these resolution types based on its defined objectives. The decision process for choosing dataset depends on balancing the resolution specifications of acquired data against processing times and storage capacity requirements for intended applications, (Catthoor & Danckaert, 2002).

Data quality and availability benefit strongly from the successful combination of information acquired from multiple data sources. Remote sensing data gains more value when scientists integrate it with auxiliary information such as meteorological data and topographic maps and socio-economic data while using it for remote sensing applications, (Gerland, 1996). The integration of sources produces enhanced analysis capabilities as well as improved decision-making processes. The main obstacle involves making sure multiple sources of data work together in a meaningful manner for effective merging. To achieve accurate results from merged datasets these issues including spatial resolution discrepancies and different data formats and time periods need resolution. New improvements in remote sensing technology through advanced data processing and machine learning methods will boost data quality and availability thus enabling remote sensing for various applications.

The Complexity of Coastal Environments

Pearl Ridge Estates and the surrounding coastal regions show continuous changes from natural physio-biological interactions alongside human activities. A diverse collection of coastal features consisting of beaches, shoreline rocks, estuaries, wetlands and coral reefs exist in these environments where wave action merges with tidal forces and sediment transport mechanisms. Coastal spaces present their operational intricacy through two principal factors: natural variations of their biodiversity along with modifications in physical chemistry effects which stem from background processes as well as human-made influences. These ecosystems exhibit permanent shifts between varying time frames from regular tidal rhythms to extended developments of sea-level increases or beachfront degradation.

The complexity of coastal zones grows higher because of human activities. The high human pressure from industrial development together with urbanization and agricultural practices and tourism operations causes environmental damage to these sensitive areas through habitat destruction and chemical pollution and physical landscape changes. The unique ecological requirements of coastal areas such as proper water quality and salinity alongside sediment makeup provide little protection against disturbances. Human activities including development drainage of wetlands result in harmed conditions for these ecosystems and coral reefs are negatively affected by pollution and ocean acidification. The modifications created by humans produce unpredictable system disturbances in coastal regions which remain challenging to manage.

The effects of human actions combine with climate change to create sensitivity problems in coastal areas, (Sánchez-Arcilla et al., 2016). The coastal regions face continuous transformation due to increasing water levels and more frequent storms along with changing meteorological conditions. The regions near coasts serve as the primary location for climate change effects because rising sea levels endanger ecosystems in lower areas while producing stronger and more frequent flood conditions. Climate change-induced ocean temperature increases along with acidification transform marine ecosystems which causes disturbances throughout the entire fish migration network as well as damaging coral reef ecosystems. The numerous interlinking factors between climate change and coastal regions make total understanding of lasting impacts difficult to predict, (Toimil et al., 2020).

Figure 5. Research priorities for coastal climate change

Thorough monitoring combined with strategic management approaches becomes necessary for protecting coastal areas because of their complex operating conditions, (Bhatti et al., 2024). Remote sensors activate vital roles for coastal area research because they deliver crucial data regarding temporal variations of ecosystem health alongside land cover and water quality changes. Organizations need to combine information from various sources including ground-based observations and climate models along with socio-economic data for complete understanding of coastal systems dynamics during management strategies. Strategies to achieve balanced conservation development of coastal areas must be further advanced since human and natural pressures keep intensifying, (Bhatti, Zhang, & Sarker, 2025).

Data Integration and Fusion

The integration and fusion of data through remote sensing requires combining various datasets to develop improved analytical results which provide a better understanding of the environment, (Bhatti, Li, & Tang, 2025). The synthesis of information resulted from combining data originating from satellite imagery ground-based sensors and aerial data enables unified analysis. Data integration aims to employ complimentary data strengths for surpassing the weaknesses of separate datasets. Animals and ground measurements demonstrate two different approaches to collection in which broad spatial data comes from satellites and accurate precise data exists through ground-based platforms. The analysis benefits from data integration which allows researchers to obtain superior findings and extensive comprehension of the examined territory, (Li, Bhatti, & Zhang, 2024).

The integration of different datasets faces the fundamental challenge to establish data compatibility between sources, (Bhatti, li, & Zhang, 2024). The integration process usually includes managing different levels of spatial precision and time frequency alongside various storage formats. Adjustments along with resampling procedures are necessary to integrate data because opposing spatial resolution levels make data from different sources incompatible. The analysis of time-sensitive phenomena including vegetation growth needs to deal with temporal discrepancies through techniques that handle images taken at different times. Data integration requires standardized formats along with proper alignment methods to protect data integrity according to successful integration and fusion principles, (Sattar et al., 2024).

The process of data fusion creates entirely new comprehensive information by uniting various datasets. Multiple fusion techniques operate at three different levels starting from pixels up to features and decision-making processes. The combination of various data layers at the pixel level generates a unified composite image that provides elaborated analytical information. The combination of high-resolution panchromatic and multispectral satellite images generates a single data collection that contains spatial precision as well as spectral information. The fusion process at feature-level combines the characteristics extracted from multiple datasets for instance merging land cover classifications with elevation data to analyze terrain features better. Decision-level fusion unites different analytical outcomes to enhance both classification performance and decision-making processes, (Yu, Zhang, & Bhatti, 2024).

Research results from data integration and fusion procedures demonstrate excellent outcomes for resource-intensive environments that include coastal regions and urban districts as well as forest ecosystems, (Zhang, Zhang, & Bhatti, 2024). Complex environmental zones need multiple sensor data including optical, radar, LiDAR and others to monitor all environmental variables and their connections. Research becomes more precise and realistic for real-world scenarios when investigators merge different data varieties which helps resolve heterogeneity issues in their studies. Successful data integration and fusion rely on advanced algorithms along with sophisticated computational tools in addition to detailed knowledge about the analyzed data sources and research questions. Remote sensing technologies now and into the future will generate enhanced capabilities through data integration and fusion which will create more effective tools for environmental surveillance together with resource supervision and disaster relief operations, (Li, Zeng, & Bhatti, 2024).

FUTURE DIRECTIONS AND EMERGING TRENDS

Advancements in Remote Sensing Technology

Remote sensing technology development has substantially increased the abilities for Earth observation as well as environmental monitoring, (Liu, Bhatti, & Zhang, 2024). The past decades have brought significant progress to sensor technologies and satellite designs and data processing algorithms which resulted in better Earth surface measurements. Scientists have created high-resolution imaging sensors to observe territories in great detail with minimal spatial limitations, (Li, Zhang, & Bhatti, 2025). Due to their capability to detect information with precision at scales ranging from a few centimeters and above these sensors prove beneficial for planning urban areas and infrastructure tracking as well as agricultural precision. The developments in multispectral and hyperspectral sensing technologies enable researchers to obtain data from multiple wavelength bands which leads to deeper measurements of environmental factors and land cover as well as vegetation health assessments, (Aamir et al., 2024).

Remote sensing technology has progressed due to the growing availability of small satellites and unmanned aerial vehicles (UAVs) according to Zhang, Xu, and Bhatti (2025). Small satellites under the name "CubeSats" function as a budget-friendly tool to capture high-frequency data which enables continuous observation of changing processes in real-time. The aircraft-driven unmanned aerial vehicles let scientists obtain detailed images across land areas which regular satellite networks struggle to reach. These data collection platforms allow researchers to perform specific observations through targeted monitoring and they can be swiftly deployed to track particular regions during events thereby providing exceptional adaptability in data collection, (Zhang, Xu, & Bhatti, 2025).

Advanced data processing emerges as a primary field of research advancement through artificial intelligence and machine learning applications. Traditional analysis models fail to provide sufficient handling of expansive remote sensing datasets since their data volume continues to increase. The complete transformation of large-scale remote sensing dataset feature classification and detection and analysis comes from machine learning algorithms particularly deep learning models, (Li, Wang, & Bhatti, 2024). Through their ability to find hidden patterns in the data remotely operating systems boost accuracy of applications like environmental monitoring, land cover classification and change detection. Through machine learning integration with remote sensing datasets have become more efficient and automatic resulting in simpler extraction of practical insights from big data sources, (Zhang, Li, & Bhatti, 2025).

Remote sensing data management has become faster and more efficient because of improved communication technology through high-bandwidth satellite data transmission. Organizations and governments and researchers can now easily access data worldwide for better worldwide data-sharing which supports informed decision-making processes, (Wu, Zhang, & Bhatti, 2025). Remote sensing technology shows promise for future real-time monitoring and improved multi-source data combination and advanced environmental assessment accuracy. Remote sensing technology shows promise to hold important positions in solving global problems (Li & Bhatti, 2024) as it provides essential instruments for environmentally conscious development, (Yang, Li, & Bhatti, 2025).

The Role of Artificial Intelligence in Coastal Monitoring

AI has become vital for coastal monitoring because it enhances productivity while providing better analysis of multidimensional coastal areas through its automated processes. The conventional process of monitoring coastal areas depends on manual data processing and analysis while providing limited scope in time. Machine learning algorithms together with deep learning algorithms from artificial intelligence automatically enhance remote sensing data interpretation allowing efficient coastal environment monitoring of extensive changing coastal zones, (Zhang, Zeng, & Bhatti, 2024). By analyzing satellite and drone imagery AI systems provide fast insights about coastal dynamics and emerging coastal problems through their rapid identification of changes in landforms and levels of vegetation health and water quality, (Zhang, Zeng, & Bhatti, 2024).

The detection and classification of land cover types represents a major application of AI in coastal monitoring which separates vegetation types from urban areas and water bodies, (Hasnain, Li, & Bhatti, 2024). High-resolution imagery gets processed through deep learning models especially convolutional neural networks which yield accurate classifications into specific categories, (Wang, Li, & Bhatti, 2024). AI demonstrates essential value for determining the well-being of environmentally susceptible coastal environments which encompass mangroves and coral reefs and seagrass beds because it enables assessments of their health from pollution and coastal modifications and climate change impacts. The automation of classification processes through Artificial Intelligence decreases time and resource costs for data analysis so monitoring of essential ecosystems becomes more frequent and detailed, (Hasnain, Hashmi, & Bhatti, 2024).

The comprehensive coastal monitoring becomes achievable through AI by means of processing various data sets together, (Yang, Li, & Bhatti, 2024). The evaluation of coastal impacts from climate change becomes more efficient through this integration because researchers can determine sea-level rise and storm intensity effects on coastal regions, (Tang, Li, & Bhatti, 2025). AI operates through modeling and predictions of environmental elements that impact coastal zones and facilities alongside communities because it provides knowledge helpful in developing disaster mitigation strategies. Large-scale coastal resource optimization becomes possible through AI systems which determine vulnerable regions for enhanced preservation work, (Zhang, Li, & Bhatti, 2024).

The assessment of coastal hazard impacts during disasters receives enhancement from AI systems which process data about hurricanes and tsunamis and oil spills, (Li, Zhang, & Bhatti, 2024). The extent of damage to infrastructure as well as habitats and human population areas gets quickly evaluated through AI algorithms analyzing satellite and drone imagery following such events. Complete damage assessments become faster through AI automation that directly identifies locations requiring immediate help, (Yuan, Li, & Bhatti, 2024; Liu, Li, & Bhatti, 2025; Gao & Bhatti, 2024). AI models enhance disaster monitoring through their ability to analyze data collected from satellite sensors thus identifying environmental conditions which suggest coastal disasters before they occur through processes of monitoring sea-level changes and atmospheric patterns. AI revolutionizes coastal monitoring through its processing of big real-time datasets for complex prediction analysis which delivers better accuracy in addition to efficiency for daily operations and disaster response activities, (Wu & Bhatti, 2024).

The Integration of Remote Sensing with Other Environmental Data

The joint use of remote sensing alongside environmental datasets creates a strong tool which helps ecological oversight and environmental management especially in challenging environments including urban locations and both coastal and forested regions. Remote sensing technology generates precise spatial data by identifying a complete array of soil and plant cover and water quality indicators and environmental air conditions data, (Asif & Bhatti, 2025). By merging with other environmental data collections such as ground-based measurements and meteorological data and climate models remote sensing provides a more thorough environmental process understanding, (Zhang & Bhatti, 2025). When scientists integrate different data sources they gain enhanced capabilities to address environmental problems better, (Li & Bhatti, 2024).

Soil samples along with vegetation health evaluations and water quality measurements supply face-to-face information that corresponds with the extensive spatial characteristics of remote sensing information, (Li, Zhang, & Bhatti, 2025). These different datasets together deliver a superior capability to determine environmental factors across multiple scales. Ground-based observations can verify specific causes of environmental changes by confirming data points such as illegal logging activities which satellites have observed but not specifically identified. The assessment of water bodies includes remote sensing detection of surface variations that become more complete when researchers conduct in-situ measurements of water quality including temperature and salinity and pH data to understand aquatic ecosystem health better. Combining remote sensing technology with natural observations produces more valid analyses which lead to complex environmental understanding of the current issues, (Zhang & Bhatti, 2024).

The connection between remote sensing methods and meteorological and climatic information delivers essential knowledge about climate change predictions, (Tang, Li, & Bhatti, 2025). Technological data from remote sensors reveals earth cover modifications together with ecosystem physical features yet meteorological measurements including wind patterns and temperature data assist in uncovering environmental changes. These datasets function together for climate change effect modeling and prediction purposes including sea-level rise and temperature changes and precipitation pattern alterations, (Li, Bhatti, & Zhang, 2025). Researchers use combined climate models with remote sensing data to understand which specific regions are most affected by climate changes then create better adaptation and mitigation plans. This method proves useful for managing coastal zones and agricultural protection and disaster readiness programs.

For sustainable development studies researchers need to combine remote sensing data with socio-economic information because this combination helps understand human-environment relationships, (Li, Zhang, & Bhatti, 2025). Remote sensing data enables assessment of human activities' impact on ecosystems and natural resources by combining information about population density along with land use and economic activity data. An integrated strategy provides data to create procedures for land-use planning together with resource management systems that balance community economic needs with natural environment protection goals, (Zeng, Li, & Bhatti, 2025; Zhang, Li, & Bhatti, 2025; Liu, Zhang, & Bhatti, 2024). Integrated environmental and remote sensing and socio-economic data enables comprehensive sustainable development approaches through improved surveillance of urban expansion and agricultural operation assessment and infrastructure development planning, (Aamir & Bhatti, 2025). Remote sensing technology integrated with environmental data will continue to be a principal instrument for addressing worldwide environmental issues given increasingly better access to improved data sets, (Bhatti & Hasnain, 2025).

CONCLUSION

Summary of Key Findings

Surging cooperation between environmental data sources with remote sensing builds increased observation capabilities in monitoring as well as protecting complex environments such as forests and urban zones and coastal areas. Efforts in remote sensing collection allow detailed spatial data collection which extends to several environmental parameters including land cover classification and vegetation distribution and water and atmospheric conditions reports. With addition of ground-based measurements and meteorological data and climate models to remote sensing data scientists acquire improved abilities to understand environmental processes and their interactions. When scientists integrate different data sources they gain enhanced capabilities to address environmental problems better.

Remote sensing data benefits from local and specific information which ground-based tests of soil and vegetation health and water condition assessments and measurements provide. These datasets become stronger together because they produce better environmental condition assessments across different scales. Ground-based observations can verify specific causes of environmental changes by confirming data points such as illegal logging activities which satellites have observed but not specifically identified. Surface-level monitoring of water bodies through remote sensing produces observational data that in-situ measurements of water quality expand with metrics including temperature, salinity and pH levels to determine water ecosystem health. Combining remote sensing technology with natural observations produces more valid analyses which lead to complex environmental understanding of the current issues.

The analysis of climate effects requires remote sensing data which should be merged with both meteorological data and climatic data for better predictions. Remote sensing enables scientists to monitor ecosystem physical characteristics as well as land cover transformations while meteorological data including temperature and precipitation patterns helps scientists explain environmental condition trends. These datasets function together for climate change effect modeling and prediction purposes including sea-level rise and temperature changes and precipitation pattern alterations. The combination of climate models with remote sensing data allows researchers to evaluate the susceptibility of individual regions to climate changes for building better adaptation approaches and mitigation responses. This method proves its worth during management of coastal zones and agricultural risk reduction and natural disaster readiness.

Sustainable development needs the combination of remote sensing data alongside socio-economic data for understanding human-environment relationships. Remote sensing data enables researchers to evaluate ecosystem and natural resource fragility through the integration of information about population distribution and economic activity regimes alongside land development patterns. An integrated strategy provides data to create procedures for land-use planning together with resource management systems that balance community economic needs with natural environment protection goals. Integrated environmental and remote sensing and socio-economic data enables comprehensive sustainable development approaches through improved surveillance of urban expansion and agricultural operation assessment and infrastructure development planning. Remote sensing technology integrated with environmental data will continue to be a principal instrument for addressing worldwide environmental issues given increasingly better access to improved data sets.

Implications for Coastal Zone Management

The combination of remote sensing technologies with other environmental data offers a strong method to detect and control ecosystem systems in complex environments including coastal regions and forests together with urban areas. Remote sensing systems collect detailed geographical data that covers the analysis of surface features together with terrestrial growth patterns together with aquatic quality measurements and atmospheric elements. Remote sensing systems produce enhanced environmental insights regarding process understanding along with interaction examination when they utilize ground-based environmental measurements and meteorological metrics and climate models. When scientists integrate different data sources they gain enhanced capabilities to address environmental problems better.

Remote sensing data benefits from ground-based information about soil samples along with vegetation health check-ups and water quality measurement which provide detailed local data points. Few environmental assessment efforts succeed when datasets are combined to offer exact measurements at different spatial levels. Ground-based observations can verify specific causes of environmental changes by confirming data points such as illegal logging activities which satellites have observed but not specifically identified. Remote sensing detects surface-level modifications of water bodies yet measurements made in-situ enable researchers to determine water quality factors including temperature, salinity and pH levels which provide detailed knowledge about aquatic ecosystem health. Combining remote sensing technology with natural observations produces more valid analyses which lead to complex environmental understanding of the current issues.

Researchers need remote sensing to integrate with meteorological and climatic data because this combination serves as a key tool for climate change impact prediction and understanding. Remote sensing reveals critical knowledge about ecosystem physical characteristics as well as land cover transformations simultaneously with meteorological information that explains environmental condition transformations. These datasets function together for climate change effect modeling and prediction purposes including sea-level rise and temperature changes and precipitation pattern alterations. Climate models work more effectively when combined with remote sensing data for understanding the specific risks specific areas face. This allows scientists to devise better adaptation approaches and mitigation plans. Such data combination delivers essential benefits to fields including coastal zone management along with agriculture and disaster preparedness.

Academic research requires the combination of remote sensing information with socio-economic information for effective examinations of human-environment relationships and sustainable development objectives. The combination of population density data and land usage and economic activity information obtained from remote sensing allows analysts to determine how human activities affect natural resources along with the health of ecosystems. An integrated strategy provides data to create procedures for land-use planning together with resource management systems that balance community economic needs with natural environment protection goals. Integrated environmental and remote sensing and socio-economic data enables comprehensive sustainable development approaches through improved surveillance of urban expansion and agricultural operation assessment and infrastructure development planning. Remote sensing technology integrated with environmental data will continue to be a principal instrument for addressing worldwide environmental issues given increasingly better access to improved data sets.

Final Thoughts and Recommendations

Combining remote sensing data with different environmental information systems creates important opportunities for enhanced coastal zone management that leads to better decisions and sustainable growth. Effective and precise monitoring of coastal regions gains extreme importance when human activity and climate changes intensify their pressures. Remote sensing technology delivers extensive spatial data which becomes more beneficial through integration with ground survey information together with meteorological and socio-economic records for fully understanding coastal ecosystems. The combination of these methods enables better environmental monitoring processes which results in superior conservation methods and leads to the development of predictive management plans.

Remote sensing technology advancement enables coastal zone management whose future success depends heavily on implementing these advancements along with small satellite and unmanned aerial vehicles (UAVs) as new sensors become mainstream. The combination of modern technologies enables efficient coastal environment monitoring through inexpensive high-quality data acquisition that allows closer and more precise observation of coastal areas. Machine learning techniques together with data fusion methods will improve integrated dataset analysis which leads to enhanced accuracy and operational efficiency of coastal monitoring systems. Matching technology requirements need frameworks and policies to ensure complete data access alongside standards and network compatibility among different data sources.

The main recommendation for coastal zone management calls for making capacity development in technology and data analysis a top priority. The application of remote sensing techniques requires coastal managers and scientists together with decision-makers to receive suitable training which enables them to utilize remote sensing and combine it successfully with available data systems. The necessary steps include training initiatives along with data accessibility improvements combined with interagency and community collaboration between local entities and governmental institutions and research centers. Data-sharing initiatives must be developed to enable valuable information exchanges between stakeholders because they improve collective decision-making processes which lead to more effective environmental challenge responses.

Adaptable coastal zone management practices will become essential because climate change intensification continues to worsen. The incorporation of remote sensing technology must be designed to handle immediate problems and understand upcoming threats and unpredictability in environmental information. Sustainable coastal ecosystem protection and community defense together with extended environmental sustainability will be achieved through this visionary method. Coastal zone management will achieve better ecological protection through human development needs by maintaining innovative technology combined with interdisciplinary collaboration.

REFERENCES

Aamir, M., & Bhatti, U. A. (2025). Remote sensing for climate monitoring and agriculture. *Computers, Materials & Continua*, *75*(3).

Aamir, M., Gao, J., Zhang, Y., Tang, H., & Bhatti, U. A. (2024). Overview of Underwater Object Detection Based on Image Processing Techniques. In *7th International Conference on Pattern Recognition and Artificial Intelligence*.

Al-Mhiqani, M. N., Ahmad, R., Zainal Abidin, Z., Yassin, W., Hassan, A., Abdulkareem, K. H., & Yunos, Z. (2020). A review of insider threat detection: Classification, machine learning techniques, datasets, open challenges, and recommendations. *Applied Sciences (Basel, Switzerland)*, *10*(15), 5208.

Asif, M., & Bhatti, U. A. (2025). Evaluating the effectiveness of green finance mechanisms, digitalization, globalization, and renewable energy consumption in advancing environmental sustainability in G20 countries. *Energy & Environment*.

Beechie, T., Pess, G., Roni, P., & Giannico, G. (2008). Setting river restoration priorities: A review of approaches and a general protocol for identifying and prioritizing actions. *North American Journal of Fisheries Management*, *28*(3), 891–905.

Bhatti, U. A., & Hasnain, A. (2025). Modeling and assessing environmental change using remote sensing and machine learning. *Computers, Materials & Continua*.

Bhatti, U. A., Li, D., & Tang, H. (2025). Oceanic biometric recognition algorithm based on generalized zero-shot learning. In *Deep Learning for Earth Observation and Climate Monitoring* (pp. 249–262). CRC Press.

Bhatti, U. A., Li, J., Huang, M., & Ahmad, R. M. (2024). Remote-sensing image classification: A comprehensive review and applications. In *Deep Learning for Multimedia Processing Applications* (pp. 18–47). CRC Press.

Bhatti, U. A., Li, J., & Zhang, Y. (2024). Enhancing lung cancer diagnosis with data fusion and mobile edge computing using DenseNet and CNN. *IEEE Transactions on Biomedical Engineering*, •••, 1–4.

Bhatti, U. A., Zhang, Y., & Asif, M. (2024). *Deep Learning for Earth Observation and Climate Monitoring*. CRC Press.

Bhatti, U. A., Zhang, Y., & Sarker, S. K. (2025). Artificial intelligence for remote sensing and climate monitoring. In *Deep Learning for Earth Observation and Climate Monitoring* (pp. 35–62). CRC Press.

Catthoor, F., & Danckaert, K. (2002). *Data access and storage management for embedded programmable processors*. Springer Science & Business Media.

Dale, V. H. (1997). The relationship between land-use change and climate change. *Ecological Applications*, *7*(3), 753–769.

Estes, J. E., & Loveland, T. R. (1999). Toward the use of remote sensing and other data to delineate functional types in terrestrial and aquatic systems. In *Developments in Atmospheric Science* (Vol. 24, pp. 125–150). Elsevier.

Fomalont, E. B. (1999). Image analysis. In Synthesis Imaging in Radio Astronomy II (Vol. 180, p. 301).

Ganesan, D., Ratnasamy, S., Wang, H., & Estrin, D. (2004). Coping with irregular spatio-temporal sampling in sensor networks. *Computer Communication Review*, *34*(1), 125–130.

Gao, M., & Bhatti, U. A. (2024). *Deep Learning Based Applications for Multimedia Processing Applications* (Vol. 1 and 2). Taylor & Francis.

Gerland, P. (1996, June). Socio-economic data and GIS: datasets, databases, indicators, and data integration issues. In UNEP/CGIAR (Consultative Group on International Agricultural Research), Arendal III Workshop on Use of GIS in Agricultural Research Management. Norway.

Guild, L. S., Cohen, W. B., & Kauffman, J. B. (2004). Detection of deforestation and land conversion in Rondonia, Brazil using change detection techniques. *International Journal of Remote Sensing*, *25*(4), 731–750.

Gupta, A. K., & Nair, S. S. (2011, May). Environmental knowledge for disaster risk management. In *Abstract Volume of International Conference* (pp. 10-11).

Hasnain, A., Hashmi, M. Z., & Bhatti, U. A. (2024). Prediction of particulate matter pollution using a long short-term memory model in Zhejiang Province, China. *International Journal of Wireless and Mobile Computing*, *23*(2), 163–172.

Hasnain, A., Li, Y., & Bhatti, U. A. (2024). Predicting the impact of change in air quality patterns due to COVID-19 lockdown policies in multiple urban cities of Henan: A deep learning approach. *Atmosphere*, *14*(5), 902.

Hirschmugl, M., Gallaun, H., Dees, M., Datta, P., Deutscher, J., Koutsias, N., & Schardt, M. (2017). Methods for mapping forest disturbance and degradation from optical earth observation data: A review. *Current Forestry Reports*, *3*, 32–45.

Hussain, S. M., Braydich-Stolle, L. K., Schrand, A. M., Murdock, R. C., Yu, K. O., Mattie, D. M., & Terrones, M. (2009). Toxicity evaluation for safe use of nanomaterials: Recent achievements and technical challenges. *Advanced Materials, 21*(16), 1549–1559.

Klemas, V. V. (2001). Remote sensing of landscape-level coastal environmental indicators. *Environmental Management, 27*, 47–57. PMID: 11083908

Krause, J. R., Watson, E. B., Wigand, C., & Maher, N. (2020). Are tidal salt marshes exposed to nutrient pollution more vulnerable to sea level rise? *Wetlands, 40*(5), 1539–1548. PMID: 35068652

Li, D., & Bhatti, U. A. (2024). A Sustainable Computing-Based Multi-Objective Feeding Scheduling Model for Low-Carbon Factory Aquaculture Using Enhanced NSGA-II in Agricultural Consumer Electronics. *IEEE Transactions on Consumer Electronics.*

Li, D., Bhatti, U. A., & Zhang, Y. (2024). MSTFNet: A Mamba and Dual Swin-Transformer Fusion Network for Remote Sensing Image Classification for Precision Agriculture Land Processing. *Environmental Earth Sciences, 83*(23), 1–16.

Li, D., Bhatti, U. A., & Zhang, Y. (2025). *Robust multi-watermarking algorithm for medical images based on SqueezeNet transfer learning*. Deep Learning for Multimedia Processing Applications.

Li, D., Zhang, Y., & Bhatti, U. A. (2024). Color correction methods for underwater image enhancement: A systematic literature review. *PLoS One, 20*(3), e0317306. PMID: 40063649

Li, D., Zhang, Y., & Bhatti, U. A. (2025). CSTFNet: A CNN and Dual Swin-Transformer Fusion Network for Remote Sensing Hyperspectral Data Fusion and Classification of Coastal Areas. *IEEE Journal of Selected Topics in Applied Earth Observations and Remote Sensing.*

Li, D., Zhang, Y., & Bhatti, U. A. (2025). *A robust color image zero-watermarking based on stationary wavelet transform and Daisy descriptor*. Deep Learning for Multimedia Processing Applications.

Li, H., Zeng, Y., & Bhatti, U. A. (2024). Bacterial Inhibition Zone Recognition Method Based on YOLO. In *7th International Conference on Pattern Recognition and Artificial Intelligence.*

Li, W., & Hsu, C. Y. (2020). Automated terrain feature identification from remote sensing imagery: A deep learning approach. *International Journal of Geographical Information Science, 34*(4), 637–660.

Li, X., Wang, S., & Bhatti, U. A. (2024). State-of-the-Art Approach: Boosting IoT Sensor Network Security with Visual Secret Sharing. *IEEE Smart World Congress (SWC).*

Li, Y., & Bhatti, U. A. (2024). A robust watermarking algorithm for medical images based on K-means and DCT. *International Journal of Wireless and Mobile Computing*, *23*(2), 163–172.

Li, Y., Zhang, M., & Bhatti, U. A. (2025). *Robust zero watermarking algorithm for encrypted medical volume data based on PJFM and 3D-DCT.* Deep Learning for Multimedia Processing Applications.

Liu, J., Bhatti, U. A., & Zhang, Y. (2024). EF-UV: Feature Enhanced fusion of U-Net and VIT Transformer for Brain Tumor MRI Image Segmentation. *IEEE Journal of Biomedical and Health Informatics.*

Liu, J., Li, J., & Bhatti, U. A. (2024). Medical Image Hybrid Watermark Algorithm Based on Frequency Domain Processing and Inception v3. *Advanced Intelligent Systems*, •••, 2400654.

Liu, K., Zhang, X., & Bhatti, U. A. (2024). Data-driven switched-gain observation and control. *Nonlinear Dynamics.*

Meng, X., Shen, H., Li, H., Zhang, L., & Fu, R. (2019). Review of the pansharpening methods for remote sensing images based on the idea of meta-analysis: Practical discussion and challenges. *Information Fusion*, *46*, 102–113.

Neal, S. (2013). Transition culture: Politics, localities and ruralities. *Journal of Rural Studies*, *32*, 60–69.

Rhind, D. (1992). Data access, charging and copyright and their implications for geographical information systems. *International Journal of Geographical Information Systems*, *6*(1), 13–30.

Sánchez-Arcilla, A., García-León, M., Gracia, V., Devoy, R., Stanica, A., & Gault, J. (2016). Managing coastal environments under climate change: Pathways to adaptation. *The Science of the Total Environment*, *572*, 1336–1352. PMID: 26875603

Sattar, A., Assam, M., Alahmadi, T. J., Bhatti, U. A., Tang, H., & Aamir, M. (2024). Remote sensing based advance image captioning improved feature attention. In *7th International Conference on Pattern Recognition and Artificial Intelligence.*

Simboura, N., Pavlidou, A., Bald, J., Tsapakis, M., Pagou, K., Zeri, C., & Panayotidis, P. (2016). Response of ecological indices to nutrient and chemical contaminant stress factors in Eastern Mediterranean coastal waters. *Ecological Indicators*, *70*, 89–105.

Srivastava, P. K., Han, D., Rico-Ramirez, M. A., Bray, M., & Islam, T. (2012). Selection of classification techniques for land use/land cover change investigation. *Advances in Space Research, 50*(9), 1250–1265.

Tang, H., Li, D., & Bhatti, U. A. (2025). Robust Zero-Watermarking Algorithm for Medical Images Based on AGAST-LATCH and DCT. In Deep Learning for Multimedia Processing Applications.

Tang, H., Li, J., & Bhatti, U. A. (2025). *Robust zero watermarking algorithm for medical images based on BRISK and DCT*. Deep Learning for Multimedia Processing Applications.

Thota, S., Chitta, S., Vangoor, V. K. R., Ravi, C. S., & Bonam, V. S. M. (2023). Few-ShotLearning in Computer Vision: Practical Applications and Techniques. *Human-Computer Interaction, 3*(1).

Toimil, A., Losada, I. J., Nicholls, R. J., Dalrymple, R. A., & Stive, M. J. (2020). Addressing the challenges of climate change risks and adaptation in coastal areas: A review. *Coastal Engineering, 156*, 103611.

Wallace, J. F., Caccetta, P. A., & Kiiveri, H. T. (2004). Recent developments in analysis of spatial and temporal data for landscape qualities and monitoring. *Austral Ecology, 29*(1), 100–107.

Wang, X., Li, J., & Bhatti, U. A. (2024). A robust watermarking algorithm for medical images based on DWT-Daisy and Discrete Cosine Transform. In *International KES Conference on Innovation in Medicine and Healthcare*.

. Williams, K. C., & Hoffman, J. C. (2020). Remediation to restoration to revitalization: Engaging communities to support ecosystem-based management and improve human wellbeing at clean-up sites. Ecosystem-based management, ecosystem services and aquatic biodiversity, 543.

Wu, H., Zhang, Y., & Bhatti, U. A. (2025). Climate change air quality monitoring using sentimental 2 dataset. In Deep Learning for Earth Observation and Climate Monitoring.

Wu, Y., & Bhatti, U. A. (2024). Applications of machine learning in medicine: Current trends and prospects. *Global Conference on Wireless and Optical Technologies (GCWOT)*.

Yang, G., Li, J., & Bhatti, U. A. (2024). A zero-watermarking algorithm for medical images based on Gabor-DCT. Cyberspace Safety and Security: 12th International Symposium.

Yang, M., Li, J., & Bhatti, U. A. (2025). A novel convolution kernel-based robust watermarking scheme applied in medical image. *International Journal of Wireless and Mobile Computing*, *22*(3-4), 290–299.

Yu, M., Zhang, Y., & Bhatti, U. A. (2024). Research on Cross-Platform Motion Control System Based on gSOAP. In *7th International Conference on Pattern Recognition and Artificial Intelligence*.

Yuan, Y., Li, J., & Bhatti, U. A. (2024). *Robust Zero Watermarking Algorithm for Medical Images Based on STAR-TEBLID-DCT*. Innovation in Medicine and Healthcare.

Zeng, Y., Li, D., & Bhatti, U. A. (2025). Enhancing lung cancer diagnosis with data fusion and mobile edge computing using DenseNet and CNN. *IEEE Transactions on Biomedical Engineering*.

Zhang, B., Wu, Y., Zhao, B., Chanussot, J., Hong, D., Yao, J., & Gao, L. (2022). Progress and challenges in intelligent remote sensing satellite systems. *IEEE Journal of Selected Topics in Applied Earth Observations and Remote Sensing*, *15*, 1814–1822.

Zhang, H., & Bhatti, U. A. (2024). *Deep learning for remote sensing and climate monitoring*. Deep Learning for Earth Observation and Climate Monitoring.

Zhang, J., Li, Y., & Bhatti, U. A. (2024). Dynamic Segmentation Enhanced Asynchronous Non-linear Federated Learning for Fault Diagnosis Using Incremental Aggregation Mechanism. *IEEE Smart World Congress (SWC)*.

Zhang, K., Zhang, Y., & Bhatti, U. A. (2024). A domain knowledge graph construction method based on joint extraction of GlobalPointer. In *7th International Conference on Pattern Recognition and Artificial Intelligence*.

Zhang, L., Zeng, Q., & Bhatti, U. A. (2024). Recommendations for Healthcare: An Interpretable Approach Using Deep Learning. In *7th International Conference on Pattern Recognition and Artificial Intelligence*.

Zhang, M., Li, J., & Bhatti, U. A. (2024). Secure data storage and processing architectures for climate Internet of Things systems. In Deep Learning for Earth Observation and Climate Monitoring.

Zhang, Q., & Bhatti, U. A. (2025). Forecasting the carbon emissions in African countries using EMD-GCN. IEEE 6th International Conference on Pattern Recognition and Artificial Intelligence.

Zhang, Y., Li, D., & Bhatti, U. A. (2025). Practical PID Protocol of T–S Fuzzy Positive Multi-Agent Systems Under Attacks. *International Journal of Fuzzy Systems*.

Zhang, Y., Xu, B., & Bhatti, U. A. (2025). Optimizing co-simulation with the age of information in intelligent cyber physical systems. *Internet of Things : Engineering Cyber Physical Human Systems, 31*, 101550.

Zhang, Z., Zeng, H., & Bhatti, U. A. (2024). A novel myocarditis detection combining deep reinforcement learning and an improved differential evolution algorithm. *CAAI Transactions on Intelligence Technology.*

Zielinski, O., Busch, J. A., Cembella, A. D., Daly, K. L., Engelbrektsson, J., Hannides, A. K., & Schmidt, H. (2009). Detecting marine hazardous substances and organisms: Sensors for pollutants, toxins, and pathogens. *Ocean Science, 5*(3), 329–349.

Chapter 3
Analytics for Health Resilience:
Managing Chronic Diseases During Natural Disasters

Afsheen Sabir
https://orcid.org/0009-0000-8257-2755
Balochistan University of Information Technology, Engineering, and Management Sciences, Pakistan

Sibghat Ullah Bazai
https://orcid.org/0000-0003-3042-5977
Balochistan University of Information Technology, Engineering, and Management Sciences, Pakistan

Alamgir Naushad
https://orcid.org/0000-0001-7009-1751
National University of Sciences and Technology, Balochistan, Pakistan

ABSTRACT

This chapter examines how big data analytics transforms the management of chronic disease during natural disasters. It shows how live research from wearable gadgets, electronic health records (EHRs) and social forums can provide early warning, remote monitoring, and smart resource deployment. Example case studies from catastrophes across the world – the 2005 Hurricane Katrina and the Haiti earthquake – and present COVID-19 pandemic prove the value of AI, IoT, and telemedicine for continuity of care. Some key obstacles like data privacy, infrastructure constraints, interoperability are also reviewed. Ethical frameworks, capacity building and technology integration that build on disaster-resilient healthcare systems are

DOI: 10.4018/979-8-3373-3501-8.ch003

highlighted by the chapter. Eventually, it calls for stakeholder cooperation in order to develop equitable, data-based solutions for vulnerable populations in the event of an emergency.

INTRODUCTION

Chronic diseases also called NCDs (Non-communicable diseases) are one of the major causes of illness leading to death globally that can be managed effectively through continuous monitoring, timely interpositions, and a functional healthcare system. It is difficult for chronic patients to access proper medical resources during natural disasters e.g. earthquake, hurricanes, wildfires, and floods as healthcare systems are disrupted during these times. These disruptions may include hospital damage, medical supply chains cut off, and struggles of healthcare professionals to reach affected areas that put patients at high risk with severe complications intensifying critical public health challenges.

During disasters big data analytics has been a game changer for improving healthcare outcomes as a response to these challenges. Big data is defined as an enormous amount of data that can be structured and unstructured data gathered from several sources such as social media platforms, wearable devices, electronic health records (EHRs), remote monitoring systems, and mobile health applications (Alam et al., 2024). These data sources are integrated to identify high risk populations, allocate critical medical resources in disaster affected regions, and enable real-time tracking of patient conditions. Internet of things (IOT), Artificial intelligence (AI), and machine learning (ML) are some advanced computational techniques that can assist in predicting disease progression, enhancing patient health patterns analysis, and optimizing healthcare response efforts, (Nabeel et al., 2024). Moreover, HCr platforms that refer to health cloud records allow doctors to save and manage patients' data on cloud so that it can be accessed remotely in case of unavailability of physical infrastructures of hospitals because of some natural disaster like floods etc.

Big data analytics manages chronic diseases through predictive insights. Wearable technology provides real-time physiological data that is analyzed by predictive analytics tools for identifying early warning signs of disease complications, (Ghazanchaei et al., 2023). For example, during a disaster if blood sugar level in diabetic patient fluctuates abnormally then an alert is generated to healthcare providers for rapid interpositions. Likewise, AI models can be used for enabling efficient resource allocation by authorities based on forecasts such as hospital admissions prediction for cardiovascular patients based on environmental data like air pollution index.

Furthermore, medical supply chains optimization and resource distribution are two key roles played by big data analytics during disasters. Essential medications required in affected areas can be predicted by organizations by observing past disaster response data, (Lokmic-Tomkins et al., 2023). It is made sure by cloud-based platforms that chronic disease patients receive medication on time despite challenges. One of these platforms is OpenMRS (Open Medical Record Systems) that is an open-source platform used by medical professionals for recording and analyzing patients' data Moreover, it offers scalable and customizable solutions for healthcare data management. It enables emergency response time, real-time coordination between hospitals, and humanitarian organizations.

Despite many benefits of integrating big data analytics with healthcare settings in disaster affected areas, there are several challenges that are encountered. During emergencies health data collection and sharing must be done considering ethical and legal standards for addressing major issues such as data privacy and security. for ensuring that patients' rights are reserved two major platforms are considered including HIPAA (Health Insurance Portability and Accountability Act) that refers to a policy defined by United states for securing health data of people to be shared without their consent and it also provides health insurance in case of loss or change of job and GDPR (General Data Protection Regulation) that is the law of the European Union to ensure privacy of data for individuals. It manages people's data transparently ensuring security and privacy.

Interoperability issues create barriers to data exchange between different data platforms and healthcare systems (Garson, 2021) where HIE (Health Information Exchange) refers to the process of mobilizing healthcare data across different organizations electronically within a country or a region ensures continuity of healthcare activities by efficient data sharing. Some of the healthcare institutes still rely on traditional systems that are incompatible to models based on big data analytics thus limiting the efficient usage of real-time analytics. Moreover, real-time processing capability is critical as timely medical interventions can be missed because of delay in data analysis.

In order to gain full benefits of big data analytics in healthcare, especially during disasters there should be collaboration between healthcare providers, policy makers, and technology developers (Abdul et al., 2024). Healthcare resilience during disasters can be improved significantly through investment in AI-driven decision support systems, secure data-sharing frameworks, and telemedicine infrastructure. Moreover, it is suggested to develop clear ethical guidelines while dealing with patient data to build public trust in healthcare solutions that are data-driven.

In this chapter the role of big data analytics in chronic disease management during natural disasters is explored in detail including real world applications, challenges, key technologies, and recommendations for future that is illustrated in *Figure 1*. In times of crisis vulnerable populations can be safe if stakeholders understand critical data-driven strategies for healthcare resilience and design effective policies and solutions based on technology. Following sections discuss the impact of natural disasters on chronic disease management, case studies discussing successful healthcare solutions, and recommendations for optimizing healthcare solutions using big data analytics.

This chapter attempts to establish the significance of big data analytics in the management of chronic diseases under natural hazards in terms of optimizing the use of resources, enhancing delivery of healthcare by monitoring them in real-time, and caring more about most vulnerable patients, as well as the mitigating factors and issues, such as ethical considerations and limitations.

To lead this exploration, there will be the following central research question: What are the potential ways of efficiently leveraging big data analytics to enhance management of chronic disease in natural disaster setting, and mitigate the dilemmas of privacy, interoperability, infrastructure constraints, and in considerations of ethical issues?

Figure 1 Conceptual framework: Big data analytics for disaster-resilient chronic disease management

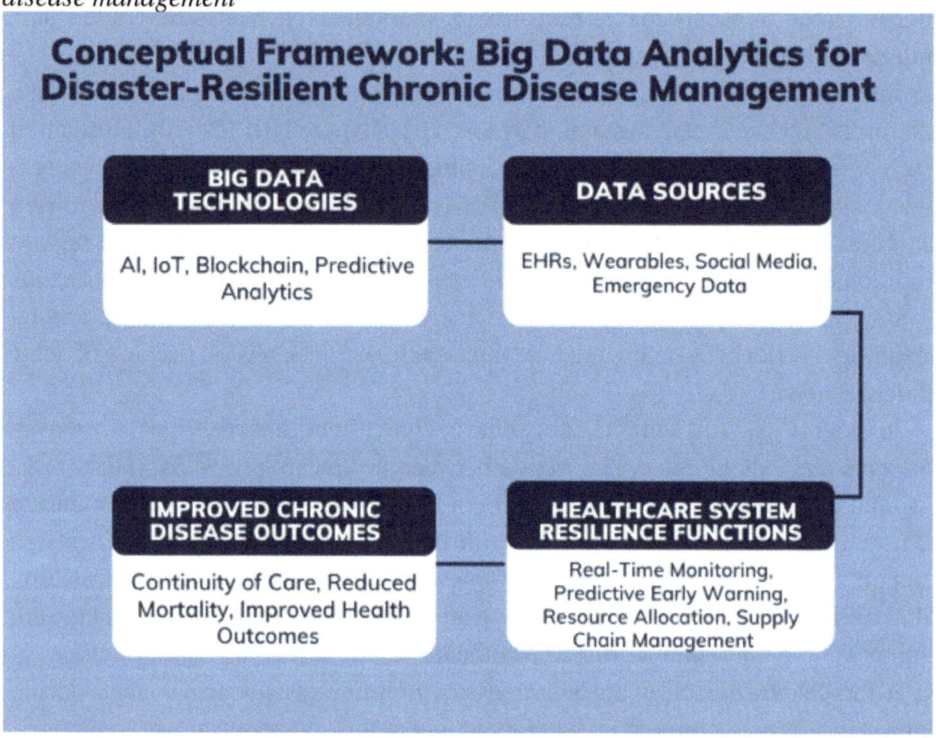

OVERVIEW OF CHRONIC DISEASE MANAGEMENT

Because these chronic diseases live in you for a long period of time or are very serious and difficult to manage, they are chronic diseases. Chronic illnesses are unlike acute illnesses, which can be treated and resolved very quickly but with chronic illnesses you have to monitor them, you adjust your lifestyle and you are required to go for regular medical treatments. This will help in the improvement of patient outcome, reduction in healthcare cost and improvement of overall quality of life.

Importance of Continuous Monitoring

Continuous monitoring is one of the essential points in chronic disease management. In timely interventions, regular tracking of vital health parameters like blood glucose levels for patients suffering from diabetes and heart rate variability for patients with cardiovascular disease becomes possible, (Ali & Thakkar, 2023). In these days of medical advancements and advancements in medical technology, wearable devices and mobile applications of health are now playing an important role in tracking and analysis of health data and any changes in the body's health in real time, (Amir et al., 2023). These tools enable the patients and healthcare providers to detect early warning signs and make early preventive measures before complications occur, (Haq et al., 2023).

Challenges in Chronic Diseases Management

In spite of the fact that continuous monitoring plays a significant role in the management of chronic diseases, chronic disease management has various challenges still confronting it. There are some outcomes that affect the management of chronic diseases regardless of the effective implementation of the modern healthcare technologies:

1. Limited Availability of Healthcare – There are very little or no healthcare facilities in remote areas or disaster-prone areas. And this accessibility gap can cause invites to diagnosis and treatment to be delayed.
2. Managing chronic diseases involves strict medication to be taken, dietary and lifestyle modification. Nevertheless, its compliance is difficult for many patients because of lack of awareness, financial limiting or psychological causes.
3. Chronic diseases are a burden to healthcare systems because many patients are repeatedly visiting hospitals, requiring long-term medication, and receiving specialized care.

4. Unfortunately, Data Fragmentation has not been addressed as the patient health records cannot flow smoothly across different platforms and healthcare professionals cannot have a panoramic view of a patient's medical history.
5. Chronic disease management involves high costs and its associated costs (medications, medical devices, routine checkups) can be overwhelming to the patients thus becoming a problem for financial reasons even in low income communities.

These challenges are addressed through a multidisciplinary approach involving healthcare policies, technological innovations as well as patient education. By applying big data analytics on large scale health data, the potential lies in transforming chronic disease management by assisting decision making, predicting what is to come in health, and allocating resources more effectively, (Albahri et al., 2024). In the subsequent sections, the implications of big data analytics on healthcare would be explored especially in situations such as healthcare systems being stressed during natural disasters.

IMPACT OF NATURAL DISASTERS ON HEALTHCARE SYSTEMS

However, natural disasters like earthquakes, hurricanes, floods and wildfires severely degrade healthcare systems, making it difficult to deliver medical care, staff resources, and maintain the status of patients. It is especially painful to those with chronic diseases that have an ointment requirement for regular maintenance, supplement or support in their life through medication and healthcare infrastructure.

Disruptions in Medical Care and Supply Chains

Widespread damage by natural disasters renders hospitals, clinics and pharmacies inoperable. Damage to transportation networks, healthcare providers' difficulty in delivering essential medical services owing to power outages and infrastructure collapse act as hindrances, (Teh & Rana, 2023). That disruption affects the availability of life saving treatments, dialysis for kidney disease patients and diabetes insulin and oxygen therapy for those with respiratory diseases.

Even more of a crisis is further supply chain breakdowns. The distribution of pharmaceuticals, medical devices and essential supplies into the clinic holds too many delays which leads to shortages putting the patient at risk. In disaster areas, where the roads are damaged owing to destruction caused due to disaster, emergency response systems fail to coordinate for the needful resources to reach at the right time.

Digital health records may also be unavailable if data centers are affected, rendering medical professionals' ability to see a patient's history and offer them proper care.

Increased Vulnerability of Chronic Disease Patients

Natural disasters are most dangerous for the patients with chronic diseases. Ongoing management of their condition becomes difficult when health care access is denied. Lack of mobility, power outages that play havoc with home medical devices, and interruption in regular medication supply could result in severe complications or even fatality, (Akindote et al., 2023). For instance, people with cardiovascular diseases might be at risk of heart attacks on account of stress and insufficient fast medical help, whereas patients with diabetes could have ill-regulated blood glucose levels on the account of lack of insulin.

Additionally, emergency shelters may not be set up for chronic disease management. These patients are overcrowded, have little medical personnel, and are exposed to a high number of infections which will make their health outcome even worse. Disasters commonly trigger or complicate conditions such as blood pressure and depression as well as psychological distress, also common following a disaster.

To address these challenges, it needs a good disaster preparedness strategy which makes use of big data analytics to monitor in real time, determine the risk of a disaster and allocate resources, (Fanelli et al., 2023). In the next section, big data analytics is found to improve disaster response and management of chronic disease during disasters.

ROLE OF BIG DATA ANALYTICS IN HEALTHCARE DURING DISASTERS

Predictive Analytics for Early Warnings

Big data analytics can offer early warnings, one of the most especially valuable contributions due to the occurrences within disaster scenarios. However, predictive analytics uses the historical disaster data, climate patterns, epidemiology models to foresee potential health crises. Big data analytics uses data from different sources including satellite imagery and social and emergency response records to analyze, thus helping governments and healthcare organizations preempt not only a disaster but also its effect (Khan et al., 2023). This capability is of high value in anticipating disease outbreak, timely evacuation plans and readiness of healthcare facilities to cope with a large number of patients.

Data-Driven Resource Allocation

In such disasters, healthcare resources typically become over stretched and it becomes critical to make good resource allocation. Big data analytics can keep track of real time medical supply levels, patient admissions, as well as emergency service availability. By looking at patterns of hospital admissions, demand for medicines and transportation disruption, providers in healthcare can make decisions on where to first shift resources. It makes sure that the more high-risk patients get their priority care and also help emergency responders get to the designated location where they are needed most. Such strategies can also reduce waste and minimize the possibility of shortage of essential medicines and equipment.

Machine Learning and AI Applications

Large amounts of healthcare data need to be analyzed in order to find trends and optimize disaster response and machine learning and artificial intelligence fit perfectly for this, (Yum & Adhikari, 2023). Electronic health records could be assessed by AI-powered systems to provide healthcare professionals with the ability to categorize folks as possible who need crisis intervention. Responding to physical barriers, chatbots and AI-based telemedicine platforms facilitate patients in disaster-hit areas to access medical guidance online. AI disease surveillance systems can also detect anomalies in health data, (Noor et al., 2023) and signal potential health outbreaks in order to take immediate measures to contain them.

SOURCES OF BIG DATA IN CHRONIC DISEASE MANAGEMENT

There are following major sources of big data also shown in *Figure 2*.

Electronic Health Records (EHRs)

Among the major contributors of big data in healthcare, Electronic Health Records (EHRs) are recognized as one of the most significant sources of big data in healthcare (Ogugua et al., 2024). A detailed history of patient admitting records is included in these records, such as medical diagnosis, treatment plan, and laboratory results, etc. EHRs enable healthcare providers to access critical patient information in times of natural disasters, so as to maintain continuity of care. Also, integrating the EHRs with the predictive analytics helps in finding at-risk patients for attention.

Wearable and Remote Monitoring Devices

Chronic disease management makes use of Wearable devices and remote technologies of monitoring. Continuous real time tracking of patient conditions can be achieved through smartwatches, glucose monitors and blood pressure sensors.

These devices in disaster situations are important because they help medical professionals to focus their efforts to care for the relevant patients(Kuh & Shlomo, 2004). However, data from wearable devices could also provide telemedicine services so that patients get proper medical advice in time regardless of barriers such as geographical distance.

Social Media and Emergency Response Data

Big data analytics is helped by social media platforms to provide real time information about disaster affected populations. Social media posts give a picture of what our health care organizations need now, and can respond to those needs by allocating their resources most efficiently. Medical helpline call logs and ambulance dispatch records are filled with emergency response data that authorities can refer to when assessing the gravity of the situation and the needed disaster relief efforts.

Figure 2. Major sources of big data

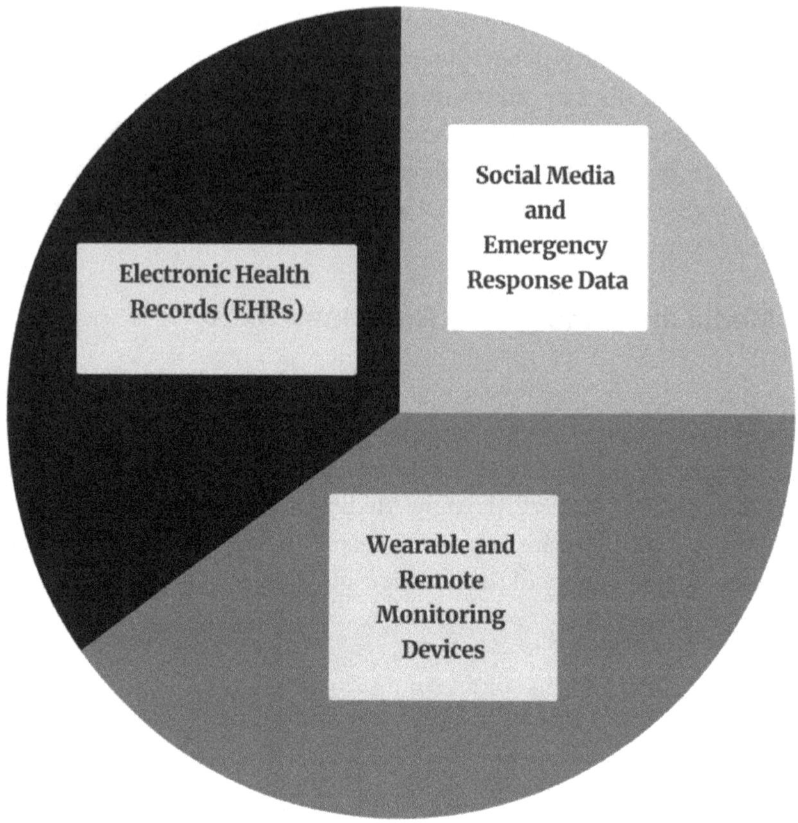

BIG DATA TECHNOLOGIES FOR DISASTER-RESILIENT HEALTHCARE

Scalable Data Storage and Access on Cloud Computing

With disasters, healthcare data management is very important using cloud computing. Therefore, traditional on-premise data storage systems can fail if there were physical damage, power outages or network failures. With cloud computing, patient record, medical history, and real time monitoring data are always available and available using a scalable and reliable solution, (Kumar et al., 2023). Hospitals and disaster response teams can securely share and retrieve information, and use it to coordinate emergency efforts in the cloud. For example, papers records in hospitals were lost in access during Hurricane Katrina and this affected patient care significantly. As of today, cloud computing takes away all such risks by providing remote access to all critical health data.

Additionally, cloud computing makes it possible for hospitals and emergency services teams to work together through interoperability of healthcare systems, even public health organizations. Data at rest (encrypted and protected by robust security protocols such as encryption and multi-factor authentication) is secure no matter what emergency occurs, (Zaland et al., 2021).

Artificial Intelligence and Machine Learning for Data Processing

The adoption of the Artificial Intelligence (AI) and Machine Learning (ML) trend in disaster resilient healthcare is crucial for the following reason: AI and ML empower handling huge amounts of data very quickly enabling it to crunch through data and discovering the patterns leading to the existence of new errors or safety risks, as well as deriving predictive models of the errors or safety risks that this data points to(Snezhana, 2023). AI powered algorithms can use a huge set of data drawn from a variety of sources including administrative databases of health claims or administrative databases for payments, administrative wage tables of offices, wearables or social media to identify emerging health threats.

AI can for instance forecast pattern to chronic disease management or spread of infectious disease in case of a disaster. In the period of the pandemic of the COVID-19, AI-based analytics made possible the prediction of the outbreak and efficient allocation of medical resources, (Haider, 2024). For natural disasters cases, AI can go through monstrous amounts of data in order to know which areas are to be attended to with priority, which patients that have to get emergency aid, as well as how to best spread resources, (Murala et al., 2023).

In addition, ML models become more efficient over time, making upcoming disaster responses more fruitful. Healthcare organizations can better manage treatments for chronic diseases in disaster situations if they can train AI based on previous disaster data.

Blockchain for Secure and Transparent Health Data Management

In such a disaster, data security and integrity are critical and blockchain technology is quickly becoming a game changer in healthcare. It is a decentralized tamper proof system that allows a patient record management that can't be lost or altered.

The most significant advantage about blockchain is that it facilitates the sharing of secure and seamless data amongst multiple stakeholders (Igwama et al., 2024). In disaster-stricken areas hospitals, NGO and government agencies are engaged to provide medical aid for the general public. Blockchain provides everyone having access to the same, accurate patient data and reliably blocking out any unauthorized modifications.

Moreover, you have transparency in medical supply chains using blockchain. Counterfeit or expired medications are a significant problem in disaster operations. Healthcare providers can be able to verify the authenticity of drugs and medical supplies using blockchain based tracking systems, thereby making sure that the patients do not receive unsafe or even ineffective treatments.

Internet of Things (IoT) for Real-Time Monitoring

The Internet of Things (IoT) is making disaster resilient healthcare real time and by enabling the real time monitoring of patients, supplies etc. IoT connects with the healthcare devices that continue to collect information, offering great results and insight for the professionals in healthcare.

In natural disasters, the IoT devices allow the tracking of patients suffering from chronic diseases which ensure they get timely intervention. For instance, if a diabetic patient's blood sugar levels go dangerously low or a heart patient's heart rhythm is abnormal, it is either this or a doctor may learn about it remotely via remote monitoring devices designed for the purpose, (Rane, 2023). This real time alerts allows the medical teams to respond immediately thereby eliminating any complications.

In addition, it manages healthcare logistics. Ambulance and medical warehouses are equipped with smart sensors to verify the correct temperature of life saving drugs and products, as well as methods to adopt, optimize and accelerate the transport of medical equipment. IoT can help healthcare providers continue on providing care to the patients even when the area is in the middle of disaster.

Big Data Analytics for Disaster Preparedness and Response

Healthcare systems use big data among other ways to analyze historical data, identify risk factors and develop proactive strategies in disaster preparedness. Big data is used by governments and public health agencies to create disaster response models that simulate different scenarios with the intent of being able to plan for numerous contingencies (Kurniawan et al., 2024).

For example, if the application of predictive analytics is to predict the most vulnerable regions to natural disasters and the high-risk people, from the point of view of natural disasters — for instance elderly people with chronic diseases. This information helps authorities in prioritizing resources and preparing to handle the situation by setting up emergency medical camps in high potential areas.

Furthermore, big data analytics is used for assessing responsiveness to a disaster. Using patient outcomes, resource utilization and emergency response timelines, healthcare organizations are able to determine where shortcomings exist and where strategy has not been effective in the past.

ETHICAL AND LEGAL IMPLICATIONS OF USING BIG DATA DURING DISASTERS

While increasing use of big data analytics in chronic disease management especially during disasters warrants consideration of a host of ethical and legal considerations associated with big data analytics, with regards to natural disasters, the historic disparity between the use of big data analytics by public officials and medical professionals heightens its significance. In the emergency scenarios however, these challenges only hamper the need for fast data use, patient consent and privacy, making these challenges even more confusing.

Data privacy is one of the main ethical issues. During a disaster, acute health systems do not always have the time to access this sensitive personal data to support vulnerable chronic disease patients; they need to access medical history, location, biometric data from wearables, behavioral data from social media, etc., all quickly. Nevertheless, such a fast utilization of data should pass the usual safeguards of privacy. For example, when healthcare providers or emergency respondents' access to Electronic Health Records (EHRs) or access to the real time data from wearable devices, this creates a conflict between wellbeing of the public and respect to the individual privacy rights.

In most cases the patients simply don't know how their data is being used. This brings up the issue of informed consent. While some jurisdictions permit relaxed rules under emergency scenarios, it is still crucial to make sure that these patients are notified — be it beforehand or retrospectively — as to how the data pertaining to these patients was utilized. Emergency driven exemptions should have clear legal frameworks, and should be documented as part of transparency.

An additional concern is data ownership and control. During the times of crisis, who actually owns health data — the patient, the government or the private healthcare platforms? In many countries though, there is still an unanswered question on this, leaving it liable for legal ambiguity. If there are multinational tech companies that are answering itself as platforms for telemedicine, data storage, or AI analysis, national privacy laws are getting enforced under a different jurisdiction.

The second is that data bias and discrimination are pressing ethical issues. Existing datasets are used by big data systems, and they may already be biased. Taking an urban data driven predictive analytics system trained on urban data alone as an example, rural and low-income areas are just as likely as any urban area to experience higher disaster vulnerabilities and the model might underperform in these areas. It can misallocate the resources and even people who are at risk.

The matter of accountability from a legal perspective notes another aspect. Imagine a scenario where a high-risk chronic disease patient is not identified by a machine learning model during a flood period and hence suffering severe consequences, who is to be blamed: the software developer, the healthcare provider or the government? There are more and more requirements on the legal determination of liability when AI driven healthcare decisions(U. A. Bhatti, Hashmi, et al., 2023).

Use of big data in a disaster needs to be guided by ethical principles such as beneficence, non-maleficence, autonomy and justice. A benefit is maximization, while nonmaleficence is the avoidance of harm. Consent is one of the most emphasized on autonomy, while justice prefers fair access to resources and technologies, (Goundar & Rayani, 2021). A big ethical challenge involves ensuring these principles happen in the face of urgency and uncertainty.

Some other recommendations may be taken in order to address these issues including:

- Data governance policies that strike the balance between the need for development of disaster-specific emergency data governance policies and privacy.
- In the case where anonymization is not possible or useful, try not to use identity as much as possible.
- Creation of ethics committees to oversee data use in crises.
- Education of ethical and legal framework to healthcare professionals and data scientists.
- Legal protocols for cross border data use between countries that are international in collaboration.

Finally, big data analytics for managing chronic diseases in the context of natural disasters has a great potential, but it should be implemented on the basis of strong ethical principles. Righting these legal and privacy problems at the outset serves to build public trust and guard against such technologies being misused and unfairly used, even in the darkest of days.

CASE STUDIES AND REAL-WORLD APPLICATIONS

In all parts of the globe, big data analytics has been critical to healthcare management during natural disasters. Emergency response teams and their healthcare organizations and government agencies have used large amounts of data to compute to predict disease outbreaks, efficiently allocate resources, and improve patient outcomes. This section looks at the successful and unsuccessful uses of big data for managing chronic diseases in disaster.

Hurricane Katrina (2005): The Role of Data in Managing Chronic Disease Patients

Disastrous Hurricane Katrina was the very real natural disaster responsible for displacing millions and destroying healthcare services. All of this meant a major challenge in continuing to provide health care for chronic disease patients such as diabetics, those with heart disease or respiratory conditions.

How Big Data Helped

- After Hurricane Katrina, the U.S. government realized the need of electronic health records for disaster preparedness and imported Electronic Health Records (EHRs) and Remote Access for that purpose. To help in such a case, big data analytics helped produce digital copies of patient records that doctors and even hospitals could use as backups if physical records were lost. For chronic disease patients, who needed medication history, allergy records or treatment plans, this was very useful.
- Public health agencies applied big data analytics to predict disease outbreaks in shelters, e.g., infections, and other health complications. They looked at past disaster trends, current health reports and were able to identify those at risk and where medical teams should be deployed.

Challenges Faced

- In many hospitals, there were different systems that were not interoperable, therefore, sharing data with each other was difficult.

- Lack of proper infrastructure in the affected areas did slow down the data collection and analysis.

The Haiti Earthquake (2010): Using Big Data for Remote Monitoring and Response

The devastating 7.0 magnitude earthquake in Haiti seriously impaired the healthcare system and hit patients who rely on regular medical supervision hard. Faced with an inadequate infrastructure to provide proper healthcare to the people, the organizations utilized big data analytics to keep track of patients' health online and to effectively allocate resources.

How Big Data Helped

- Organizations such as the World Health Organization (WHO) and the Red Cross used mobile health applications to track patient conditions using the application. Mobile apps that collected chronic disease patients' health data and transmitted them to central databases were tracked for chronic disease patients, including those with diabetes and hypertension, (Han et al., 2018).
- Prediction of Disease Prevention: Where disease prevention predictions were made using satellite imagery, patient records, and social media trends. This meant that medical resources were quickly deployed and mortality rates amongst vulnerable people were lowered.

Challenges Faced

- It was difficult to share real time data due to lack of stable internet connectivity.
- This use of patient data in emergency situations raised ethical concerns related to without explicit consent.

The COVID-19 Pandemic: The Power of Big Data in Chronic Disease Management

The COVID-19 pandemic was a global crisis that also demonstrated its potential in the sector of healthcare brought to light, (Nawaz et al., 2021). Innovative solutions needed to be found for the health management of millions of chronic disease patients at high risk of severe complications while becoming isolated in lockdowns with overwhelmed hospitals.

How Big Data Helped

- Large Datasets and Predictive Models: The predictive models analyzed large datasets to identify high-risk patients so they could be placed at the front of medical care lines.
- Big Data Analytics on Social Media: Governments and health organizations mined big data on social media sites to spot, monitor and neutralize misinformation as well as to predict outbreaks of mental health and to locate areas that require the most healthcare resources(Aamir et al., 2023).
- Remote Monitoring Devices: Smartwatches and glucose monitors were devices that healthcare providers could use to monitor vital signs and could then sound the alarm in the event of a decline in the patient's health before turning into a critical situation.

Challenges Faced

- As governments and institutions gathered a great volume of personal health data, data privacy concerns rose.
- A digital divide existed, poor communities and other underprivileged communities lacked access to digital health solutions.

Indian Floods (2018): Big Data for Emergency Response and Chronic Disease Care

In 2018, during the severe flooding in Kerala, India, million were left stranded, without their medications and healthcare access.

How Big Data Helped

- Application: Machine Learning for Resource Allocation: AI-based systems studied the patterns of flood and population density to find out the most affected areas so that the medical resources could be delivered to those people in need timely.
- Geo-Tracking for Patient Location: Mobile tracking and geo-mapping were done in mobile to locate stranded patients which required urgent medical assistance.
- EHR Integration: Cloud based HCr helped doctors continue in the treatment of patients when the local hospitals were flooded.

Challenges Faced

- The rural areas were not well covered with the internet making the digital interventions less effective.
- One big hurdle had to do with integration of data from various healthcare providers.

The Role of Big Data in Wildfire Disasters (California, 2020)

Wildfires in California have become more common and have serious health risks including to people with chronic respiratory disease such as asthma and chronic obstructive pulmonary disease (COPD).

How Big Data Helped

- Predictive Analytics for Air Quality Monitoring: Using satellite data and sensor readings, big data models were used to predict air quality degradation and provided at-risk people a chance to safety precaution prior to events.
- Virtual Assistants Offering Real Time Health Advice: AI Based Chatbot to provide real time health guidance impacting the sufferers due to smoke inhalation, directing what to do, according to the severity of health condition when one is affected by smoke inhalation, great for individuals with less exposure to medication, and information of any sort(Huang et al., 2018).
- Telemedicine platforms helped chronic disease patients to consult doctors remotely compared them to the necessity of visiting the hospitals in hazardous conditions.

Challenges Faced

- Because of power outages, real time data gathering was adversely affected.
- Better coordination was urgently needed between government agencies and health care providers.

Lessons from These Case Studies

- Big data Saves Lives: Big data analytics turns out to be a game changer in speed of response time in emergencies, better allocation of resources and continuity of care for chronic disease patients across different natural disasters.

- Seamless Data Sharing: Case studies show that it is extremely critical to seamless data sharing among the healthcare organizations, government agencies and humanitarian groups.
- Despite that, big data remains a considerable challenge: Technology Access Remains a Challenge, especially in developing countries and rural areas(Nawaz et al., 2022).
- Emergencies Require Urgent Intervention While Privacy and Ethical Issues Must Be Addressed: The fact of the matter is that in the case of emergencies, the urgency of collecting and using health data requires balance against ethical concerns.

INTEGRATION OF EMERGING TECHNOLOGIES

The management of healthcare in disasters during blogs changed with the emergence of such technologies as artificial intelligence (AI), the Internet of Things (IoT), blockchain, and advanced predictive analytics.

Artificial Intelligence (AI) for Predictive Analytics and Early Warnings

AI models can absorb huge volumes of data in order to establish patterns and foresee future public health problems. Future advancements should focus on: How AI can help to improvc the Disaster Prediction Models: The AI can be used for analyzing the climate pattern, hospital data, social media trends to predict the disease outbreak after natural disasters(Iftikhar et al., 2023) (U. A. Bhatti et al., 2019). Virtual assistants can give real time health advice to patients of chronic disease, reducing the burden on healthcare professionals. Machine learning algorithms: Patient symptoms can be analyzed by machine learning algorithms and a suitable treatment can be recommended even in remote areas(U. A. Bhatti et al., 2022).

Internet of Things (IoT) for Remote Health Monitoring

They can support continuous patient health data collection and transmission with low effort. Future developments should include:

AI Systems must include Advanced smart watches, glucose monitors, heart rate tracker and they should be integrated into wearable health sensors to send real time health alerts(U. A. Bhatti et al., 2018). Early warning signs: IoT devices can certainly give some great early warning signs to healthcare providers or emergency services(Singh et al., 2024). IoT Power ambulances: These ambulances are connected to be able to send patient information to the hospital before arrival to improve their treatment efficiency.

Blockchain for Secure Health Data Management

During disasters, blockchain technology can improve the security as well as the transparency of health data. Key applications include:

Secure and Tamper-Proof Health Records

The use of blockchain ensures that the data regarding patients remains safe and not tampered during use by a variety of providers in health.

Data Storage

The data that a particular patient is stored can be done in a decentralized way to protect records even if hospital servers are compromised, (Feng et al., 2022).

Resource Allocation through Smart Contracts on Blockchain

Blockchain can be used to enable smart agreements for delivering of medical supplies smartly based on the live analysis.

5G and High-Speed Connectivity for Telemedicine Expansion

Real time health monitoring and telemedicine consultations need fast and reliable internet connection. Future advancements should focus on: High speed internet Network ensures that even in case of disaster telemedicine services do not become inaccessible (Avalon-Cullen et al., 2023). Improving Real-time Transmission of Data: In this way, 5G will make remote healthcare much more effective by improving video consultations quality, (Hamza et al.,2023). Faster networks improve connectivity for IoT devices so that there are no more delays in data transmission as a result of points 'hopping around' in the network(U. Bhatti et al., 2023).

COMPARATIVE ANALYSIS: TRADITIONAL VS DATA-DRIVEN CHRONIC DISEASE MANAGEMENT IN CRISES

Despite constant challenges, managing chronic diseases such as diabetes, cardiovascular ailments, asthma and kidney disorders have always been cumbersome during natural disasters. The disaster response frameworks have always opted for manual procedures, paper records, as well as basic forms of communication(Topbaş et al., 2021). On the contrary, the modern ways now more and more use the big data analytics and the smart technologies to make these more precise and proactive interventions(U. A. Bhatti, Masud, et al., 2023). This section compares the traditional and data driven models and its advantages of incorporating advanced data tools in Emergency healthcare delivery through *Table 1*.

Table 1. Traditional vs data-driven chronic disease management

Chronic Disease Management	**Traditional**	**Data-Driven**
Decision-Making Processes	Decisions were taken on the basis of experience, with basic triage principles, or with a limited patient record. In situations of disasters, patient information was seldom accessible, resulting in the general ways of patient care.	Predictive analytics and real time dashboards are healthcare provider's data driven multipliers to making decisions. For instance, AI can notify teams when a patient with a chronic obstructive pulmonary disease (COPD) goes into respiratory decline in order to respond more quickly. Predictive models also prognosticate patient surges and placement of resources based on a need.
Patient Monitoring and Continuity of Care	Evacuation, separation from care facilities and lost prescriptions or repeated antibiotic use increased mortality among patients with chronic illnesses. Usually, there was no way of tracing patients' history or continued treatment due to manual recordkeeping.	Remote Patient Monitoring using Wearable devices and Cloud based EHRs: This data driven model provides easy accessibility of latest data to the healthcare providers, irrespective of his/her location. It maintains continuity of care between devices and locations through syncing data across devices and during displacement or power outage.
Resource Allocation and Supply Chain Management	Medicine and equipment distribution during disasters was traditionally slow, reactive and uneven, because of a guidance on real time visibility on the required inventory and demand.	Real time analytics tools use real time data such as medical supply level, forecast shortages and optimize delivery routes. In a crisis, machine learning algorithms can run algorithms to better simulate different distribution strategies to supply crucial critical supplies to the most vulnerable regions.
Risk Identification and Population Targeting	Risk Identification-Local knowledge, manual surveys, and obsolete registry. Sometimes there was a lack of data or data was incomplete; they were sometimes overlooked and could not receive the same level of morbidity or mortality data as there was available.	Big data of EHRs, geolocation services and census databases can be used by public health officials to create data driven model for placing the high-risk populations. Moreover, additional support of emerging hotspot identification and population movements comes from social media analysis and mobile phone data.

continued on following page

Table 1. Continued

Chronic Disease Management	Traditional	Data-Driven
Communication and Coordination	Agencies and responders communicated via phones or in person meetings, with the telephone increasing to conversations that could at times be delayed or disrupted due to collapsed infrastructure.	Integrated communication systems, real-time alerts and shared platforms Coordination. Emergency Management Platforms offer the ability for all stakeholders (including government agencies, hospital administrators and humanitarian response), who are affected by an emergency event, to have centralized dashboards(Mayer-Schönberger & Cukier, 2013).
Health Outcomes and Response Speed	The lack of sufficient personalization, delayed response and misallocation of resources, make the health outcomes during disasters, particularly for patients with NCDs (non-communicable diseases), are often poor.	An enhanced surveillance, timely interventions and personalized care plans configuration have led to improved measurables outcomes. When predictive analytics are used during crises, studies show that reductions in hospital readmissions and mortality rates are to be expected.
Documentation and Post-Disaster Evaluation	Post crisis Evaluations: These relied on manual reports or anecdotal evidence and it was therefore difficult to assess effectiveness and replicate successful interventions, despite being traditional.	The collection of data logs, automated reports, and metrics allows for complete review of the work and the work that did not. Such insights serve as the driving forces to always improve disaster preparedness and healthcare policy(Minelli et al., 2012).
Limitations of Both Models	Prevents creation of good Model: Inexperienced developers cannot maintain precision, and cannot be sure whether your Model even makes sense at first glance. Document poorly, which means the Model will never manage to meet your expectations. Care fragmented, information is not utilized effectively.	Infrastructure, internet access, a skilled workforce and may not work well efficiently in such a low resource or remote area.

Blending the Two Models

Given that in many disaster scenarios, especially in low- and middle-income countries, a hybrid approach may be necessary, it is important to recognize this. Consisting of local knowledge, community health workers, traditional outreach and the smart data tools, the strategy for including response is adaptive, scalable and inclusive.

CHALLENGES AND LIMITATIONS OF USING BIG DATA IN DISASTER HEALTHCARE

However, despite the great potential of big data analytics for chronic disease management in natural disasters, a number of challenges can hinder its achievement. Data privacy concern, interoperability issue, infrastructure limitation and ethical considerations are such challenges faced in the use of radio frequency identification.

Data Privacy and Security Concerns

The biggest challenge in using big data for healthcare in cases of emergencies is keeping patient privacy and data security. EHRs and wearable device data that contain sensitive health information have to be protected from accidental as well as cyber threats, (Bazai et al., 2019). Protecting the confidentiality of a patient's medical information during a natural disaster is often as much of an emergency as there needs to be for patient data if not for primary care data, (Bazai et al., 2021), and exposing it to viral infections is another risk of infection unless the proper security measures have been put in place, (Stawicki et al., 2021).

Cloud-based storage and sharing of data platforms are liable to hacking and that can result in patient information handed onto malicious actors(Han et al., 2024). Furthermore, unlike encryption, blockchain technology is secure; however, it immensely requires computational power and infrastructures to be implemented efficiently, hence not available during a disaster.

Therefore, to avoid these risks, healthcare organizations should follow strict encryption routines, use of a multi-factor authentication and comply with global health data regulations like HIPAA (Health Insurance Portability and Accountability Act) and GDPR (General Data Protection Regulation).

Interoperability Issues Between Healthcare Systems

It is possible to change interoperability to the ability of different healthcare systems and devices to interact with and make data changes with ease. During disasters, a major problem is that hospitals, emergency response agencies, and government agencies all tend to use diverse software applications that may be non-compatible, (Bazai & Jang-Jaccard, 2020).

For instance, if a patient is transferred from one hospital to another in case of a flood or an earthquake, their medical records will not be available to be seen without access to the information from the two hospitals having different data handling systems, (Bhatti et al., 2024). The lack of integration can delay the treatment and cause a medical error.

In light of this challenge, it is necessary to develop standardized health information exchange (HIE) protocols that enable diverse healthcare institutions to provide and obtain the data on their patients quickly and efficiently, (Noor et al., 2024). A cloud-based set of interoperability platforms and APIs (Application Programming Interfaces) can make data integration a smoother process, so that medical people can get the patient record at any location.

Infrastructure Limitations in Disaster-Affected Areas

Power supply, internet connectivity and data storage facilities of big data analytics can severely be affected by natural disasters. Unfortunately, many disasters struck regions, and worse, developing countries do not have technological infrastructure to process and analyze large healthcare during.

For example, during the 2010 Haiti earthquake, many of the healthcare facilities collapsed and medical teams were deprived of the records of patients(Bazai et al., 2017). Likewise, in remote flood-affected regions, the internet connectivity may be disrupted thus preventing the real time data transmission from wearable devices and remote monitoring systems.

To defeat these challenges, governments and healthcare companies should allot investment in resilient digital infrastructure like the internet through satellite, mobile health (mHealth) solutions, and portable energy sources. Decentralized data centers and backup systems can also be established so that continuity of healthcare services can be ensured during crises.

CONCLUSION

The increasing number of big data analytics in chronic disease management is due to increased frequency and severity of natural disasters across the globe, especially those caused by climate change. Therefore, by strengthening healthcare infrastructure, promulgating ethical policies and implementing advanced technologies, the healthcare system could be made more resilient and able to provide uninterrupted care to chronic disease patients during an emergency.

A game changer if the incorporation of big data analytics into chronic disease management during a natural disaster time. The need to collect, analyze, and act on real time health data as disasters continue to disrupt almost every community across the globe has become very essential. This chapter looks at a range of features of big data in disaster resilient healthcare, including how big data can (and can't) impact healthcare systems, predict and make early warnings, allocate health data and resources, and present real world case studies of successful interventions.

One of the most pressing needs in disasters is to ensure continuity of care to chronic disease patients. There is an inherent risk of disruptions in medical supply chains and healthcare infrastructure damage along with lack of access to healthcare providers. Big data analytics tackle these challenges with its provision of real time decision making, and also in remote patient monitoring and resource distribution. Healthcare providers use predictive analytics to know when disease outbreaks are coming, and when they are on the increase in medical demand. Machine learning and artificial intelligence (AI) invents new ways of diagnosing and monitoring chronic diseases even in the times of disaster.

The chapter also identified the different sources of big data in healthcare e.g. electronic health records (EHRs), wearable devices, social trends in media, and emergency response data. The data from these sources will give a broad view of a patient's health status and will help health care professionals in the field of personalized care even during tough times. Additionally, blockchain, Internet of things (IoT), high speed connectivity like 5G can go a long way to increase the security, efficiency as well as reach of healthcare service in a natural disaster.

However, big data analytics have their own set of challenges in implementation. It’s also thought that ethical issues of patient privacy, data security and algorithmic bias need to be thoroughly considered. The responsibility of big data should be standardized data governance policies and have stronger regulations to protect patient’s rights. Furthermore, an equitable approach to access to digital health technologies, especially in the low income and disaster-prone regions, is essential to the whole inclusive and a reliable healthcare system.

Thus, across all regions, health care infrastructure would need to be strengthened, policy reforms progressed and emerging technologies integrated, in order to strengthen disaster preparedness and response. To maximize the potential of big data in managing chronic diseases, governments, healthcare institutions, technology developers and policymakers have to work together on creating a robust framework. Investments in telemedicine, diagnoses and real-time health monitoring systems using AI will make it easier for patients and reduce the number of disruptions to medical care during emergencies.

Finally, big data analytics is the future of healthcare, especially in minimizing the impact to chronic patients of the devastation caused by natural disasters. It involves extracting meaning from data and using it in health systems in order to provide quick interventions to patients, ameliorate health outcomes, and make health systems stronger. However, as more and more research is done, innovation and big data analytics are evolving disaster resilient healthcare and saving lives and improving global health security.

FUTURE RESEARCH DIRECTIONS AND RECOMMENDATIONS

The evaluation of the capability of big data analytics for the management of chronic diseases during natural disasters has been already implemented. There are however a few things that need to be improved to fully unleash its potential. This section discusses future directions and recommendations of improving the healthcare infrastructure, transitioning policy changes, ethics of taking care of chronic diseases in disasters, and incorporating new technologies for better chronic disease management in disasters.

Strengthening Healthcare Infrastructure

For natural disasters, it is crucial that big data analytics be used effectively and this is possible only when a resilient healthcare infrastructure exists(Simon, 2013). To strengthen this architecture, it is about data collection systems, digital healthcare facilities, and robust networks that can share information.

Expanding Digital Health Infrastructure

In order to reap all the benefits big data offers, governments and healthcare organizations have to build on expanding digital health systems including:

- *Cloud Technology:* The advantage of Cloud Technology is that if physical infrastructure is damaged, medical records are still available. Cloud based systems also provide healthcare professionals with the ability to do real time updates and get them from remote places.
- *Strengthening Telemedicine and Remote Monitoring:*

In the midst of disruption of healthcare facilities, telemedicine and wearable devices have become critical in the management of chronic diseases. Strengthening this aspect includes:

Government and private organizations should invest more in expanding availability of telehealth platforms to people in the natural disaster prone areas.

Future of Improving Remote Monitoring Technology: Devices such as blood glucose, heart rate and oxygen tracking should be integrated into a single healthcare system that will enable health care providers to monitor patients in real time.

Implementing AI based health assistants to enable the virtual assistants to provide personalized recommendations to the chronic disease patients in times of disasters.

- *Building Disaster-Resilient Health Facilities:* Therefore, hospitals and health care centers must be designed to withstand the natural disasters. Recommendations include:
- *Implement Big Data Tech Medical Care in the Disaster Area:* Deployable health units with big data technology can create a health unit for disaster area mobile health unit.

Having alternatives to digital health services can provide Healthcare Facilities with securing backup power sources that prevent disruptions.

- *Making Supply Chain Management stronger:* Big data can predict demand and keep stocks and transfer of essential medical supplies including insulin, oxygen tanks and medicine used for cardiovascular disease among others in order.

Policy Changes and Ethical Considerations

With increasingly emerging data in the health industry, the government and the policymakers must set certain rules for ethical data collection; privacy protection, and proper use of patient data. To ensure that the Patients have control on how the data is being gathered, stored and used, even in emergencies. Sensitive medical information in your computer should be protected and not accessible by anyone in the public except the person named. The data comprising sensitive medical information must be stored securely to keep it away from potential access and cyber-attacks. Governments need to develop policies that mandate the same data governance framework for all healthcare service providers in order to handle their patient's data.

Ethical Use of AI and Machine Learning in Disaster Management

The ethical consideration is coming up because AI is getting more involved helping with healthcare decisions making. The use of AI should be set according to guidelines in order to ensure it is used responsibly. To avoid Algorithmic Bias, they have to train their AI models on diverse datasets in order to avoid such biases that might impact some people disproportionately. Patients and healthcare providers should be clear about how AI driven recommendations are made. Periodic Auditing of the AI Systems: The AI tools that are employed in the critical analysis of the disasters should go through continuous surveillance processes to validate the veracity and the fairness of their functionality.

Addressing Health Inequality in Disaster Response

Different communities are not all accessing digital healthcare solutions. Policymakers should focus on:

Closing Digital Division

Governments should financially support underprivileged portions with digital health technology.

Special Consideration

Must be given to elderly patients, the disabled and the people living in remote areas in relation to healthcare access enhancement for vulnerable groups.

Data Sharing Agreements

Data sharing between countries can help in faster global disaster response and ensure that the Best practices worldwide.

REFERENCES

Aamir, M., Bazai, S., Bhatti, U. A., Dayo, Z. A., Liu, J., & Zhang, K. (2023). Applications of machine learning in medicine: Current trends and prospects. In *2023 Global Conference on Wireless and Optical Technologies (GCWOT)* (pp. 1-4). IEEE.

Aamir, M., Rahman, Z., Abro, W. A., Bhatti, U. A., Dayo, Z. A., & Ishfaq, M. (2023). Brain tumor classification utilizing deep features derived from high-quality regions in MRI images. *Biomedical Signal Processing and Control*, *85*, 104988.

Abdul, S., Adeghe, E. P., Adegoke, B. O., Adegoke, A. A., & Udedeh, E. H. (2024). AI-enhanced healthcare management during natural disasters: Conceptual insights. *Engineering Science & Technology Journal*, *5*(5), 1794–1816.

Akindote, O. J., Adegbite, A. O., Dawodu, S. O., Omotosho, A., Anyanwu, A., & Maduka, C. P. (2023). Comparative review of big data analytics and GIS in healthcare decision-making. *World Journal of Advanced Research and Reviews*, *20*(3), 1293–1302.

Alam, M. A., Sohel, A., Uddin, M. M., & Siddiki, A. (2024). Big Data and Chronic Disease Management Through Patient Monitoring And Treatment With Data Analytics. Academic Journal on Artificial Intelligence, Machine Learning. *Data Science and Management Information Systems*, *1*(01), 77–94.

Albahri, A. S., Khaleel, Y. L., Habeeb, M. A., Ismael, R. D., Hameed, Q. A., Deveci, M., Homod, R. Z., Albahri, O. S., Alamoodi, A. H., & Alzubaidi, L. (2024). A systematic review of trustworthy artificial intelligence applications in natural disasters. *Computers & Electrical Engineering*, *118*, 109409.

Ali, A. H., & Thakkar, R. (2023). Climate changes through data science: Understanding and mitigating environmental crisis. *Mesopotamian Journal of Big Data*, *2023*, 125–137.

Avalon-Cullen, C., Caudill, C., Newlands, N. K., & Enenkel, M. (2023). Big data, small island: Earth observations for improving flood and landslide risk assessment in Jamaica. *Geosciences*, *13*(3), 64.

Bazai, S. U., & Jang-Jaccard, J. (2019). SparkDA: RDD-based high-performance data anonymization technique for Spark platform. In International conference on network and system security (pp. 646-662). Cham: Springer International Publishing.

Bazai, S. U., & Jang-Jaccard, J. (2020). In-memory data anonymization using scalable and high performance rdd design. *Electronics (Basel)*, *9*(10), 1732.

Bazai, S. U., Jang-Jaccard, J., & Alavizadeh, H. (2021). A novel hybrid approach for multi-dimensional data anonymization for apache spark. *ACM Transactions on Privacy and Security*, *25*(1), 1–25.

Bazai, S. U., Jang-Jaccard, J., & Wang, R. (2017). Anonymizing k-NN classification on MapReduce. *International Conference on Mobile Networks and Management*, 364–377.

Bhatti, U., Bazai, S., Hussain, S., Fakhar, S., Ku, C., Marjan, S., Por, Y., & Jing, L. (2023). Deep learning-based trees disease recognition and classification using hyperspectral data. *Computers, Materials & Continua*, *77*, 681–697.

Bhatti, U. A., Hashmi, M. Z., Sun, Y., Masud, M., & Nizamani, M. M. (2023). Artificial intelligence applications in reduction of carbon emissions: Step towards sustainable environment. [). Frontiers Media SA.]. *Frontiers in Environmental Science*, *11*, 1183620.

Bhatti, U. A., Huang, M., Wang, H., Zhang, Y., Mehmood, A., & Di, W. (2018). Recommendation system for immunization coverage and monitoring. *Human Vaccines & Immunotherapeutics*, *14*(1), 165–171. PMID: 29068748

Bhatti, U. A., Huang, M., Wu, D., Zhang, Y., Mehmood, A., & Han, H. (2019). Recommendation system using feature extraction and pattern recognition in clinical care systems. *Enterprise Information Systems*, *13*(3), 329–351.

Bhatti, U. A., Masud, M., Bazai, S. U., & Tang, H. (2023). Investigating AI-based smart precision agriculture techniques. [). Frontiers Media SA.]. *Frontiers in Plant Science*, *14*, 1237783. PMID: 37496855

Bhatti, U. A., Mengxing, H., Li, J., Bazai, S. U., & Aamir, M. (2024). *Deep learning for multimedia processing applications: Volume two: Signal processing and pattern recognition*. CRC Press.

Bhatti, U. A., Zeeshan, Z., Nizamani, M. M., Bazai, S., Yu, Z., & Yuan, L. (2022). Assessing the change of ambient air quality patterns in Jiangsu Province of China pre-to post-COVID-19. *Chemosphere*, *288*, 132569. PMID: 34655644

Fanelli, S., Pratici, L., Salvatore, F. P., Donelli, C. C., & Zangrandi, A. (2023). Big data analysis for decision-making processes: Challenges and opportunities for the management of health-care organizations. *Management Research Review*, *46*(3), 369–389.

Feng, S., Liu, Q., Patel, A., Bazai, S. U., Jin, C., Kim, J. S., Sarrafzadeh, M., Azzollini, D., Yeoh, J., & Kim, E. (2022). Automated pneumothorax triaging in chest X-rays in the New Zealand population using deep-learning algorithms. *Journal of Medical Imaging and Radiation Oncology*, *66*(8), 1035–1043. PMID: 35224858

Garson, G. D. (2021). *Data analytics for the social sciences: Applications in R.* Routledge.

Ghazanchaei, E., Allahbakhshi, K., Khorasani-Zavareh, D., Aghazadeh-Attari, J., & Mohebbi, I. (2023). Challenges in providing care for patients with chronic diseases during disasters: A qualitative study with focus on diabetes and chronic respiratory diseases in Iran. *Tanaffos*, *22*(1), 83. PMID: 37920319

Goundar, S., & Rayani, P. K. (2021). *Applications of Big Data in Large-and Small-Scale Systems. IGI Global.Kuh, D., & Shlomo, Y. Ben. (2004). A life course approach to chronic disease epidemiology (Issue 2).* Oxford university press.

Haider, A., Hussain, L., Tareen, A. W., Bazai, S. U., Aslam, S., Neo, M., & Amphawan, A. (2024). A comparative study of machine learning techniques for accurate disease prediction using symptom-based diagnosis. In AIP Conference Proceedings (Vol. 3153, No. 1). AIP Publishing.

Hamza, M., Bazai, S. U., Ghafoor, M. I., Ullah, S., Akram, S., & Khan, M. S. (2023). Generative adversarial networks (gans) video framework: A systematic literature review. In *2023 International Conference on Energy, Power, Environment, Control, and Computing (ICEPECC)* (pp. 1-5). IEEE.

Han, H., Bazai, S. U., Bhatti, M. A., Basit, A., Wahid, A., Bhatti, U. A., Ghadi, Y. Y., & Algarni, A. (2024). Hybrid Climate Forecasting: Variational Mode Decomposition and Convolutional Neural Network with Long-Term Short Memory. *Polish Journal of Environmental Studies*, *33*(2).

Han, H., Huang, M., Zhang, Y., & Bhatti, U. A. (2018). An architecture of secure health information storage system based on blockchain technology. Cloud Computing and Security: 4th International Conference, ICCCS 2018, Haikou, China, June 8-10, 2018, Revised Selected Papers, Part II 4, 578–588.

Haq, S. U., Bazai, S. U., Fatima, A., Marjan, S., Yang, J., Yee, L., & Ku, C. S. (2023). Reseek-arrhythmia: Empirical evaluation of ResNet architecture for detection of arrhythmia. *Diagnostics (Basel)*, *13*(18), 2867. PMID: 37761234

Huang, M., Han, H., Wang, H., Li, L., Zhang, Y., & Bhatti, U. A. (2018). A clinical decision support framework for heterogeneous data sources. *IEEE Journal of Biomedical and Health Informatics*, *22*(6), 1824–1833. PMID: 29994279

Iftikhar, A., Purvis, L., Giannoccaro, I., & Wang, Y. (2023). The impact of supply chain complexities on supply chain resilience: The mediating effect of big data analytics. *Production Planning and Control, 34*(16), 1562–1582.

Igwama, G. T., Olaboye, J. A., Maha, C. C., Ajegbile, M. D., & Abdul, S. (2024). Big data analytics for epidemic forecasting: Policy Frameworks and technical approaches. *International Journal of Applied Research in Social Sciences, 6*(7), 1449–1460.

Khan, S. M., Shafi, I., Butt, W. H., Diez, I. de la T., Flores, M. A. L., Galán, J. C., & Ashraf, I. (2023). A systematic review of disaster management systems: Approaches, challenges, and future directions. *Land (Basel), 12*(8), 1514.

Kumar, A., Jain, R., Gupta, M., & Islam, S. M. N. (2023). *6G-enabled IoT and AI for smart healthcare: challenges, impact, and analysis.* CRC Press.

Kurniawan, M. H., Handiyani, H., Nuraini, T., Hariyati, R. T. S., & Sutrisno, S. (2024). A systematic review of artificial intelligence-powered (AI-powered) chatbot intervention for managing chronic illness. *Annals of Medicine, 56*(1), 2302980. PMID: 38466897

Lokmic-Tomkins, Z., Bhandari, D., Bain, C., Borda, A., Kariotis, T. C., & Reser, D. (2023). Lessons learned from natural disasters around digital health technologies and delivering quality healthcare. *International Journal of Environmental Research and Public Health, 20*(5), 4542. PMID: 36901559

Mayer-Schönberger, V., & Cukier, K. (2013). *Big data: A revolution that will transform how we live, work, and think.* Houghton Mifflin Harcourt.

Minelli, M., Chambers, M., & Dhiraj, A. (2012). *Big data, big analytics: emerging business intelligence and analytic trends for today's businesses.* John Wiley & Sons.

Murala, D. K., Panda, S. K., & Dash, S. P. (2023). MedMetaverse: Medical care of chronic disease patients and managing data using artificial intelligence, blockchain, and wearable devices state-of-the-art methodology. *IEEE Access : Practical Innovations, Open Solutions, 11*, 138954–138985.

Nabeel, S. M., Bazai, S. U., Alasbali, N., Liu, Y., Ghafoor, M. I., Khan, R., & Por, L. Y. (2024). Optimizing lung cancer classification through hyperparameter tuning. *Digital Health, 10*, 20552076241249661. PMID: 38698834

Nawaz, S. A., Li, J., Bhatti, U. A., Bazai, S. U., Zafar, A., Bhatti, M. A., Mehmood, A., & Ain, Q. ul, & Shoukat, M. U. (2021). A hybrid approach to forecast the COVID-19 epidemic trend. *PLoS One, 16*(10), e0256971. PMID: 34606503

Nawaz, S. A., Li, J., Bhatti, U. A., Shoukat, M. U., & Ahmad, R. M. (2022). AI-based object detection latest trends in remote sensing, multimedia and agriculture applications. *Frontiers in Plant Science*, *13*, 1041514. PMID: 37082514

Noor, R., Wahid, A., Bazai, S. U., Khan, A., Fang, M., Bhatti, U. A., & Ghadi, Y. Y. (2024). DLGAN: Undersampled MRI reconstruction using deep learning based generative adversarial network. *Biomedical Signal Processing and Control*, *93*, 106218.

Noor, S., Bazai, S. U., Ghafoor, M. I., Marjan, S., Akram, S., & Ali, F. (2023). Generative adversarial networks for anomaly detection: a systematic literature review. In 2023 4th International Conference on Computing, Mathematics and Engineering Technologies (iCoMET) (pp. 1-6). IEEE.

Ogugua, J. O., Onwumere, C., Arowoogun, J. O., Anyanwu, E. C., Odilibe, I. P., & Akomolafe, O. (2024). Data science in public health: A review of predictive analytics for disease control in the USA and Africa. *World Journal of Advanced Research and Reviews*, *21*(1), 2753–2769.

Rane, N. (2023). Integrating leading-edge artificial intelligence (AI), internet of things (IOT), and big data technologies for smart and sustainable architecture, engineering and construction (AEC) industry: Challenges and future directions. Engineering and Construction (AEC) Industry: Challenges and Future Directions (September 24, 2023).

Simon, P. (2013). *Too big to ignore: the business case for big data* (Vol. 72). John Wiley & Sons.

Singh, S., Sharma, P., Pal, N., Sarma, D. K., Tiwari, R., & Kumar, M. (2024). Holistic one health surveillance framework: Synergizing environmental, animal, and human determinants for enhanced infectious disease management. *ACS Infectious Diseases*, *10*(3), 808–826. PMID: 38415654

Snezhana, D. (2023). Applying artificial intelligence (AI) for mitigation climate change consequences of the natural disasters. Dineva, S.(2023). Applying Artificial Intelligence (AI) for Mitigation Climate Change Consequences of the Natural Disasters. *Research Journal of Ecology and Environmental Sciences*, *3*(1), 1–8.

Stawicki, S. P., Firstenberg, M. S., Galwankar, S. C., Izurieta, R., & Papadimos, T. (2021). *Contemporary Developments and Perspectives in International Health Security* (Vol. 1). BoD–Books on Demand.

Teh, D., & Rana, T. (2023). The use of Internet of Things, Big Data analytics and artificial intelligence for attaining UN's SDGs. In *Handbook of big data and analytics in accounting and auditing* (pp. 235–253). Springer.

Topbaş, A., Jamil, A., Hameed, A. A., Ali, S. M., Bazai, S., & Shah, S. A. (2021). Sentiment analysis for covid-19 tweets using recurrent neural network (rnn) and bidirectional encoder representations (bert) models. 2021 International Conference on Computing, Electronic and Electrical Engineering (ICE Cube), 1–6.

Yum, S.-G., & Das Adhikari, M. (2023). Suitable site selection for the development of solar based smart hydrogen energy plant in the Gangwon-do region, South Korea using big data: A geospatial approach. *International Journal of Hydrogen Energy*, *48*(93), 36295–36313.

Zaland, Z., Bazai, S. U., Marjan, S., & Ashraf, M. (2021). Three-tier password security algorithm for online databases. In 2021 2nd International Informatics and Software Engineering Conference (IISEC) (pp. 1-6). IEEE.

Chapter 4

Applications of Remote Sensing in Coastal Zone Land Cover Mapping: Challenges, Advances, and Future Directions

Huang Mengxing
Hainan University, China

Mughair Aslam Bhatti
https://orcid.org/0009-0000-8039-0173
Department of Robotics and Aritificial Intelligence, SZABIST University, Karachi, China

Mohammad Asif
Hunan University, China

Gongsong Peng
https://orcid.org/0009-0001-6597-1997
Hainan University, China

Yangmeng Yao
Hainan University, China

ABSTRACT

Studying land cover through remote sensing technologies has become essential for coastal zone management since these approaches let users make important environmental observations regarding habitat degeneration while monitoring ecosystem changes. The paper investigates current developments together with implementa-

DOI: 10.4018/979-8-3373-3501-8.ch004

tion areas and remote sensing obstacles in coastal zone land cover cartography. The evaluation reviews diverse remote sensing systems which begin with satellite systems then advance to flying sensors and aerial LiDAR and uncrewed aerial vehicles showing their tracking functionality for shoreline identification and their uses in monitoring vegetation and analyzing wetland conditions and urban coastal development. This paper examines classification algorithms with detection methods for changes alongside a discussion of data fusion techniques as well as evaluation of machine learning technologies and deep learning approaches to enhance map classification qualities.

INTRODUCTION

All across the globe coastal areas remain the most vibrant position of diverse ecological habitat which exists on the planet. Multiple species choose their critical habitats among these geographical areas where economic operations take place alongside healthy ecosystem conditions. The zones hosting ecological sensitivity persist vulnerable to climate change alongside pollution issues excessive fishing and rapid urban developments. Scientists have established these areas as hazardous to such actions. Caring for both coastal zone management and protection demands urgent attention from all areas. The strategic development requires an understanding of present Coastal changes to formulate sustainable defensive methods for vulnerable ecosystems.

Remote sensing technologies function as advanced coastal zone monitoring instruments through their capability to deliver extensive recent data about extensive geographic regions. Remote sensing reveals information about coastal features using different sensors that contain optical and radar data as well as LiDAR technology and hyperspectral imaging instrumentation. Scientists can track minimal environmental changes in coastal areas by using this technology for long-term repetitive observations.

Tracking the coast together with visualizing land cover serves as the necessary fundamental foundation. Overlapping systems for land cover mapping provide vital statistics about the placement of different ecosystems as well as measurements of urban expansion along with coastal environment assessments. Remote sensing tools are essential instruments for observing modifications such as shoreline movements as well as environmental deterioration effects and climate change impacts. Remote sensing systems generate high-resolution geographic information which utility serves policy advisors during their environmental conservation and sustainable coastal zone management implementations.

Remote sensing technology advancements open new possibilities for its uses to create increased effectiveness in coastal zone management. Higher accuracy and precision exist in coastal land cover studies as scientists combine specialized spaceborne observation systems with unmanned aerial vehicles and artificial intelligence frameworks. The detection capabilities of circular innovations have surpassed traditional limits to analyze coastal changes at previously impossible scales leading to advanced environmental protection possibilities against rising threats. This paper investigates remote sensing technology for coastal zone land cover mapping by analyzing current obstacles as well as research guidelines for this essential domain.

REMOTE SENSING TECHNOLOGIES FOR COASTAL ZONE MAPPING

Coastal areas throughout the planet remain the most active and diverse ecological areas worldwide. The primary habitats for many species exist in coastal areas and these zones additionally support the economy and preserve planetary life systems. Geological regions experience a growing set of perils stemming from polluting agents coupled with altering climate conditions and excessive fishing activities and urban population advancements. Coastal protection together with efficient zone management need to become urgent priorities. Strategic development requires full comprehension of modern changes in these regions to develop enduring protective systems for vulnerable ecosystems.

The combination of remote sensing tools serves as an advanced instrument for coastal zone monitoring because they provide detailed up-to-date information about extensive geographic areas. Primitive sensing systems help establish coastal measurements from multiple instrument types working together with equipment for optical and radar and LiDAR instruments alongside hyperspectral imaging systems. Through this technology scientists can perform continuous monitoring of coastal regions while identifying small environmental changes that develop through time periods.

The tracking procedure together with coastal land cover visualization stands as an essential fundamental feature for this endeavor. Maps which precisely classify land surfaces generate information related to habitat distribution together with urban area expansion statistics and coastal ecosystem health status. Remote sensing tools serve as basic monitoring devices to identify shoreline alterations and habitat demise simultaneously with climate change alterations. Remote sensing capabilities built with sophisticated technology provide public officials complete maps to create pollution-free coastal conservation strategies as well as sustainable management frameworks.

The development of remote sensing technology enables coastal zone projects to implement diverse solutions that create more comprehensive effects in their applications. New technological developments consisting of satellite imaging drones and machine learning algorithms produce better precision levels during coastal land cover assessment procedures. The modernized systems provide detailed coastal observation data allowing for better proactive actions to repair endangered ecosystems threatened by progressive threats. The research discusses remote sensing applications for coastal zone land cover mapping by assessing existing limitations and establishing research suggestions in this essential field.

Figure 1. Applications of remote sensing in met-ocean studies and climate monitoring

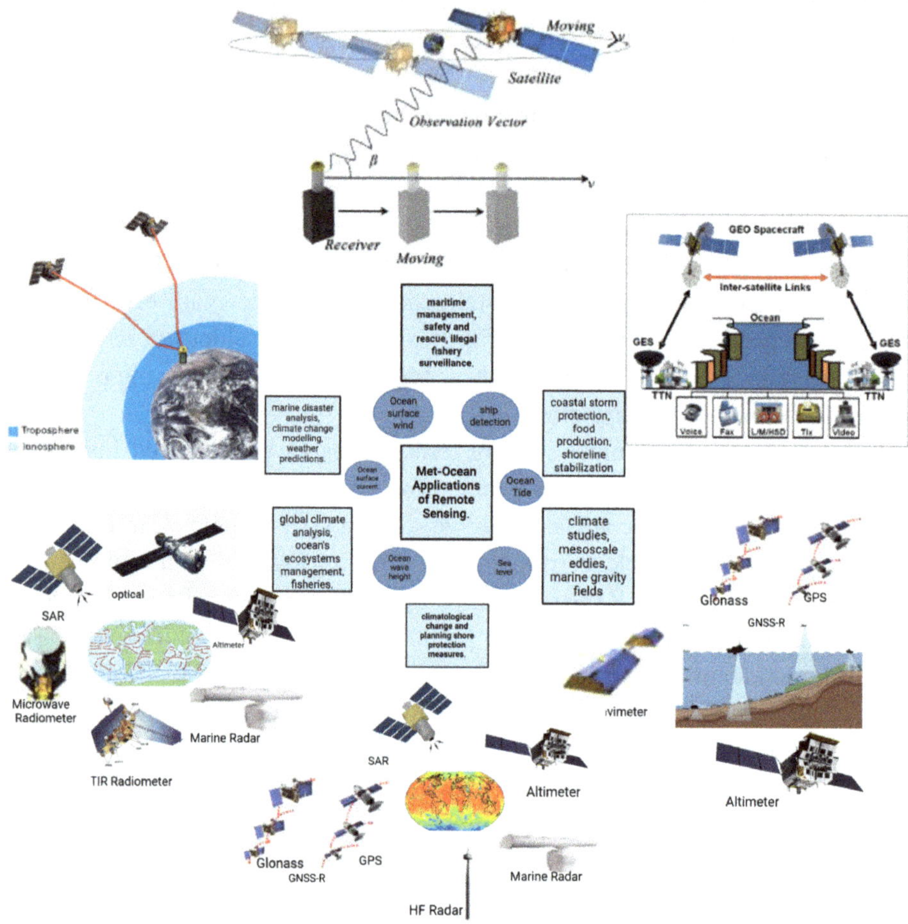

SAR plays a vital role in coastal zone mapping through remote sensing by providing capabilities beyond optical and LiDAR technologies according to Joyce et al. (2014). Synthetic aperture radar detects data in all weather conditions and throughout daytime and night hours making it the perfect choice for maintaining coastal area surveillance wherever weather patterns include clouds, (Brown, Hambidge, & Brownett, 2016). SAR proves most effective when identifying coastal erosion and following water body displacements and evaluating flood situations. Surface characteristics of coastal zones change detectably through radar backscatter analysis using SAR imagery allowing scientists to track landform activities which include sediment movements and shoreline positions alongside storm surge effects, (Lapetina & Sheng, 2015). Coastal zone monitoring benefits extensively from the joint application of SAR technology with optical and LiDAR data sources, (Sun et al., 2024).

LAND COVER MAPPING TECHNIQUES

Scientists require land cover mapping techniques for coastal zones because they need both detailed spatial knowledge about different land surfaces and tracking their long-term changes in distribution, (Mitsuda & Ito, 2011). The main technique in land cover mapping utilizes image classification because analysts assign satellite and aerial image pixels to predefined categories by analyzing their spectral characteristics, (Khatami, Mountrakis, & Stehman, 2016). Analytical selection of training data takes place under supervised classification methods for identifying known land cover types, (Pal & Mather, 2003). The models undergo training to learn image patterns that allow the identification of features leading to classification of water, vegetation, urban areas and bare land categories. Various land-cover classification algorithms include Support Vector Machines (SVM), Random Forests and Maximum Likelihood based on the data complexity level along with resolution, (Otukei & Blaschke, 2010; Hansen et al., 2000).

The process of unsupervised classification operates without training data so pixels are grouped through spectral similarity examinations. The classification technique provides useful results when investigators have restricted knowledge about the study site or interact with large areas, (Rittle-Johnson, Star, & Durkin, 2009). The data analysis makes use of clustering algorithms that incorporate k-means or ISODATA. The final product generated from unsupervised classification exists as groups of clusters that represent various land cover categories. Additional post-processing steps using manual labeling or validation become necessary to guarantee the accuracy and relevance of the produced map after running this method, (Li, Liu, & Pfeifer, 2019).

Another complex approach for coastal land cover mapping is Object-Based Image Analysis (OBIA). The analysis method known as Object-Based Image Analysis (OBIA) incorporates spectral data along with spatial data elements of objects found inside images, (Ouchra, Belangour, & Erraissi, 2022). The methodology divides images into significant objects including vegetation patterns and coastal elements and after that assigns categorizations to these elements according to characteristics such as size and texture and form, (Li & Wan, 2015). Mapping complex coastal areas succeeds best with OBIA because this method shows superior results compared to standard pixel-based operations in environments with mixed pixel mixing such as coastal wetlands and urbanized shorelines and estuarine regions, (Lee et al., 2006). Spatial contextual analysis through OBIA produces superior classification results in environment domains where coastal features are highly dynamic, (Sreekesh, Kaur, & Sreerama, 2020).

The adoption of deep learning technologies has improved coastal land cover mapping through Convolutional Neural Networks (CNNs) specifically. CNNs advance image understanding since they perform similar operations to the human visual system to autonomously acquire hierarchical patterns directly from raw image data. The application of CNNs to high-resolution remote sensing images leads to powerful performance because they succeed at recognizing complicated objects alongside noise and hidden objects. The application of deep learning techniques proves successful for coastal zone classification at different scales because they can distinguish vegetation features along with detecting complicated coastal shapes, (Belcore et al., 2024). Deep learning models excel at combining several data sources such as optical images and SAR and LiDAR data which enhances both mapping precision and final product detail level. The combination of these processing requirements with substantial labeled datasets does not diminish their value as future coastal land cover mapping tools because they can recognize intricate patterns, (Datta & Deb, 2012).

APPLICATIONS OF REMOTE SENSING IN COASTAL LAND COVER MAPPING

Another complex approach for coastal land cover mapping is Object-Based Image Analysis (OBIA). The analysis method known as Object-Based Image Analysis (OBIA) incorporates spectral data along with spatial data elements of objects found inside images, (Ouchra, Belangour, & Erraaissi, 2022). The methodology divides images into significant objects including vegetation patterns and coastal elements and after that assigns categorizations to these elements according to characteristics such as size and texture and form, (Li & Wan, 2015). Mapping complex coastal areas succeeds best with OBIA because this method shows superior results compared to standard pixel-based operations in environments with mixed pixel mixing such as coastal wetlands and urbanized shorelines and estuarine regions, (Lee et al., 2006). Spatial contextual analysis through OBIA produces superior classification results in environment domains where coastal features are highly dynamic, (Sreekesh, Kaur, & Sreerama, 2020).

Deep learning research has increased land cover mapping accuracy in coastal regions because Convolutional Neural Networks (CNNs) became widely used for this purpose. CNNs advance image understanding since they perform similar operations to the human visual system to autonomously acquire hierarchical patterns directly from raw image data. The application of CNNs to high-resolution remote sensing images leads to powerful performance because they succeed at recognizing complicated objects alongside noise and hidden objects. The ability of deep learning algorithms has proven capable of precise coastal zone identification with their achievement at both fine vegetation identification and advanced coastal landform detection, (Belcore et al., 2024). Deep learning models implement the integration of optical and SAR and LiDAR datasets which enhances classification precision resulting in advanced land cover maps. Future land cover mapping operations in coastal areas will benefit from these techniques because they solve complex patterns yet demand large computational power alongside extensive labeled datasets, (Klemas, 2001).

Figure 2. Framework for complex coastal surface interpretation and land use classification

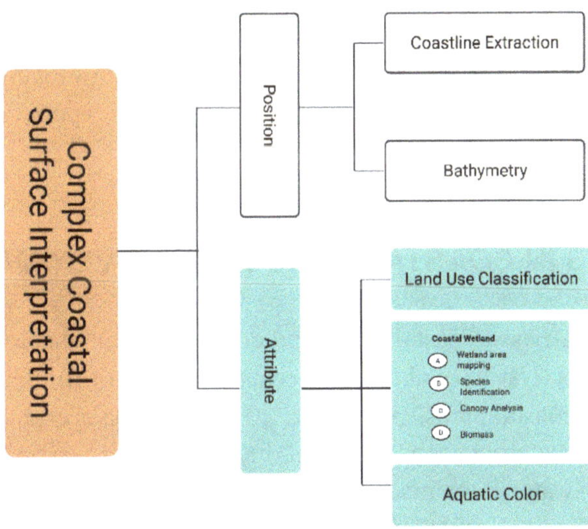

Another complex approach for coastal land cover mapping is Object-Based Image Analysis (OBIA). The analysis method known as Object-Based Image Analysis

(OBIA) incorporates spectral data along with spatial data elements of objects found inside images, (Ouchra, Belangour, & Erraissi, 2022). The methodology divides images into significant objects including vegetation patterns and coastal elements and after that assigns categorizations to these elements according to characteristics such as size and texture and form, (Li & Wan, 2015). Mapping complex coastal areas succeeds best with OBIA because this method shows superior results compared to standard pixel-based operations in environments with mixed pixel mixing such as coastal wetlands and urbanized shorelines and estuarine regions, (Lee et al., 2006). Spatial contextual analysis through OBIA produces superior classification results in environment domains where coastal features are highly dynamic, (Sreekesh, Kaur, & Sreerama, 2020).

Deep learning technology has improved coastal zone land cover mapping because scientists now employ Convolutional Neural Networks (CNNs) for this purpose. CNNs advance image understanding since they perform similar operations to the human visual system to autonomously acquire hierarchical patterns directly from raw image data. The application of CNNs to high-resolution remote sensing images leads to powerful performance because they succeed at recognizing complicated objects alongside noise and hidden objects. Studies have demonstrated the effectiveness of deep learning approaches in identifying coastal areas through their applications in small-scale vegetation detection as well as intricate terrain recognition, (Belcore et al., 2024). Multi-source data integration using deep learning models allows them to process optical and SAR and LiDAR information which boosts classification performance and generates detailed land cover maps. Complex pattern processing capabilities of these methods make them a beneficial technology for upcoming coastal land cover mapping operations although they need vast computational processing power and large labeled datasets, (Klemas, 2001).

CHALLENGES IN COASTAL ZONE LAND COVER MAPPING

Remote sensing-based coastal zone land cover mapping encounters numerous important obstacles (Nawaz et al., 2023) because of the hard-to-predict and mobile characteristics of coastal areas, (Liu et al., 2019). The main obstacle arises from coastal land cover's heterogeneous nature that includes mixed characteristics between land features with water elements such as beaches and wetlands and mangroves and urban areas, (Huang et al., 2018). Different environments cause classification problems through spectral interactions between land cover types that produce blended pixel data. Accurate classification and distinction between neighboring land cover classes becomes complex when vegetation meets water bodies closely or complex landforms exist in the area, (Aamir et al., 2021; Galvan et al., 2022; Li et al., 2022).

Environmental factors together with atmospheric disturbances generate serious obstacles for mapping coastal land surface features. The combined effects of coastal fog and common cloud cover and unstable atmospheric conditions reduce the quality of optical satellite images in such areas, (Hasnain et al., 2022). Though synthetic aperture radar (SAR) and LiDAR technologies endure these conditions, synthetic remote sensing continues being fundamental in coastal research for vegetation monitoring and land cover detection. Dense clouds across satellite images create difficulties for researchers who try to monitor extended alterations in coastal environments because the data's reliability decreases, (Bhatti et al., 2021).

The main hurdle when combining remote sensing data sources is the fact that their resolutions do not match up well, (Fan et al., 2021). Federally grounded imagery from aerial systems combined with UAV data provides enhanced details about smaller coastal regions although satellite-platform observations edge superior in extending full-area coverage, (Gao et al., 2022). The different resolutions in spatial, spectral and temporal domains create integration obstacles for creating detailed coastal maps across extended regions and multi-time analyses. The implementation of data fusion methods addresses this issue although dataset calibration and alignment demands time-consuming and complex procedures, (Wu et al., 2024).

Remote sensing data limitations for coastal land cover mapping stem from their high purchase cost and limited accessibility that mainly affects researchers and practitioners with scarce resources in various regions, (Bhatti et al., 2023; Wu et al., 2022). The free data provided by satellite systems including Landsat and Sentinel does not stop many institutions from having to pay exorbitant costs for high-resolution commercial satellite imagery according to Aamir et al. (2019). The implementation of LiDAR systems alongside UAV data collection methods requires sophisticated equipment together with knowledgeable operators which limits accessibility for general coastal management agencies, (Fang et al., 2023). The processing requirements along with data analysis for remote sensing creates significant obstacles for wide-scale implementation because these tasks need complex software and computing equipment which remains out of reach in some coastal regions, (Nawaz et al., 2021). The successful implementation of remote sensing technology for coastal zone management depends on advancing remote sensing systems as well as providing economical access to data and designing intuitive tools for coastal zone administrators, (Zhang & Bhatti, 2024).

RECENT ADVANCES IN REMOTE SENSING FOR COASTAL LAND COVER MAPPING

Remote sensing technologies have recently achieved major progress which enhances coastal land cover mapping capabilities, (Li, Bhatti, & Zhang, 2022). Convolutional Neural Networks (CNNs) join machine learning algorithms in a significant development because they prove highly effective for automated classification tasks, (Sheng, Zhang, & Bhatti, 2023). The analytical techniques can extract hierarchical components directly from unprocessed remote sensing data so the process becomes more precise and time-efficient for mapping complex coastal systems such as wetlands and mangroves and urban environments, (Wu, Tang, & Bhatti, 2024). Small datasets no longer impede deep learning models from operating effectively due to their processing capability of substantial datasets which benefits classification detection in heterogeneous coastal regions that traditional techniques fail to manage, (Xu, He, & Bhatti, 2023).

Coastal land cover mapping receives improved benefits from satellite and airborne platforms through recent advancements in resolution capabilities, (Long et al., 2023). The WorldView or Pleiades satellites provide high-resolution imagery which permits precise observation of subtle coastal features since such precision is vital for tracking sensitive environmental areas such as coral reefs or seagrass beds, (Alhatami et al., 2024). Advanced hyperspectral imaging allows scientists to make increasingly precise spectral measurements between different forms of land cover thus enhancing their ability to separate features with similar appearances such as distinct vegetations or soils, (Peng, Liu, & Bhatti, 2023). Better coastal land cover representation emerges from these advancements thus facilitating precise management decisions and improved conservation activities.

Remote sensing technologies have recently achieved major progress which enhances coastal land cover mapping capabilities, (Liu, Bhatti, & Zhang, 2022). Convolutional Neural Networks (CNNs) join machine learning algorithms in a significant development because they prove highly effective for automated classification tasks, (Sheng, Zhang, & Bhatti, 2023). The analytical techniques can extract hierarchical components directly from unprocessed remote sensing data so the process becomes more precise and time-efficient for mapping complex coastal systems such as wetlands and mangroves and urban environments, (Wu, Tang, & Bhatti, 2024). Large datasets enable deep learning models to manage mixed pixels because they improve classification precision specifically in heterogeneous coastal regions resistant to traditional mapping approaches, (Xu, He, & Bhatti, 2023).

Coastal land cover mapping receives improved benefits from satellite and airborne platforms through recent advancements in resolution capabilities, (Long et al., 2023). The combination of WorldView and Pleiades satellite imagery provides high-resolution assessments of coastal zones because it reveals small details about sensitive areas such as seagrass beds and coral reefs, (Alhatami et al., 2024). Advanced hyperspectral imaging allows scientists to make increasingly precise spectral measurements between different forms of land cover thus enhancing their ability to separate features with similar appearances such as distinct vegetations or soils, (Peng, Liu, & Bhatti, 2023). Better management and conservation efforts receive enhanced support because these developments enable precise coastal land cover representation.

Figure 3. Land surface reflectance detected by satellite

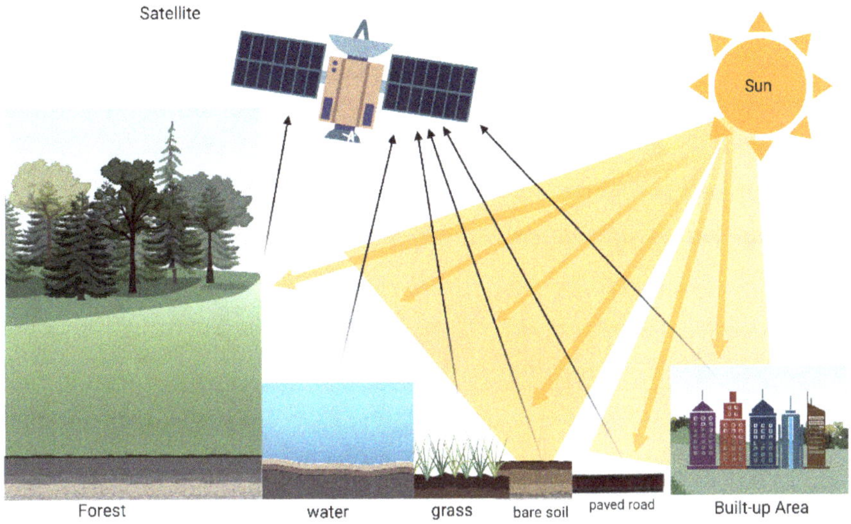

FUTURE DIRECTIONS IN REMOTE SENSING FOR COASTAL ZONE MAPPING

Remote sensing technology for coastal zone mapping will experience substantial progress because of recent developments in satellite systems together with data processing strategies, (Sun, Khan, & Bhatti, 2024). The future of remote sensing for coastal zone mapping advances because next-generation satellites offer improved specifications in resolution at space, temporal and spectral levels. Procedures for acquiring satellite imagery will increase frequency and resolution detection of coastal regions thus providing advanced monitoring capabilities for modifying shoreline features and changing wetlands together with urban infrastructure structures. With growing satellite constellation networks global coastal monitoring will establish better reliability and consistency for real-time coastal observation of erosion and habitat destruction and sea-level rise effects, (Zhang, Xu, & Bhatti, 2023).

Remote sensing technology will progress thanks to artificial intelligence (AI) and machine learning integration for data analysis purposes. Virtual technologies automate large and complex datasets by enabling superior interpretation of these volumes which results in higher accuracy of land cover identification and shift detection without requiring human interaction. Deep learning algorithms demonstrate excellent capabilities in automated feature recognition and pattern detection within coastal data networks that no human observer could identify. Continued advancement of AI-based techniques should create more efficient mapping processes which require less professional experience for their operation. AI systems will significantly contribute to data combination through multiple sources by resolving contradictory patterns between resolution and time frequency and sensor attributes, (Wu, Liu, & Bhatti, 2024).

Figure 4. Applications of remote sensing technologies in coastal and terrain analysis

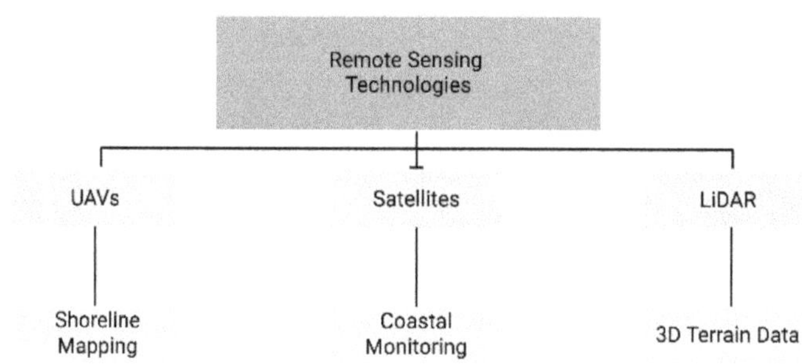

The future of coastal zone mapping will significantly increase the importance of unmanned aerial vehicles (UAVs). UAVs can now capture detailed images along with LiDAR data because drone technology has advanced to make these vehicles more accessible combined with their high-resolution abilities. UAV deployment provides instant access to perform precise mapping operations on small coastal areas when aerial and satellite sensors cannot reach these regions. Operational monitoring systems will integrate UAVs in the future to watch coastal land cover changes occur in near real-time, (Zeeshan, Li, & Bhatti, 2024). Onboard data processing through edge computing partnerships with UAVs will lead to shorter delays between information collection and assessment periods.

Remote sensing shall achieve advanced coastal zone mapping through its combination with citizen science and crowdsourced technologies. Remote sensing technology benefits from public and local knowledge integration when data from public sources validates remote sensing output thus producing enhanced comprehensive maps, (Fan, Zeng, & Bhatti, 2021). The method of citizen science enables the collection of current coastal condition data thereby supporting a management system that relies on community involvement. Open-platform developments along with data-sharing initiatives will enhance academic researcher collaboration with coastal managers and community members to distribute remote sensing technology benefits across all coastal management stakeholders, (Peng, Liu, & Bhatti, 2024).

CONCLUSION

The essential oversight of coastlines becomes possible through sustainable management systems because remote sensing operations heavily depend on modern technology. The tools receive high-resolution spatial data throughout multiple years allowing scientists to determine important information about coastal land cover changes and environmental patterns and human-induced impacts. Remote sensing technologies have progressed through satellite camera connections with UAV systems and LiDAR teamed up with hyperspectral equipment to give improved observation capabilities for coastal domains which helps detection of problems and develops solutions for habitat destruction and coastal erosion and development. The advancement of technology will strongly support coastal zone management through improved precise monitoring programs and complete solution development.

Emerging information technology processing methods like artificial intelligence and machine learning show enormous promise to improve coastal land cover mapping operations. Enhanced automated procedures that boost remote sensing data interpretation produce improved accuracy rates along with accelerated environmental change detection capabilities. Data fusion development involving multiple sensing systems enables well-planned coordination between optical imagery and SAR and LiDAR sources to generate whole understanding of coastal process dynamics. Advanced technological advancements provide coastal administrators improved solutions to overcome mounting environmental problems they face.

Multiple ongoing obstacles limit the use of remote sensing technology when it is applied to programs monitoring coastal regions. The management of current coastal land cover representation challenges and improved understanding require modern technological advancements and latest data processing methods such as deep learning to overcome regexional analysis barriers including dataset diversity as well as resolution conflicts and environmental factors like pixel mosaicking and time-dependent and cloud-shrouded conditions.

Remote sensing technology used for coastal zone mapping will show advancement from better technological progress and expanded data availability with improved analytical methods. Modern ecological management enhancements combined with environmental real-time tracking operations will be achieved through satellite expansion together with rising UAV use and AI enhancements. Better cooperation between coastal stakeholders using available remote sensing data enables adaptive management practices to become more effective. The entire array of developments available to us enables better understanding and robust protection of coastal zones to ensure sustainability for future society.

REFERENCES

Aamir, M., Li, Z., Bazai, S., Wagan, R. A., Bhatti, U. A., Nizamani, M. M., & Akram, S. (2021). Spatiotemporal change of air-quality patterns in Hubei province—A pre-to post-COVID-19 analysis using path analysis and regression. *Atmosphere*, *12*(10), 1338.

Aamir, M., Rahman, Z., Abro, W. A., Bhatti, U. A., Dayo, Z. A., & Ishfaq, M. (2019). Brain tumor classification utilizing deep features derived from high-quality regions in MRI images. *Biomedical Signal Processing and Control*, *85*, 104988.

Alhatami, E., Bhatti, U. A., Mengxing, H., & Feng, S. (2024). Advanced fuzzy denoising technique for agricultural remote sensing: Modified partition filter for suppressing impulsive noise. *IEEE Access : Practical Innovations, Open Solutions*, *12*, 6721–6735.

Belcore, E., Latella, M., Piras, M., & Camporeale, C. (2024). Enhancing precision in coastal dunes vegetation mapping: Ultra-high resolution hierarchical classification at the individual plant level. *International Journal of Remote Sensing*, *45*(13), 4527–4552.

Bhatti, U. A., Marjan, S., Wahid, A., Syam, M. S., Huang, M., Tang, H., & Hasnain, A. (2023). The effects of socioeconomic factors on particulate matter concentration in China's: New evidence from spatial econometric model. *Journal of Cleaner Production*, *417*, 137969.

Bhatti, U. A., Mengxing, H., Neira-Molin, H., Marjan, S., Baryalai, M., Hao, T., & Wu, G. (2023). MFFCG–Multi feature fusion for hyperspectral image classification using graph attention network. *Expert Systems with Applications*, ●●●, 120496.

Bhatti, U. A., Ming-Quan, Z., Qing-Song, H., Ali, S., Hussain, A., Yuhuan, Y., & Yu, Z. (2021). Advanced color edge detection using Clifford algebra in satellite images. *IEEE Photonics Journal*, *13*(2), 1–20.

Bhatti, U. A., Tang, H., Wu, G., Marjan, S., & Hussain, A. (2023). Deep learning with graph convolutional networks: An overview and latest applications in computational intelligence. *International Journal of Intelligent Systems*, *2023*, 278.

Bhatti, U. A., Yan, Y., Zhou, M., Ali, S., Hussain, A., Qingsong, H., & Yu, Z. (2021). Time series analysis and forecasting of air pollution particulate matter (PM2.5): An SARIMA and factor analysis approach. *IEEE Access : Practical Innovations, Open Solutions*, *9*, 41019–41031.

Bhatti, U. A., Yu, Z., Chanussot, J., Zeeshan, Z., Yuan, L., Luo, W., & Nawaz, S. A.. (2021). Local similarity-based spatial–spectral fusion hyperspectral image classification with deep CNN and Gabor filtering. *IEEE Transactions on Geoscience and Remote Sensing*, *60*, 1–15.

Bhatti, U. A., Yu, Z., Yuan, L., Zeeshan, Z., Nawaz, S. A., Bhatti, M., & Mehmood, A. (2020). Geometric algebra applications in geospatial artificial intelligence and remote sensing image processing. *IEEE Access : Practical Innovations, Open Solutions*, *8*, 155783–155796.

Brown, K. M., Hambidge, C. H., & Brownett, J. M. (2016). Progress in operational flood mapping using satellite synthetic aperture radar (SAR) and airborne light detection and ranging (LiDAR) data. *Progress in Physical Geography*, *40*(2), 196–214.

Datta, D., & Deb, S. (2012). Analysis of coastal land use/land cover changes in the Indian Sunderbans using remotely sensed data. *Geo-Spatial Information Science*, *15*(4), 241–250.

Fan, Y., Li, J., Bhatti, U. A., Shao, C., Gong, C., Cheng, J., & Chen, Y. (2021). A Multi-Watermarking Algorithm for Medical Images Using Inception V3 and DCT. *Computers, Materials & Continua*, *74*(1), 1–13.

Fan, Y., Zeng, C., & Bhatti, U. A. (2021). Robust zero-watermarking algorithm for medical images based on dual-tree complex wavelet transform and discrete cosine transform. *Journal of Medical Imaging and Health Informatics*, *9*(1), 188–194.

Fang, Y., Liu, J., Li, J., Cheng, J., Hu, J., Yi, D., Xiao, X., & Bhatti, U. A. (2023). Robust zero-watermarking algorithm for medical images based on SIFT and Bandelet-DCT. *Multimedia Tools and Applications*, *81*(12), 16863–16879.

Galvan, L. P. C., Bhatti, U. A., Campo, C. C., & Stanojevic, S. (2022). Toward sustainable development: The nexus between CO2 emission, economic growth, trade openness, and the middle-income trap in Latin American countries. *Frontiers in Environmental Science*, •••, 1003.

Gao, J., Zhang, Y., Geng, X., Tang, H., & Bhatti, U. A. (2022). PE-Transformer: Path enhanced transformer for improving underwater object detection. *Expert Systems with Applications*, •••, 123253.

Hansen, M. C., DeFries, R. S., Townshend, J. R., & Sohlberg, R. (2000). Global land cover classification at 1 km spatial resolution using a classification tree approach. *International Journal of Remote Sensing*, *21*(6-7), 1331–1364.

Hasnain, A., Sheng, Y., Hashmi, M. Z., Bhatti, U. A., Hussain, A., Hameed, M., & Yu, Z. (2022). Time series analysis and forecasting of air pollutants based on Prophet forecasting model in Jiangsu province, China. *Frontiers in Environmental Science*, *10*, 945628.

Huang, M., Han, H., Wang, H., Li, L., Zhang, Y., & Bhatti, U. A. (2018). A clinical decision support framework for heterogeneous data sources. *IEEE Journal of Biomedical and Health Informatics*, *22*(6), 1824–1833. PMID: 29994279

Joyce, K. E., Samsonov, S. V., Levick, S. R., Engelbrecht, J., & Belliss, S. (2014). Mapping and monitoring geological hazards using optical, LiDAR, and synthetic aperture RADAR image data. *Natural Hazards*, *73*, 137–163.

Khatami, R., Mountrakis, G., & Stehman, S. V. (2016). A meta-analysis of remote sensing research on supervised pixel-based land-cover image classification processes: General guidelines for practitioners and future research. *Remote Sensing of Environment*, *177*, 89–100.

Klemas, V. V. (2001). Remote sensing of landscape-level coastal environmental indicators. *Environmental Management*, *27*, 47–57. PMID: 11083908

Lapetina, A., & Sheng, Y. P. (2015). Simulating complex storm surge dynamics: Three-dimensionality, vegetation effect, and onshore sediment transport. *Journal of Geophysical Research. Oceans*, *120*(11), 7363–7380.

Lee, S. Y., Dunn, R. J. K., Young, R. A., Connolly, R. M., Dale, P. E. R., Dehayr, R., & Welsh, D. T. (2006). Impact of urbanization on coastal wetland structure and function. *Austral Ecology*, *31*(2), 149–163.

Li, D., Zhang, H., & Bhatti, U. A. (2022). A robust zero-watermarking algorithm based on PHTs-DCT for encrypted medical images. *Innovation in Medicine and Healthcare*, *5*(2), 102–113.

. Li, G., & Wan, Y. (2015). A new combination classification of pixel-and object-based methods. international journal of remote sensing, 36(23), 5842-5868.

Li, N., Liu, C., & Pfeifer, N. (2019). Improving LiDAR classification accuracy by contextual label smoothing in post-processing. *ISPRS Journal of Photogrammetry and Remote Sensing*, *148*, 13–31.

Li, T., Li, J., Liu, J., Huang, M., Chen, Y. W., & Bhatti, U. A. (2022). Robust watermarking algorithm for medical images based on log-polar transform. *EURASIP Journal on Wireless Communications and Networking*, *2022*(1), 24.

Liu, J., Li, J., Ma, J., Sadiq, N., Bhatti, U. A., & Ai, Y. (2019). A robust multi-watermarking algorithm for medical images based on DTCWT-DCT and Henon map. *Applied Sciences (Basel, Switzerland)*, *9*(4), 700.

Liu, J., Li, J., Zhang, K., Bhatti, U. A., & Ai, Y. (2019). Zero-watermarking algorithm for medical images based on dual-tree complex wavelet transform and discrete cosine transform. *Journal of Medical Imaging and Health Informatics*, *9*(1), 188–194.

. Liu, M., Bhatti, U. A., & Zhang, S. (2022). Remote sensing techniques for crop monitoring and environmental analysis. Big Data Research, 100478.

Long, K., Shen, C., Tian, C., Zhang, K., Bhatti, U. A., Kong, D. F., Feng, S., & Cheng, H. (2023). Single UWB anchor aided PDR heading and step length correcting indoor localization system. *IEEE Access : Practical Innovations, Open Solutions*, *9*, 11511–11522.

Mitsuda, Y., & Ito, S. (2011). A review of spatial-explicit factors determining spatial distribution of land use/land-use change. *Landscape and Ecological Engineering*, *7*, 117–125.

Nawaz, S. A., Li, J., & Bhatti, U. A., SU, B., & A Mehmood. (2021). Hybrid medical image zero watermarking via discrete wavelet transform-ResNet101 and discrete cosine transform. *Computers & Electrical Engineering*, *112*, 108985.

Nawaz, S. A., Li, J., Bhatti, U. A., Shoukat, M. U., Ahmad, R. M., & Hussain, S. (2023). AI-based object detection latest trends in remote sensing, multimedia, and agriculture applications. *Frontiers in Plant Science*, •••, 4273. PMID: 37082514

Otukei, J. R., & Blaschke, T. (2010). Land cover change assessment using decision trees, support vector machines and maximum likelihood classification algorithms. *International Journal of Applied Earth Observation and Geoinformation*, *12*, S27–S31.

Ouchra, H., Belangour, A., & Erraissi, A. (2022). A comparative study on pixel-based classification and object-oriented classification of satellite image. *International Journal of Engineering Trends and Technology*, *70*(8), 206–215.

Pal, M., & Mather, P. M. (2003). An assessment of the effectiveness of decision tree methods for land cover classification. *Remote Sensing of Environment*, *86*(4), 554–565.

Peng, M., Liu, Y., & Bhatti, U. A. (2023). Crop monitoring using remote sensing land use and land change data: Comparative analysis of deep learning methods using pre-trained CNN models. *Big Data Research*, *36*, 100448.

Peng, M., Liu, Y., & Bhatti, U. A. (2024). Deep learning-based tree disease recognition and classification using hyperspectral data. *Computers, Materials & Continua, 77*(1), 112–121.

Rittle-Johnson, B., Star, J. R., & Durkin, K. (2009). The importance of prior knowledge when comparing examples: Influences on conceptual and procedural knowledge of equation solving. *Journal of Educational Psychology, 101*(4), 836.

. Sheng, D., Zhang, J., & Bhatti, U. A. (2023). Precision agriculture applications using machine learning and remote sensing technology. Environmental Systems, 1017.

Sreekesh, S., Kaur, N., & Sreerama Naik, S. R. (2020). An Obia and rule algorithm for coastline extraction from high-and medium-resolution multispectral remote sensing images. *Remote Sensing in Earth Systems Sciences, 3*, 24–34.

Sun, S., Xue, Q., Xing, X., Zhao, H., & Zhang, F. (2024). Remote Sensing Image Interpretation for Coastal Zones: A Review. *Remote Sensing, 16*(24), 4701.

Sun, Y., Khan, M. A., & Bhatti, U. A. (2024). Advanced plant disease segmentation in precision agriculture using optimal dimensionality reduction with fuzzy C-means clustering and deep learning. *IEEE Journal of Selected Topics in Applied Earth Observations and Remote Sensing, 17*, 121–131.

. Wu, C., Liu, M., & Bhatti, U. A. (2024). Deep learning techniques for enhanced mangrove land use and land change from remote sensing imagery: A blue carbon perspective. Big Data Research, 100478.

Wu, G., Tang, H., & Bhatti, U. A. (2024). Remote sensing and GIS-based applications in coastal zone management and environmental monitoring. *Environmental Monitoring and Assessment, 207*, 142.

Wu, X., Li, J., Tu, R., Cheng, J., Bhatti, U. A., & Ma, J. (2022). Contourlet-DCT based multiple robust watermarkings for medical images. *Multimedia Tools and Applications, 78*, 8463–8480.

Wu, Y., Zhang, L., Bhatti, U. A., & Huang, M. (2024). Interpretable machine learning for personalized medical recommendations: A LIME-based approach. *Diagnostics (Basel), 13*(16), 2681. PMID: 37627940

Xu, B., He, B., & Bhatti, U. A. (2023). AI-based smart agriculture systems for remote sensing applications in coastal regions. *Frontiers in Plant Science, 14*, 124455.

Zeeshan, Z., Li, J., & Bhatti, U. A. (2024). Robust multi-watermarking algorithm for medical images based on GoogLeNet and Henon map. *Computers, Materials & Continua, 75*(1), 565–586.

. Zhang, S., & U. A. Bhatti. (2024). Image fusion techniques and applications for remote sensing and medical images. Deep Learning for Multimedia Processing Applications, 154-175.

Zhang, S., Xu, H., & Bhatti, U. A. (2023). Hybrid climate forecasting: Variational mode decomposition and convolutional neural network with long-term short memory. *Polish Journal of Environmental Studies*, *32*(1), 15–25.

Chapter 5
Big Data Analytics and AI-Driven Approaches for Air Pollution Trend Analysis

Urooj Khalid
https://orcid.org/0009-0002-2691-4081
Balochistan University of Information Technology, Engineering, and Management Sciences, Pakistan

Sibghat Ullah Bazai
https://orcid.org/0000-0003-3042-5977
Balochistan University of Information Technology, Engineering, and Management Sciences, Pakistan

Alamgir Naushad
https://orcid.org/0000-0001-7009-1751
National University of Sciences and Technology, Balochistan, Pakistan

ABSTRACT

To enhance the air quality monitoring and decision-making, this research investigates the combination of Big Data technologies, Machine learning, Deep learning models, and Artificial Intelligence. Using IOT sensor networks, satellite imaging, and meteorological data will be beneficial to meet the volume, velocity, and variety of big data characteristics. Advanced models like CNNs, LSTMs, and CNN-LSTM hybrid models are employed for accurate and precise spatiotemporal forecasting. Such models are evaluated using metrics such as RMSE, MAE, and R^2. Domain adaptation and transfer learning are used in regions where there is limited data. AI-powered public health analysis, wearable sensors for real-time exposure and

DOI: 10.4018/979-8-3373-3501-8.ch005

pollution detection in various countries like China and India, are some real-world applications. Despite the limitations in the model's performance and data dependability, this research outlines a future path integrating ethical AI, citizen science, and integration with smart cities for sustainable pollution management.

INTRODUCTION

Background of Air Pollution Analysis

Air pollution has one of the most major environmental challenges overall the world which affecting human health, ecosystems and economy. The growth in industrial development, urbanization and harmful emissions has degraded air quality, climate change, triggering respiratory problems and minimize survival rates (Gómez et al., 2021; Noor et al., 2023, Huang et al., 2024; Zhang et al., 2024a; Peng et al., 2024a; Bhatti et al., 2024a; Li et al., 2024; Han et al., 2024a; Noor et al., 2024). The world Health Organization (WHO) estimates that air pollution causes millions of untimely deaths per year, making it difficult area of research and policy intervention (Sebestyén et al., 2021; Han et al., 2024b; Bhatti et al., 2024b; Bhatti et al., 2024c;)

Understanding air pollutions trends are important for decision makers, environment bodies, and investigators to implement effective risk management approaches. Classical approaches of air quality monitoring, such as ground-based sensors have disadvantages over areas and time periods.

Figure 1. Air pollution (Shutterstock March 2025)

Significance of Air Pollution Analysis

The terms "Big data", "Machine learning" (ML), and "Artificial intelligence" (AI) have redefined the way air pollution is monitored, examined and predicted (Rautela & Goyal, 2024). The ability of big data is to process large amount of information which includes sensor data, satellite imagery etc. When applying machine learning algorithms, which can find complex hidden patterns in large dataset, we can gain deep valuable insights into the sources and behavior of population (Kang et al., 2018). Artificial intelligence in the form of Neural Network and deep learning enables further developments by improving models and predictions based on new data (Dubey et al., 2022). In this chapter, we will explore the technical details of these tools which provides a detailed information of how these tools are implemented in air pollution monitoring and trend analysis(Aamir et al., 2023). We will also discuss the role of data collection techniques, integration of different data sources, and the application of predictive models to help understand the future trends of air quality This chapter highlights the challenges and opportunities presented by big data, Machine learning and artificial intelligence addressing this important issue of our time.

Understanding Air Pollution and Its Impact

Air pollution indicates to the presence of dangerous material in the air which can be solid particles or gases that have harmful consequences for human health, the environment, and climate. These pollutants may emerge from various sources, natural sources which include wildfire, pollen etc. and human interventions which include vehicle emissions, large industry activities (Chen et al., 2024). The significance of air pollution is extensive and can effect various element of life. For instance, it is recognized to play a role in respiratory conditions including asthma, Chronic Obstructive Pulmonary Disease (COPD), and bronchitis. Air pollution has the potential to harm wildlife, disturb plant growth, and destroy ecosystems (Subramaniam et al., 2022; Chen et al., 2024; Bhatti et al., 2024d; Peng et al., 2024b; Han et al., 2024c; Zhang et al., 2024b). The World Health Organization (WHO) predicts that air pollution is responsible for seven million deaths annually. This chapter will explore the various methods for analyzing air pollution trends and how big data, ML, and AI are being used to track, predict, and mitigate these impacts.

Figure 2. Big data-based decision support for air quality

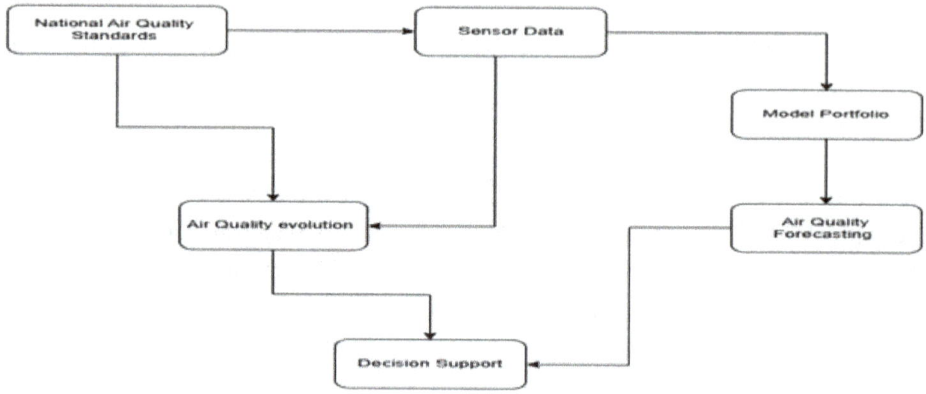

Common Air Pollutants and Their Sources

Air pollution can be classified into two major groups one is primary pollutants and secondary pollutants. Primary pollutants are directly affected from sources while secondary pollutants are created in the atmosphere as an outcome of chemical interactions between main pollutants. Particulate matter especially PM2.5 is one of the most dangerous forms of air pollution and its capability to enter and spread through the lungs and bloodstream (Lees et al., 2023). These particles may come from industrial pollutants, construction dust, wildfires, and automobile exhaust. Fossil fuel burning in automobiles and power plants is the main source of sulfur dioxide and nitrogen oxides (Kommineni et al, 2024).

Figure 3. Common air pollutants (Depositphotos.com)

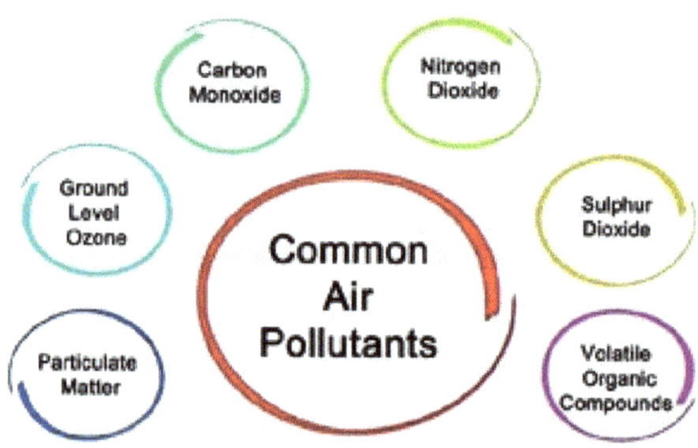

When nitrogen oxides and volatile organic molecules combine with sunlight, ground-level ozone is created. VOCs (Volatile Organic Compounds) come from a variety of natural sources, including plants, automobile fumes, and industrial solvents (Sebestyén et al., 2021; He & Bhatti, 2024; Zhang et al., 2024c; Asif et al., 2024; Chen et al., 2024; Hasnain et al., 2024, Hamza et al. 2023; Han et al. 2024, Haq et al. 2023). By examining the sources of these pollutants is important for efficient pollution management. Big data analytics, merged with machine learning (ML) approaches, allows us to determine the sources and monitor the movement of pollutants between regions (Sebestyén et al., 2021). Artificial intelligence (AI) can help model the pollutant circulation in the atmosphere and predict their real-time pollutant levels resulting in mitigation and management techniques that are more successful.

Health and Environmental Impact

The health and environmental impacts of air pollution are extensive and diverse. Air pollution is a one of the major cause of controllable health conditions and early global deaths. It is connected to elevated respiratory illness such as "Asthma, Lung cancer, and chronic obstructive pulmonary disease (COPD)." Sustained exposure to air pollution can contribute to heart dieses risk, such as "Heart attack and Strokes" (Lee et al., 2023). In addition to human health, air pollution also has detrimental effects on the environment. Acid rain, which can harm soil, water bodies, and plants, can result from high concentrations of sulfur dioxide and nitrogen oxides. Fish and other marine species may suffer when pollutants are deposited on the surface of lakes and oceans, upsetting aquatic ecosystems. Additionally, contaminants like ground-level ozone can hinder agricultural productivity and plant growth, lowering crop yields and endangering food security (Zaland, 2021; Subramaniam et al., 2022).. The reduction of biodiversity, can therefore have a domino effect on entire ecosystems, diminishing their capacity to deliver vital functions like food production, climate management, and water purification (Kommineni et al., 2024).

Figure 4. Air pollution impacts on human body (GlobalActionPlan.org.uk)

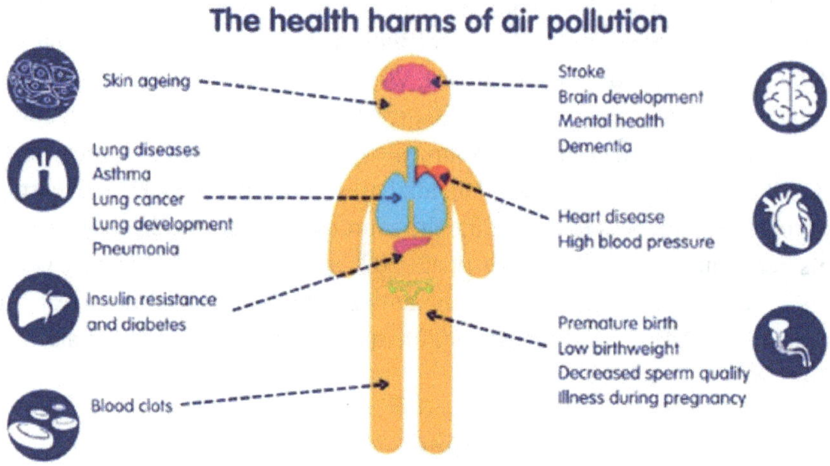

Analyzing the health and environmental effects of air pollution is essential for developing polices and methods aimed to minimize exposure and militating its effects. The integration of big data, Machine learning (ML) an Artificial intelligence method into air quality monitoring system allows for more accurate pollutant tracking, early detection of harmful trends, and more effective decisions-making to protect both health and environment (Chen et al., 2024).

Government sectors, big industries and environmental departments are working together to apply rules and regulations aimed at controlling vehicle exhaust output, factories and power plants (Gómez et al., 2021). These rules and regulations combined with the support of cleaner methods and non-polluting energy sources, can minimize the amount of polluting particles expelled into the environment. Technology enhancements in air quality monitoring and pollution control system are also important in handling the barriers caused by air pollution (Subramaniam et al., 2022).

Economic and Social Consequences of Air Pollution

The economic hardship of air pollution is extensive, affecting productivity, healthcare expense and infrastructure maintenance. Billions are invested by governments on handling health issues caused by pollution and setting up environmental methods. Poor air quality leads to productivity declining efficiency and work absences resulting in financial losses for businesses (Gómez et al., 2021). Socially undeserved communities often bear worst consequences of pollution because they live nearby to industrial areas and high traffic zones contributing to broader social and health injustices

Policy Measures and Regulations for Air Pollution Control

Governments authorities and non-profit organizations frequently conduct campaigns to raise awareness among cities about the health impacts of air pollution and promote actions that decrease emissions such as carpooling, minimize energy consumption and supporting policies that protect the environment. In addition to safeguarding the environment and public health, effective laws and policies also support long-term sustainable development objectives (Chen et al., 2024).

THE ROLE OF BIG DATA IN AIR POLLUTION ANALYSIS

Big data indicates datasets that are so huge or complex that traditional data processing applications and tools are inefficient to manage them. To extract insights from these datasets, sophisticated data management, storage, and analysis technologies are usually needed.

The three V's of big data analytics refers to the key characteristics that define big data which are Volume, Velocity and Variety.

Volume

It indicates the amount of data generated every second. In real-world, data is produced in large volume from multiple sources like social media, mobile devices sensors and more. The scale can range from terabytes (TB) to petabytes (PB) and even exabytes (EB) for example every day, Facebook processes more than 4 petabytes of data. Hundreds of gigabytes of data can be produced during a single aircraft flight.

Velocity

It refers the speed at which data is produced, examined and processed in real-time air quality monitoring. Data is being created and broadcast frequently and must be processed in real-time to extract value like fraud detection, stock trading, or sensor monitoring. For example, credit card firms must identify fraudulent transactions as soon as they happen, not hours or days later.

Variety

It refers the different types of data format. Big data is not just about numbers and database in includes structured data, unstructured data, semi-structured data. Unstructured text, semi-structured hashtags, and even photos or videos can all be included in a tweet; each requires a separate processing method.

To address air pollution issue 3v's of big data plays an important role. Large volume of data from various resources such as satellite imaging and real time sensor networks will be analyzed for efficient processing. Similarly, effective processing is necessary to enable prompt responses to such pollution events due to the speed at which the data is generated from the real time sensors. Lastly, variety of data sources that includes various contaminants and their origins, enables thorough knowledge and focused mitigation techniques.

Big data play a crucial role in air pollution examination by giving essential tools and resources to execute huge amount of data in real-time. Air quality data, gathered from different sources like stellates, sensors and other sources (Kang et al., 2018) usually expands across various locations and timeframes. However, big data approaches, like "Hadoop" and "Spark" both are efficient for processing large amount of datasets and allow for faster insights and better decision making.

DATA SOURCES FOR AIR POLLUTION ANALYSIS

Satellite Data

Satellites have altered the way we notice and analyze on a global scale. Remote sensing capabilities, satellites detect and measure multiple pollutants in the environment using imaging and spectroscopic methods, often from high attitudes above the earth. This allows experts and policymakers to monitor air quality over biggest area which includes sites with no ground level data collection (Sebestyén et al., 2021).

Important Satellites for Monitoring Air Pollution

NASA's Aqua and Terra (MODIS – Moderate Resolution Imaging Sepctroradiometer). NASA MODIS (Moderate Resolution Imaging Spectroradiometer) is involves in the "Terra and Aqua" satellites, MODIS **measures** Aerosol Optical Depth (AOD) which is too much important for calculating the amount of airborne particles in the atmosphere. The data given by MODIS is crucial for analyzing and understanding global air quality patterns, as it revels valuable information regarding pollution from natural sources as well as human caused emission (Sebestyén et al.,

2021). This data is accessed through NASA's LAADS DAAC, where participants can download aerosol data products for further analysis.

Figure 5. Aqua (Satellite meteorology)

European Space Agency – ESA (Sentinel-5P). The tropospheric Strategies observe detector, determining for observe the Earth's atmosphere, increase Sentinel-5p, a lead by the European Space Agency (ESA). TROPOMI is particularly suited to measure a range of pollutants, like carbon uniquely on oxide (CO), methane (CH4), ozone (O3), sulfur dioxide (SO2), and nitrogen dioxide (NO2) (Gómez et al., 2021). These chemicals are analytic for studying air quality and its effect on the environment and human health. Air quality data at grater resolution, specifically in this cities, can be obtain through Sentinel-5P (Sebestyén et al., 2021). Satellite monitoring can be worked by researchers, environmental organizations, and decision-makers to observe air pollution trends and make data-driven decisions about environmental management and public health (Sebestyén et al., 2021). Sentinel-5P offers a thorough, detailed, and real-time assessment of atmospheric composition, which crucially provide to pollution monitoring and air quality enhancement worldwide.

Figure 6. Introducing sentinel-5P (The European Space Agency)

OMI (Ozone Monitoring Instrument) on NASA's Aura Satellite. The Ozone monitoring instruments (OMI) on NASA's aura satellite is important tool give to monitor the earth's atmosphere. It evaluation significant pollutants such as NO2, SO2, and aerosols. Long-term air quality measurements, which carried the information about patterns and variability in air pollution, are essential in such case (Zayed, 2024). OMI is also a critical tool to measure the transport of contaminants between continents. It will help out scientists and policy makers understand the effects of emissions from one region on other regions by monitoring the global flow of pollutants, therefore it is simpler to develop international pollution control legislation (Sebestyén et al., 2021). The Geostationary Environmental Monitoring Spectrometer (GEMS), a South Korean technology, is the first geostationary satellite with a sole focus on monitoring continuously air quality across Asia (Kılınç, 2024).

Geostationary Environmental Monitoring Spectrometer (GEMS). GEMS offers high-resolution, hourly information on key air pollutants like aerosols, volatile organic compounds (VOCs), ozone (O_3), sulfur dioxide (SO_2), and nitrogen dioxide (NO_2). The near-real-time facility express in regional environmental policies and public health intervention by giving information on both short-lived pollution events and long-term atmospheric patterns, explored air quality forecasts (Kılınç, 2024). HIMAWARI-8, a geostationary weather Satellite operated by the Japan Meteorological Agency (JMA), is also critical for atmospheric monitoring.

Figure 7. Process for GEMS (European Geo Sciences Union)

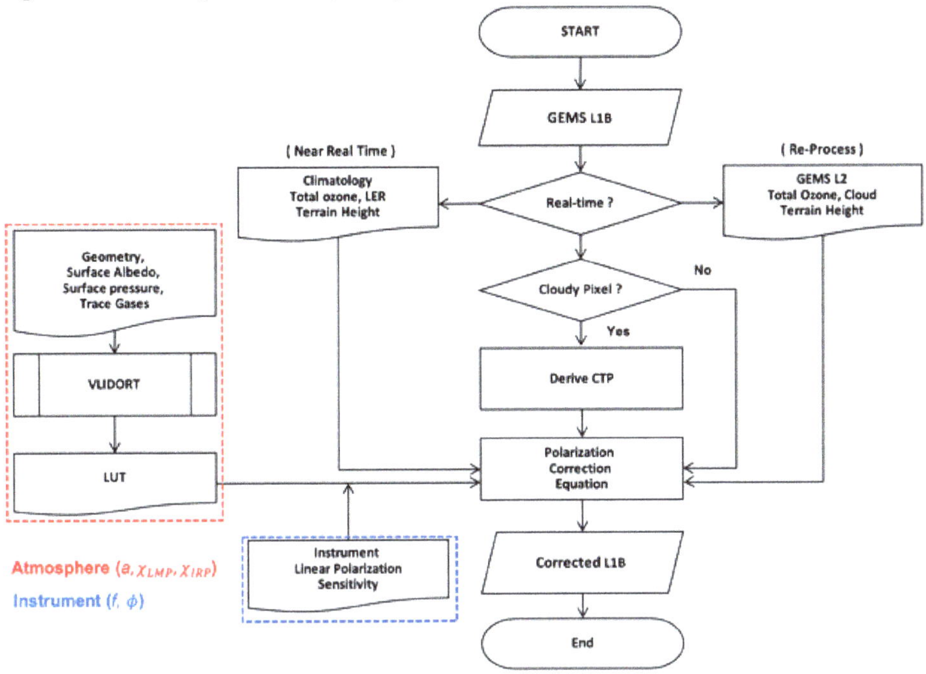

HIMAWARI-8 Satellite. It is discovered in 2014, it monitors the aerosol flow across the Asia-Pacific region, such as dust storms, volcanic ash, wild fire smoke, and industrial pollution, and gives real-time data every ten minutes. HIMAWARI pollution and their cross-border transportation with precision, which is vital for early warning systems, research on the environment, and disaster relief. To enhance understanding and control of local air pollution, GEMS and HIMAWARI-8 integrate real-time aerosol observation with chemical pollutant monitoring to provide a complete picture of air quality.

Advantages of Satellite Data

- Global monitoring gives extensive conservation which allowing regular monitoring of air pollution around the world.
- Satellite can monitor huge and remote sites that are difficult to access on the earth, which makes them unique for studying global phenomena like climate change, loss of forests, and oceans flows.
- Many satellites provide real-time data which offering immediate information after gathering data. This is important for disaster response, providing assistance during events like floods, wildfires, and hurricanes.

- Satellite sensors can capture high accurate and detailed data. It can determine even the few changes on the surface of earth due to their high quality frequency and spatial resolution.
- Satellites are capable of detecting pollution sources in remote or hard-to-reach areas, where ground-based monitoring stations may not be available or feasible.

Disadvantages of Satellite Data

- Launching and keeping satellites is very expensive. Satellite data can be a bit costly to acquire because the technology involved in satellite construction as well as ground systems is very expensive.
- Although satellite photography has improved over the years, certain applications cannot utilize the fine resolution of the data. High-resolution data may only be available from specific satellites or cost too much.
- Some areas can be particularly difficult to monitor using satellites such as the Polar Regions and areas with extreme conditions depending on the satellites orbit and type.
- When interpreted satellite data without ground-truth validation is found to be possibly incorrect, satellite photo analysis mistakes may result due to wrong satellite photo interpretations.
- There is increasing concern about space debris, which can damage operating satellites and compromise data quality, as more satellites are launched.

ENVIRONMENTAL SENSORS (GROUND-BASED MONITORING STATIONS)

There is a need for environmental sensors to obtain real-time ground-level measures of air pollution and these sensors are typically placed in urban, industrial, and rural settings. Typically, government agencies, research institutes, and environmental protection agencies run these sensors in a bid to monitor and manage pollution(Sebestyén et al., 2021). Fixed monitoring stations operated by organizations such as the EPA in the USA and the CPCB in India are used to quantify major pollutants such as PM2.5, PM10, NO2, SO2, CO, and O3 (Rautela & Goyal, 2024). These stations yield credible information on pollution levels, e.g., Air Quality Index (AQI) monitoring networks in major cities worldwide. Pollution levels at different points are dynamically tracked using mobile air quality monitors, which tend to be fitted on cars, drones, or bicycles (Gómez et al., 2021). Through the direct observation of pollutant emissions from power plants and factories, industrial emission sensors which contains sustained Emission Monitoring Systems (CEMS) are applied for legal compliance.

Figure 8. Ground-based monitoring stations (Research paper name: "Balancing Greenhouse Gas Budget")

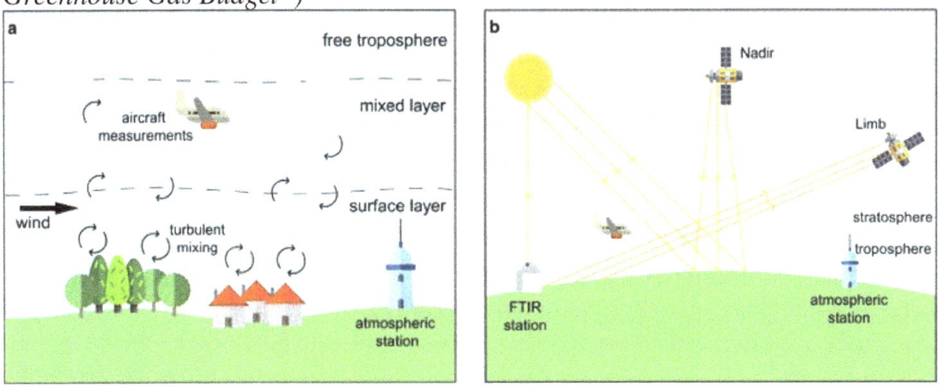

Advantages of Environmental Sensors

They make real-time data collection, enabling fast responses to pollution levels and public health issues. Environmental sensors give a complete picture of pollution sources and trends therefore they are efficient for monitoring urban and industrial zones.

The sensors are versatile, and there are different versions to meet multiple requirements and functions, such as fixed, mobile, and low-cost. They play a vital role in enhancing environmental policies and environmental regulations by providing precise information for tactical choices.

The sensors allow for crowd sourced data collection, improving databases on air quality and increasing awareness of pollution in society.

Limitations

It is important to remember that environmental sensors requires to routine maintenance, in addition to calibration in order to provide accurate data and optimal performance. Satellite data provides a border scale for logging pollution where the sensors have few geographical areas.

The deployment of these sensors around the universe would carry a cost that may preclude some regions or smaller companies from deploying them.

Environmental conditions, such as extreme weather, can have an impact on the sensors, thereby affecting their accuracy and performance.

INTERNET OF THINGS (IOT) FOR AIR POLLUTION MONITORING

The Internet of Things (IoT) plays a crucial role in air pollution monitoring by connecting networks of linked sensors that gathered and move real-time air quality data. These sensors are applied in different locations such as urban areas, residential homes, and industrial sites which enabling continuous monitoring of pollution levels (Sebestyén et al., 2021). By providing real-time data on pollutants, IoT helps official authorities and individuals to monitor air quality, find pollution sources, and take actions timely to save public health (Taj & Navi, 2020).

Figure 9. Internet of Things (IoT) for Air pollution monitoring ("An IoT System for Air Pollution Monitoring with Safe Data Transmission")

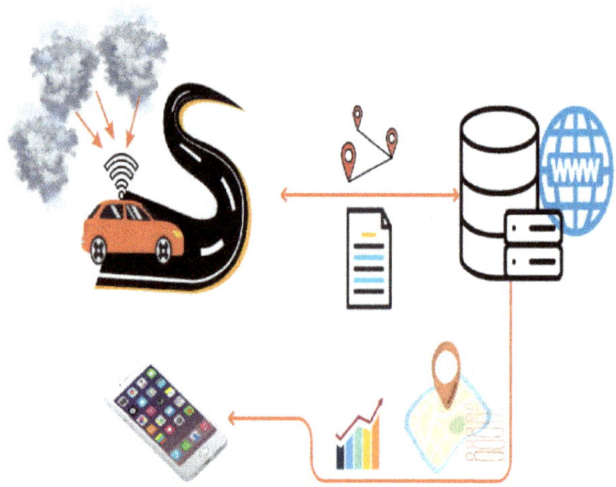

IoT Applications in Air Quality Monitoring

Smart City Air Monitoring Networks

To continue monitoring the quality of the air, several cities are adopting Internet of Things-based on pollution sensors into their networks. At traffic crossings, for instance, Barcelona's IoT- The permitted air quality network monitors contaminants such as NO_2, PM2.5, and O_3 (Essamlali et al., 2024). City planners can implement this information to make decisions on urban development and traffic control.

Wearable Air Quality Sensors

IoT devices which includes small sensors linked into digital wearables like smartwatches or masks which allows individuals to monitor pollution in real-time data collection. Plume Labs' Flow is a personal air pollution tracker that provides users with real-time air quality data based on their location (Gómez et al., 2021).

Building Indoor Air Quality

IoT devices can monitor harmful indoor pollutants such as CO_2, volatile organic compounds (VOCs), and humidity levels. Products like Aware and Robot smart air monitors help improve air quality in homes and offices by giving real-time information into air conditions (Zayed, 2024) (Wu et al., 2022.).

Advantages of IoT-based Monitoring

Real-time, localized air quality information is made available through IoT-based sensors, enabling accurate air quality assessment in specific locations and immediate response to pollution events.

These monitoring systems are easily scalable within cities, neighborhoods, or even homes, and tend to be less costly compared to traditional air quality monitoring stations.

Limitations of IoT-based Monitoring

The precision of the air quality data is based on the performance of sensors. Low expensive sensors may provide inferior data quality to advanced monitoring solutions.

Real time communication of air quality data depends continuous internet availability (Topbaş et al., 2021).

CROWD-SOURCED DATA FOR AIR POLLUTION ANALYSIS

Crowd-sourced data has established as a powerful tool for air pollution analysis by allowing individuals to participate data using personal sensors and mobile applications. This approach improves standard monitoring networks, specifically in undeveloped countries by addressing data inefficiencies and providing contextual insights(Yu et al., 2024). Most common platforms like OpenAQ, Breezometer, and Air Visual are play an important role in crowd sourced air quality monitoring. Air-Visual gather data from public platforms and personal sensors, offering real-time AQI maps for international cities Breezometer merge satellite data, sensor measurements and pollution levels using machine learning methods to generate customized personal data. OpenAQ an open-source platform, collect data from official sources and citizen sensors which gives free access to historical pollution data.

Table 1.

AQI	Air Pollution level
0-50	Excellent
51-100	Good
101-150	Lightly polluted
151-200	Moderately polluted
201-300	Heavily polluted
300+	Severely polluted (Air quality Classification)

Source: (Research gate)

Figure 10. Air quality classification

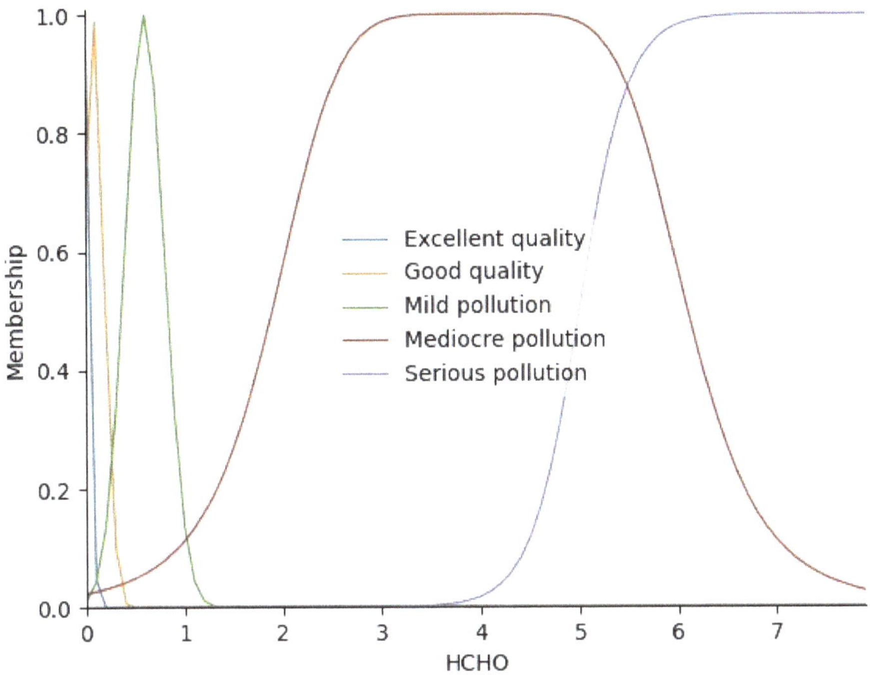

Advantages of Crowd-Sourced Data

The benefits of crowd sourced datasets are abundant. It improves air quality tracking scope, actions community take part in environmental surveillance, and gives localized data where official sensors are not present. It gives real-time information which allowing for quick actions, response and developing community awareness.

Disadvantages of Crowd-Sourced Data

The data quality can be poor, inefficient and low dimension, as it frequently depends on sensors that are not widely categorized which causing potential errors which are difficult to handle and solved. Furthermore, the need for community volunteers means the quantity and precision, fitness and accuracy of data may fluctuate, and there is no guarantee that all submissions meets basic protocols.

METEOROLOGICAL AND CLIMATIC DATA

Crucial and major meteorological elements like wind speed and directions, temperature and humidity, air-fall and air pressure are most essential in identifying and analyzing the movement and level of pollution in the atmosphere. Big data meteorological sectors, investigators can gain useful and in-depth information into the relationship between climate factors and pollution intensity which helps in predicting and investigating air quality and enhance pollution control methods effectively (Sebestyén et al., 2021).

Wind speed and directions are basic element in the movement and circulation of pollutants. High wind flow can move pollutants across large regions which decrease their presence in a pre-defined location while low wind flow allow pollutants to assemble in s particular location, leading to high densities(Zayed, 2024). The direction of the wind also highlights the movement of pollutants in air which affects the air quality. High temperatures can promote chemical changes that generate secondary pollutants like ozone, sulfur and more chemicals. These reactions are more affective in urban sites during summer weather, causing a decline in air quality(Gómez et al., 2021). On the other hand, high humidity levels can improve the creation of airborne particles and secondary pollutants. For the most part, high-pressure systems are associated with flat, stable weather that can trap pollutants near the ground, especially in urban areas (Gómez et al., 2021). Low-pressure systems facilitate easier passage of air, which can disperse contaminants more effectively. Meteorological information is a useful instrument in environmental studies and policy formulation since, in most cases, an understanding of how these meteorological variables affect air pollution is required in order to develop effective strategies for managing and improving air quality.

APPLICATIONS OF BIG DATA IN AIR POLLUTION ANALYSIS

Big data changed the way air pollution is examined by facilitating the gathering, strong and analysis huge amount of environmental dataset. With modifications in machine learning (ML), artificial intelligence (AI), and the Internet of Things (IoT), current air pollution tracking and forecasting methods have become more effectives and efficient.

Real-Time Air Quality Monitoring

The classical approaches of air quality monitoring based on permeant public facilities which gives few geographical areas and do not detect the full pollution intensity in multiple locations. But the emergence of big data radically altered this technology by merging several data sources, improving more comprehensive and current air quality monitoring(Kang et al., 2018).

One such modification is the use of IoT based sensor networks. The frequent use of these sensors enables more unambiguous air quality monitoring in urban regions and rural regions on pollution levels (Essamlali et al., 2024) (Sebestyén et al., 2021).

Air quality data is continuously collected by smart devices, wearables and remote sensors, increasing the spatial range across multiple locations (Zayed, 2024). For suppose, Google's Air View project gives dynamic, real-time air quality information's from urban streets to different places by mapping the air quality throughout cities using Street View automobiles fitted with air pollution sensors. Satellite-based Remote Sensing approaches plays an important role in large-scale air quality monitoring. Satellites like NASA's MODIS and Sentinel-5P gather expansive pollution data covering wide areas, enabling researchers to monitor pollutants across countries. These technologies develop a multi-dimensional view of air quality, enabling official authorities, organizations, and people to fight pollution quickly and intelligently, improving the environment in the process(Gómez et al., 2021).

Figure 11. Sensor data monitoring system ("Sensor Data Management for IOT, June 2018)

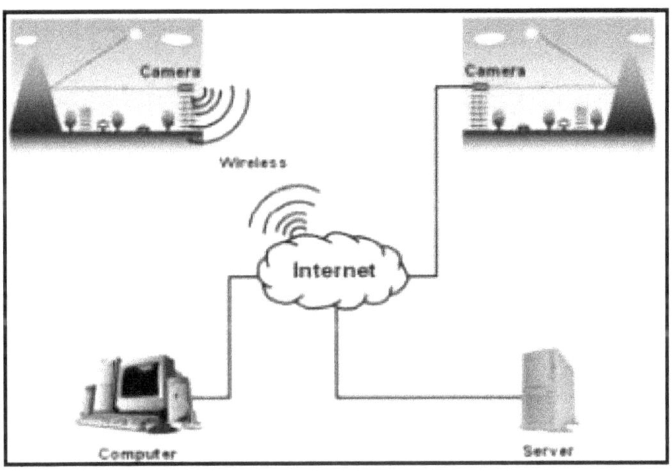

Predictive Analytics and forecasting

Predictive analytics and forecasting with Big Data enormously enhanced air pollution level forecasts and enabled the authorities to take preventive measures to mitigate the health impacts of air pollution. Random Forest, Support Vector Machines (SVM), and Long Short-Term Memory (LSTM) networks are few common and famous algorithms that analyze historical air quality data to predict upcoming pollution patterns (D. D. Wu et al., 2022). By integrating weather data, such as temperature, wind speed, and humidity, with remote data, machine learning approaches are capable to predict shifts in pollution concentrations (Gómez et al., 2021)(Kang et al., 2018). For example, IBM's Environmental Intelligence Suite integrates Big Data from satellite, weather stations, and sensor data to provide highly accurate forecasts of air quality, enabling people to prepare their communities better to face pollution bursts and improve the response mechanisms. Predictive analytics and forecasting allow officials to not only recognize real-time pollution levels but also participate future trends, improving their ability to save public health and handle environmental policies effectively (Lee et al., 2023).

Figure 12. Air quality forecasting system ("A visual study of Wind impact on PM2.5 Concentration" November 2015)

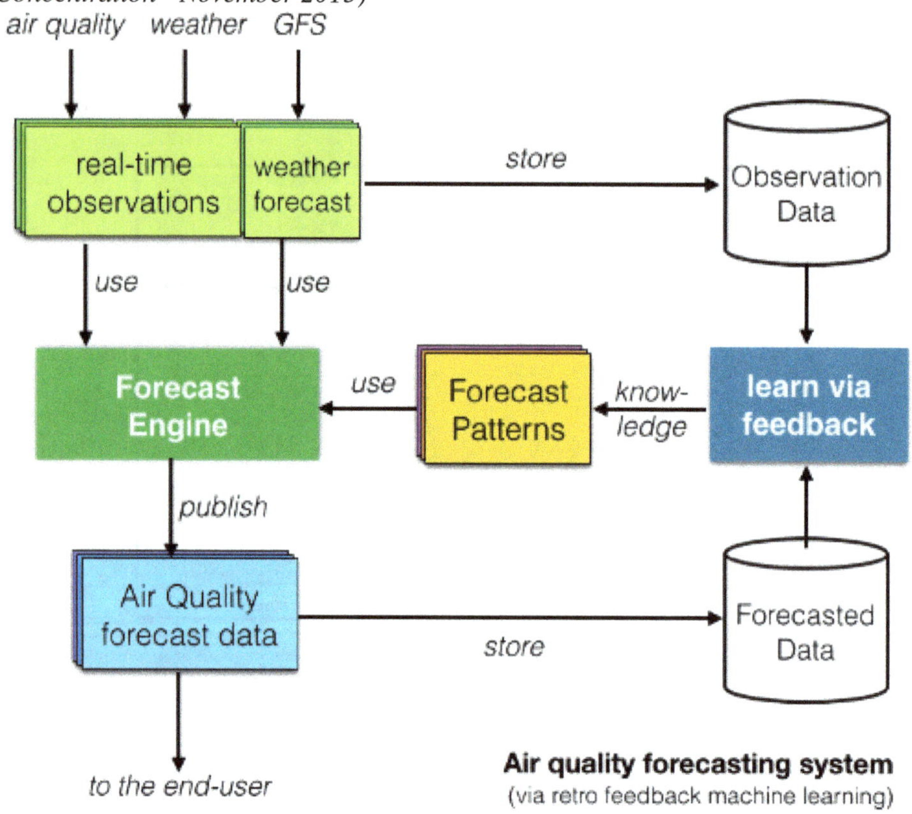

Identification and Pollution Attribution

By applying big data solutions, the current authorities can target interventions more sharply and pinpoint concrete sources of emissions (Yu et al., 2024). One key technique is Emission Source Profiling, which relies on AI-based analysis to identify and separate sources of pollution such as farm emissions, transport emissions, and industrial emissions. By intricate big data analysis scientists concluded that coal burning was the primary cause of Delhi's winter smog (Rautela & Goyal, 2024). Another effective technique for source finding is source mapping with satellites and GIS.

GIS (Geographic Information System) software combines pollution information with geospatial analysis to explore the correlations between trends in pollution and certain geographic conditions, e.g., industrial estates, highways, and urbanized development areas (Essamlali et al., 2024; Subramaniam et al., 2022; Bhatti et al. 2022; Bhatti et al. 2023a) . Through mapping such patterns, governments are better able to understand the spatial pattern of pollution sources and the consequent effects on people in the area. NASA's Ozone Monitoring Instrument (OMI) had already been able to detect NO_2 hotspots over China's industrial areas in order to demonstrate the prevalence of industrial contamination (Uzair A.B et al, 2020). With this knowledge, interventions in locations of highest contamination could improve the efficiency of air pollution control measures (Essamlali et al., 2024).

Health Impact Assessment

Possibly the most essential of these steps for understanding, analyzing and assessing the potential challenges of environmental elements like air pollution, on the public's health and it is health impact assessment (Rautela & Goyal, 2024). The inclusion of Big Data allows for a better grasp of the ways in which pollution harms human health by making it possible for researchers to analyze vast data sets that consolidate health data with levels of air pollution (Chen et al., 2024).

AI-Driven Epidemiological Studies

Artificial intelligence (AI) is redefining our epidemiological research completely, specifically in finding how air pollution impacts multiple health conditions. Artificial intelligence (AI) methods examine large amounts of data from hospitals, sensors, and public health databases to detect patterns (Haq et al., 2023) and relationships between pollution exposure and diseases like asthma, chronic obstructive pulmonary disease (COPD), and cardiovascular disease(Rautela & Goyal, 2024).

For suppose, a famous Harvard University research investigated the correlation with COVID-19 fatalities and air pollution. Based on Big Data from hospital reports (Feng et al. 2022; Haider et al. 2024) and pollution sensor networks, the authors concluded that those regions with greater air pollution had significantly higher COVID-19 fatality rates (D. D. Wu et al., 2022).

Real-Time Exposure Monitoring

With technological developments, real-time exposure monitoring has revolutionized the fight against air pollution. These offer instant feedback on air quality and may recommend alternative routes or less polluted areas, enabling users to steer clear of high-risk environments (Chen et al., 2024) (Zayed, 2024).

For example, intelligent masks with air pollution sensors not only track the air quality but also alert people to possible hazards and offer recommendations on the optimal course of action. This immediate information empowers people to make well-informed decisions regarding their health, mitigating the ill effects of extended exposure to polluted air (Lee et al., 2023).. In public health research, the combination of big data and artificial intelligence providing cutting-edge information into the health impacts of environmental factors like air pollution (Yu et al., 2024) (Essamlali et al., 2024). By merging health risk studies and real-time monitoring we can better predict to save public health and develop connections that can minimize the harmful effects of air pollution.

Figure 13. Real-time exposure monitoring

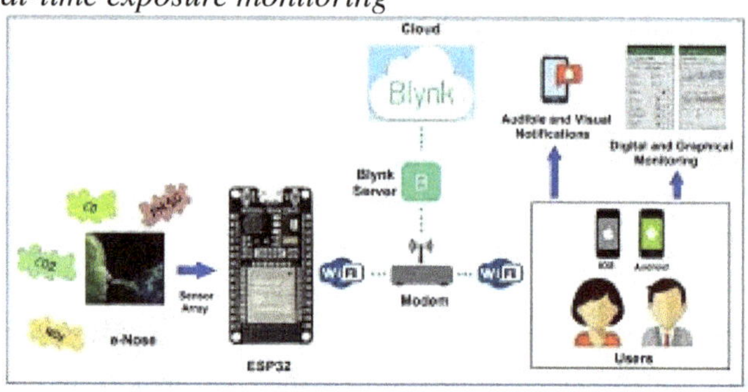

Smart City and Urban Planning

Big Data is crucial in assisting urban planners in creating sustainable cities with better air quality (Bhatti et al., 2023). In traffic optimization, AI-powered traffic management systems aid in easing congestion and car emissions, making it more efficient and environmentally friendly (Essamlali et al., 2024). A good example of this is the Ultra-Low Emission Zone (ULEZ) in London, which was identified along Big Data processing to control vehicle pollution beneficially and efficiently. Constantly the implementation of Geographic Information Systems (GIS) and satellite imagery, urban analyzer is ability to pinpoint locations that need urban forests, parks, and green belts, which help combat pollution (Chen et al., 2024).

Policy and Regulatory Compliance

Governments are gradually using Big Data to impose environmental regulations and observe compliance. In the case of industrial emission tracking, AI-powered compliance systems track emissions data from industries, facilitating the identification of violations in real time (Zayed, 2024). Big Data enables policymakers to take decisions about public policy through the utilization of AI simulations to test the likely effects of different pollution control calculate prior to implementing them (Zayed, 2024).

REMOTE SENSING APPLICATIONS

Big Data collect satellite images with AI to give a useful information into pollution patterns and natural events. A major use is monitoring pollution around the world, with satellites such as NASA's AERONET and Sentinel-5P tracking PM2.5 and NO_2 levels in universe. The example is the rapid decrease in air pollution across big cities during COVID-19 lockdowns, which was picked up with satellite evidence (D. D. Wu et al., 2022). In addition, Big Data is key to monitoring wildfire and dust storm events, Artificial intelligence algorithms decode satellite data to track and predict the flow of smoke and dust clouds supporting disaster preparedness and response (Kang et al., 2018). Predictive power increase the capacity to minimize the effect of these natural disasters on public health and protection (Subramaniam et al., 2022). Citizen Science and Community Participation Public participation in air quality monitoring has much greater via Big Data technologies, enabling specific to actively contribute to environmental monitoring and policy making (Essamlali et al., 2024). The suitable example is crowdsourced air quality data, where mobile applications allow users to report pollution and contribute data. Big Data reinforces policy advocacy based on communities by exposing environmental data to the public, this transparency energizes environmental activism and informs policy demands (Kang et al., 2018).

Figure 14. **Remote sensing applications *(Researchgate.net)***

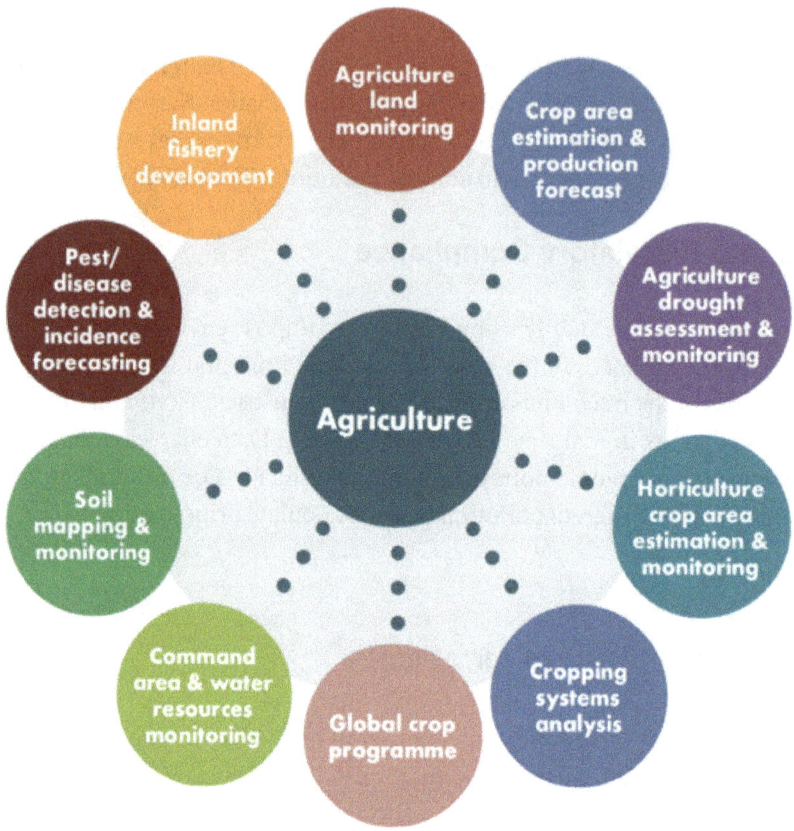

Citizen Science and Community Engagement

Public participation in air quality monitoring has much larger through Big Data than normally. Applications that enable individuals specifically to support environmental monitoring and legislation started with great scope to act (D. D. Wu et al., 2022). Public policies suggested one renowned illustration is crowdsourced air quality data since mobile applications let users share information and let know of pollution levels (Chen et al., 2024). The European Citizen Science started exemplifies this, as it uses crowdsourced air quality data to influence government policies, ensuring that public input plays a critical role in shaping environmental regulations.

MACHINE LEARNING, DEEP LEARNING, AND AI TECHNIQUES IN AIR POLLUTION ANALYSIS

Air pollution - the most popular issue that affecting not only our environment but also the public health around over all the world. It contributes to respiratory diseases and impacts our cardiovascular conditions(Rautela & Goyal, 2024). Not only that, air pollution cultivates issues like climate change which is hindering our day to day life. The traditional methods used to detect the air quality are outdated as they rely on statistical models(G. Wu et al., 2022). The models do not perform well when dealing with large scale, high dimensional and nonlinear data (Kang et al., 2018). These limitations hinder the accuracy and effectiveness of the decision making process to control and monitor the environment.

With new AI tools and technologies, it has become easier and precise to detect and monitor the air quality. They allow more exact real-time analysis of the air quality, strong predictive monitoring, and quick decision making(Subramaniam et al., 2022). By applying machine learning and deep learning approaches researchers gaining right and useful information into the air quality situations, predict pollutants events and required activities mitigate their adverse effects (Kang et al., 2018).

Artificial intelligence air quality monitoring network link data from various sources like sensor systems, satellites, meteorological data. Machine learning and Deep learning methods analyses and process these data to achieve a complete data of air quality (Kang et al., 2018). These elaborated models allow developers and environmental companies can develop effective action plans to reduce air pollution and reducing its effects on ecosystem and human health.

Analyzing and predicting air pollution using machine learning (ML) methods has become a powerful approach. They are also widely used in Air Quality monitoring, pollutant level classification, air quality data processing, and feature selection (Yu et al., 2024; Nabeel et al. 2024 ;Nawaz et al. 2021) . With ML's modified capabilities of identifying patterns among diverse environmental data these models gives a more accurate and timely evaluation of air quality as compared to conventional standard techniques (Lee et al., 2023). Supervised learning approaches, such as decision trees, random forests, support vector machines (SVM), and deep neural networks, have been effectively implemented to predict short- and long-term air quality. Unsupervised methods like clustering and principal component analysis (PCA) are also used for data preprocessing, anomaly detection, and discovering hidden patterns in air pollution (Gómez et al., 2021) (Lee et al., 2023).

Regression Models

Supervised learning models are trained on datasets with labels, where the values of both the input variables (e.g. meteorological data, traffic patterns, etc.) and the target variable (pollutant concentration) are known (Subramaniam et al., 2022). These models analyze past data to learn about it and then are used to consider future levels of air pollution or to classify types of pollution (Sebestyén et al., 2021). The most popular approaches are regression models but other supervised learning techniques are also applied in air pollution analysis.

Convolutional Neural Networks (CNN's) and Artificial Neural Networks (ANN's) can play a significant role in predicting air quality. As, CNN's can be used for identifying spatial patterns such as high smog locations from satellite imagery, whereas ANN's are capable for modeling patterns such as forecasting PM2.5 concentrations overtime.

Regression Models and Air Quality Prediction

Regression models are one of the most basic method to monitor continuous air quality factors such as the air pollution intensity. Some of the methods used to predict the air quality are described below:

Linear Regression. Linear regression (LR) is one of the simplest and clear regression technique which is based on assumption of a linear relationship between pollution levels and input parameters(Essamlali et al., 2024).. The main drawback of linear regression is that it ignores the non-linear relationship that are frequently present in air pollution dataset although it is both rapid and easy to apply (Subramaniam et al., 2022).

Support Vector Regression (SVR). Support Vector Regression (SVR) is a common type of the Support Vector Machine (SVM), which is used for regression purposes. Since SVR can determine nonlinear relationships between input factors and output elements(Essamlali et al., 2024). It is most powerful and unique tool for air quality prediction. For example, SVR may be applied for identifying nitrogen dioxide particularly in cities where a variety of factors influence the quality of the air, such as weather, traffic volume, and industrial emissions, because SVR can easily manage nonlinear data, it is a powerful technique for air quality forecasting, (Gómez et al., 2021).

CLASSIFICATION MODELS

Support Vector Machines

Support Vector Machines (SVM) are one of the most powerful machine learning classification approach. By creating a hyperplane in a high-dimensional feature space, SVM classifies many data classes. In air pollution monitoring, SVM is used to classify air quality as "Good," "Moderate," "Unhealthy," and "Hazardous" based on pollutant levels such as PM10 (particulate matter with diameters less than or equal to 10 micrometers) (Rautela & Goyal, 2024). SVM works extremely well if there is a perceptible difference between the classes. The air quality classification in relation to PM10 levels, for example, would be capable of determining whether air is suitable for outdoor activities or whether exposure has to be limited due to health concerns (D. D. Wu et al., 2022).

Random Forest (RF)

Random Forest is an ensemble learning approach popular for solving classification and regression issues in air quality analysis. The end categorization outcome is obtained by combining the results of several decision trees, each trained on a different part of the data (Kang et al., 2018). Choosing traits for an air quality study depends on identifying environmental factors that have the most influence on air quality—that is, on traffic, industrial emissions, or weather—via random forest techniques (D. D. Wu et al., 2022). Random Forest, for instance, can assist in determining if the main sources of pollution in a given region are local weather patterns, industrial operations, or traffic congestion.

Naive Bayes

Naïve Bayes is a technique that predict outcomes by estimating probabilities. It uses Naïve Bayes theorem to measure how identifying the probability of and occurrence of data (Essamlali et al., 2024). For suppose, Naïve Bayes theorem can be used to monitor pollution levels across different cities. Given the pollution level theses variables works independently which makes the prediction process quicker in simpler(D. D. Wu et al., 2022). Various real-time air quality monitoring system uses this method because it gives fast and clear prediction regarding pollution level.

Table 2. Extended comparative analysis of ML techniques for air pollution trend analysis

Technique	Strengths	Weaknesses	Suitability for Air Pollution Data	Interpretability	Training Time	Common Use Cases
Support Vector Machine (SVM)	High accuracy for high-dimensional data- Effective for small to medium datasets- Robust to overfitting	Computationally intensive for large datasets- Sensitive to parameter tuning- Requires large memory for large datasets	Suitable for detecting complex patterns in air pollution data, especially for classification tasks	Moderate: Can be hard to interpret decision boundaries in high dimensions	Moderate: Training can be slow for large datasets	Classifying air quality levels, detecting anomalies in pollution trends
Random Forest (RF)	Handles large datasets well- Robust to outliers and noise- Feature importance ranking- Versatile with both regression and classification tasks	Can be slow to train with many trees- May overfit on noisy data- Less interpretable with many trees	Excellent for analyzing AQI over time, identifying influential features (like wind speed, temperature) on pollution levels	Low: Harder to interpret due to the large number of trees	Fast to moderate: Faster compared to SVM for large datasets	Predicting air quality levels, feature importance analysis, trend analysis
Naive Bayes (NB)	- Fast and simple- Works well with large datasets- Effective for categorical input- Handles missing data well	- Assumes feature independence (rare in real data)- Lower accuracy than other methods when features are highly correlated	Useful for real-time predictions, generating probability-based pollution forecasts (e.g., "chance of high pollution")	High: Provides simple, interpretable probabilities for predictions	Fast: Quick to train, even on large datasets	Real-time pollution alerts, predicting pollution likelihood based on historical data

UNSUPERVISED LEARNING

Unsupervised learning algorithm uses data without labels or classify that have been established. The program determines hidden patterns, or groupings in the autonomous collection due to environmental datasets collected from multiple sources such as satellite sensors, Internet of Things devices, or weather systems are frequently large and unlabeled. Unsupervised learning provides researchers to examine this data without needing prior knowledge or examples with annotations in order to identifying pollution hotspots, discover new patterns, and understand how different factors influence air quality.

K-Means Clustering

K-mean clustering is one of the most suitable and famous unsupervised learning algorithm that helps Categorize data points within a given figure of clusters depends on similarity. In air pollution analysis, K-mean can be implemented in group of multiple regions or time stamp based on pollution risk categories (Sebestyén et al., 2021). For example, one cluster may include cities with comparable PM2.5 and NOx measurements, whereas another cluster may include cleaner regions. This categorization helps in determining which regions are at high risk and need quick action. K- mean clustering is very unique because it is fast and scalable which makes it better for managing huge amount of datasets (Gómez et al., 2021). Policymakers can highlight particular areas for strategic planning, resource allocation, and pollution handling by splitting the data into useful clusters.

Hierarchical Clustering

Hierarchical grouping is another method that produces a node. A treelike organization of layered clusters. Since it does not require a defined number of clusters, this strategy is more suited for data investigation. Hierarchical clustering can be used to analyze trends in air pollution and uncover how pollution patterns change over time and space (Sebestyén et al., 2021). Knowledge of this sort is valuable for urban development since it may help to guide legislation about industrial emissions, green buffer zones, traffic management, and so on (Gómez et al., 2021). Hierarchical clustering's visual organization also helps us to more easily contrast pollution patterns around several sites and infer connections.

Table 3. Comparison of unsupervised learning techniques for air pollution trend analysis

Technique	Strengths	Weaknesses	Suitability for Air Pollution Data	Interpretability	Training Time	Common Use Cases
K-Means Clustering	Fast and scalable- Easy to implement	Requires predefined number of clusters- Sensitive to outliers	Good for identifying pollution hotspots and segmenting regions based on pollution levels	Moderate (depends on cluster meaning)	Fast (especially with small k)	Pollution source classification- Regional segmentation
Hierarchical Clustering	No need to specify cluster number. Dendrogram gives rich insights	High computational cost- Not scalable for large datasets	Useful for small-scale pollution datasets and observing nested relationships	High (visual dendrogram aids understanding)	Slow (especially for large datasets)	Pollution level grouping in small areas- Pattern discovery

DEEP LEARNING FOR AIR POLLUTION ANALYSIS

Air pollution is a difficult problem affected by several fluctuating variables which includes weather situations, traffic patterns, and industrial activities (Kang et al., 2018). Standard machine learning (ML) algorithms frequently fails to note important details of correlations between time and space data due to limitations in Data transformation and scalability challenges, deep learning methods like Convolutional Neural Networks (CNNs), Recurrent Neural Networks (RNNs), and Long Short-Term Memory (LSTM) networks, have displayed outstanding proficiency in modeling these difficult patterns (Yu et al., 2024). These models automatically extract uncover patterns from raw data, minimizing the need for non-automated attribute selection and allowing more precise air quality predictions.

Figure 15. Deep learning for air pollution analysis ("Deep Learning for Air Quality Forecasts: a Review")

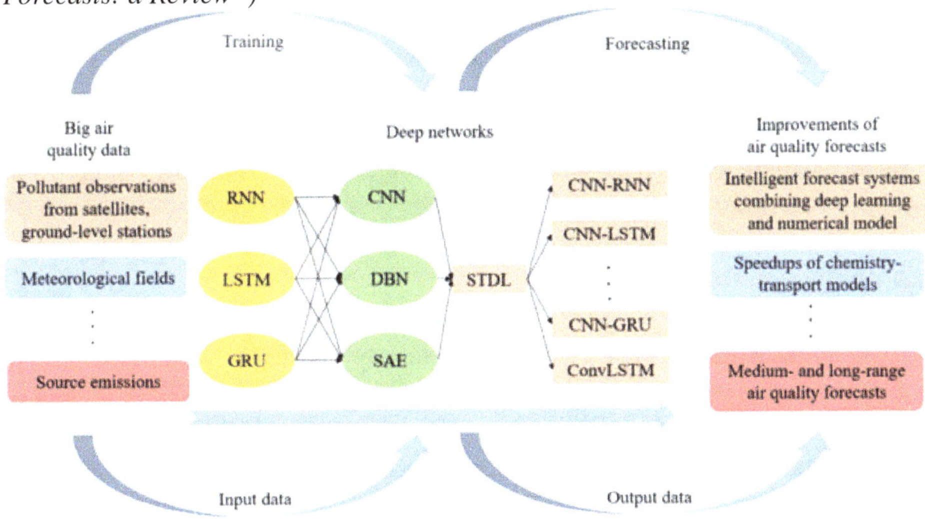

Deep learning models are suitable for connecting multiple data sources which exist satellite imagery, weather forecasts, sensor data, and geographic data, to provide a more detailed understanding and analysis of pollution patterns (Yu et al., 2024). Deep learning methods improve the prediction of pollution hotspots, monitor pollution levels, and support timely decision-making for environmental management. Deep learning is becoming an essential asset in the field of air pollution analysis, cutting-edge research and its practical use (Zayed, 2024).

Artificial Neural Networks (ANNs)

The major use of Artificial Neural Networks (ANNs) in mainly in predicting air pollution levels using previous data. These methods are specified in executing large amount of dataset and learning difficult data patterns which makes it easier and simpler to predict pollutant with high accuracy (Gómez et al., 2021). Their capabilities in determining precious pollutants levels, ANN's can detect future outcomes which gives more effective monitoring and control air quality(Gómez et al., 2021). For example, The ANN is commonly use in the area of predicting future outcomes of particulate matter expected concentration levels over the next 24 hours. They process patterns and make predictions with high precision which helps regulators should take proactive measures to lower air pollution (Subramaniam et al., 2022).

Convolutional Neural Networks (CNNs)

CNN is an image based analysis model and processes image based data while ANNs and RNNs focused on numerical data. These models can even process satellite images or drone-captured photos, and thus can monitor air pollution in real time (Rautela & Goyal, 2024). By analyzing these images, CNNs can detect changes in air quality indicators, such as smog or industrial emissions, that are visible from aerial perspectives.

Example

CNNs models are applied to visual data mostly to satellite or drone imagery so that they can better detect smog levels or industrial emissions. Due to this high feature these models can be extensively used in providing valuable information for environmental agencies to target pollution (Guo et al., 2022)

Hybrid Deep Learning Models

A hybrid deep learning model is a powerful computer program that combines different tools to solve complex problems more effectively and efficiently. It uses multiple types of algorithms to learn from data and achieve better outcomes than using a single approach.

CNN-LSTM

The idea of merging the powerful feature of CNN models that is to capture features from spatial data and LSTM that is to capture temporal dependencies, we can lead to a more powerful approach and can make more accurate and detailed predictions about air pollution data (Rautela & Goyal, 2024). CNN's can provide useful information about the location of pollution sources in graphical data while LSTMs highly effective at recognizing temporal relationships which are important for understanding and analysis how pollution levels change over time (Essamlali et al., 2024)big data. By merging both architectures, the precision in monitoring level can be improved. For example, this hybrid model can be used to monitor levels in unique locations depends on previous data patterns and real-time data from sensors and satellites (Essamlali et al., 2024).

Figure 16. DL (a) Original structure of self-attention; (b) improved structure of self-attention ("A Hybrid Deep Learning Model for Air Quality Prediction Based on the Time–Frequency Domain Relationship")

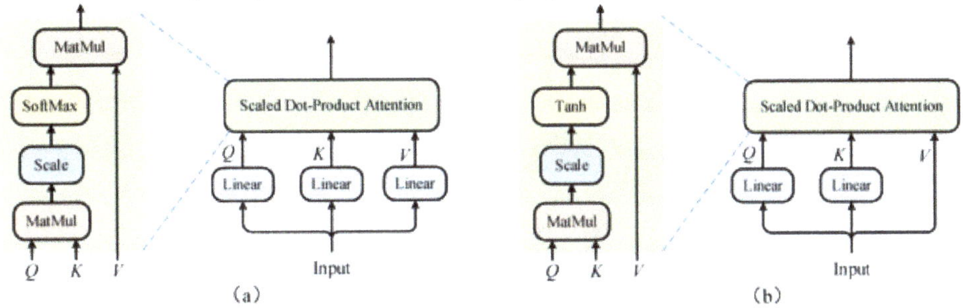

AI TECHNIQUES IN AIR POLLUTION MONITORING

Today, a wide range of sensors ranging from imaging and non-imaging types, from distant to in-contact sensors can produce massive amount of data and serves as an important part of data providers. Such data can be analyzed by AI algorithms for extracting crucial information from the sensor readings and historical data.

IoT and AI for Smart Pollution Monitoring

The connection of Internet of Things (IoT) sensors with Artificial Intelligence (AI) is innovating pollution monitoring techniques (Subramaniam et al., 2022). IoT sensors take current data which depends on several pollutants present in air like PM2.5, NOx and CO2 which can be used in further processing and analyzing by artificial intelligence methods (Sebestyén et al., 2021). These artificial and big data methods predict patterns, trends which give in-depth details into different pollution sources and their effect on public health (Baloch et al., 2023). For Example: In smart cities, IoT sensors provide frequent pollution data, which is analyzed by AI-driven systems to optimize pollution control strategies, such as adjusting traffic patterns or regulating industrial emissions in real-time (Essamlali et al., 2024) (Sebestyén et al., 2021).

Figure 17. IoT smart monitoring system ("Leveraging Machine Learning for Fault-Tolerant Air Pollutants Monitoring for a Smart City Design")

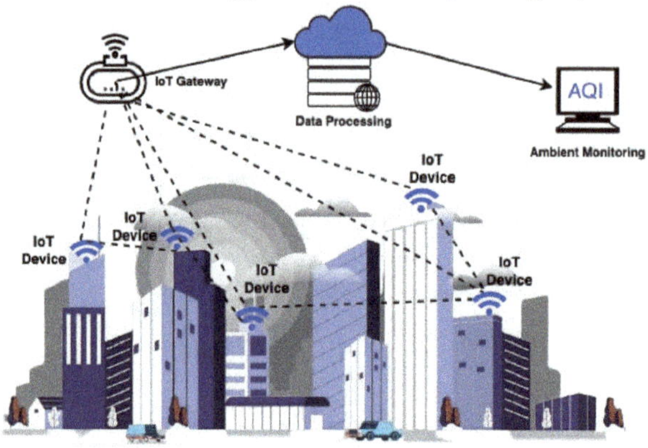

Geographic Information Systems (GIS) and AI

As the name implies, this technique makes spatial analysis of pollution data across geographic regions. AI models generally map pollutant levels in specific regions and then AI can check the spatial distribution of pollution identifying hotspots and trends (Subramaniam et al., 2022). Thus we can put more targeted efforts in pollution control and environmental planning by the use of this technique Example: AI-driven GIS models can map the concentration of PM2.5 across urban areas, helping local governments identify the most polluted zones and prioritize mitigation efforts in those areas (Sebestyén et al., 2021).

Figure 18. GIS system (SatelliteDataAcquisition.com)

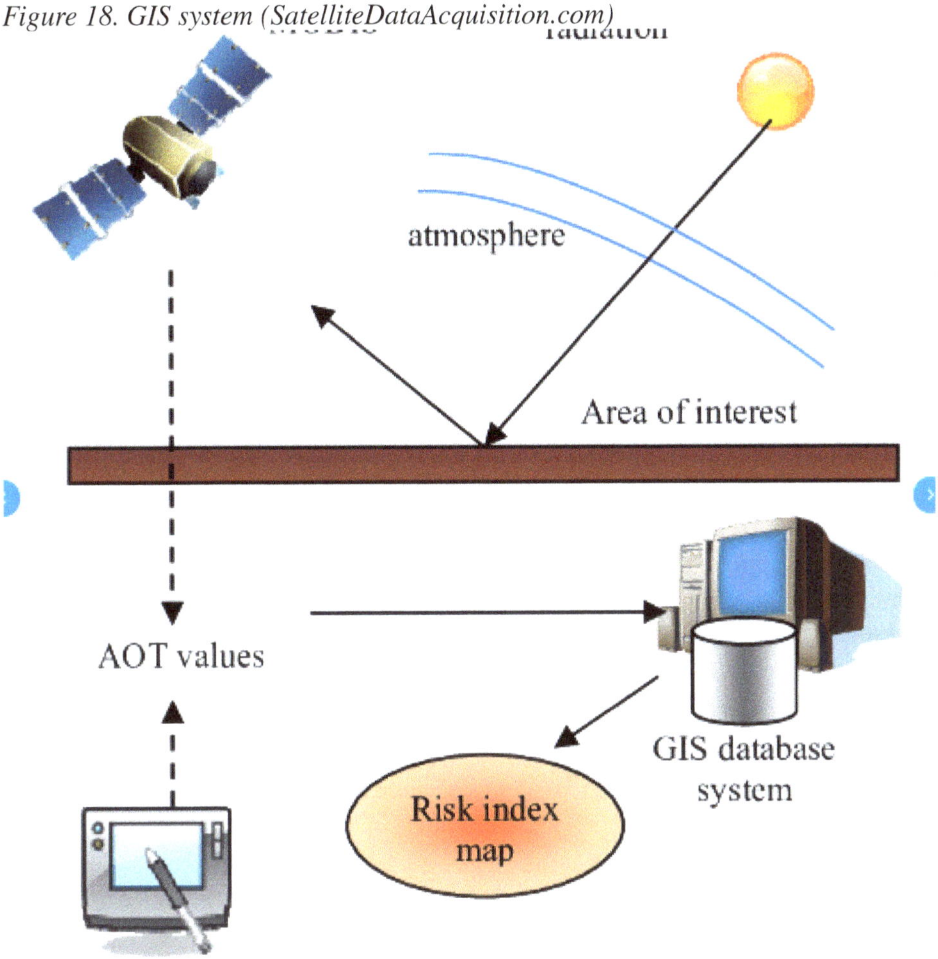

HYBRID AI SYSTEMS FOR POLLUTION CONTROL

Fuzzy Logic and AI

Fuzzy logic deals with uncertainty and imprecision and thus addresses complex decision making in pollution control (Subramaniam et al., 2022). When combined with AI, fuzzy logic can enhance pollution mitigation strategies by considering multiple variables and making decisions based on imprecise data (Subramaniam et al., 2022). Example: Fuzzy logic can help AI and issue prior warnings even having uncertain data at hand.

Expert Systems

As the name implies these systems help human in making informed decisions timely as they have large volume of data and predefined rules and thus can provide support for policymakers in air pollution management. Example: An AI-powered expert system can analyze air quality data, historical trends, and environmental policies to recommend optimal strategies for pollution mitigation, helping decision-makers prioritize actions effectively.

EQUATIONS

In order to determine the Air Quality Index to assess whether the air quality is either good or bad or worst. To achieve this, we must have done some research on the existing literature to develop the basic concepts. In this chapter, we have identified and applied equations from the already existing body of knowledge, with examples of our own through appropriate approach and use it for calculating Air Quality Index correctly.

Equation 1: Air Quality Index (AQI) Calculation

The Air Quality Index helps to understand and analysis the level of air pollution. It can be measured

based on the pollution level in the air.

Air quality index Formula (Mintz, 2012):

Where
Ip = the index for pollutant p
Cp = the rounded concentration of pollutant p
BPHi = the breakpoint that is greater than or equal to Cp
BPLo = the breakpoint that is less than or equal to Cp
IHi = the AQI value corresponding to BPHi
ILo = the AQI value corresponding to BPLo

Example

Cp=30.0
BPHi=40.4
BPLo=22.1

IHi=110

ILo=71

Calculate

Ip = (30.4−22.1) / (40.0−22.1) × (110−71) + 71

Ip = 18.37.9 × 39 + 71

Ip ≈ 0.4317 × 39 + 71 ≈ 16.82 + 71 = 87.82

Result = 88(AQI) which lies in Moderate Category

Equation 2: Regression Model for Pollution Forecasting

The regression model is used to predict pollution levels based on factors like temperature, traffic, and other environmental parameters.

Regression Formula (Loganathan et al., 2022):

Y=β0+β1X1+β2X2+⋯+βnXn+ϵY = \beta_0 + \beta_1 X_1 + \beta_2 X_2 + \dots + \beta_n X_n + \epsilon

Where:

Y = Predicted pollution level (e.g., PM2.5 concentration)

β_0 = Intercept (baseline pollution level when all factors are zero)

β_1, β_2, ..., βn = Coefficients (weights showing how much each factor influences pollution)

X_1, X_2, ..., Xn = Independent variables (factors affecting pollution)

ε = Error term (captures the model's prediction errors)

Dependent Variable (Y): The value we want to predict (e.g., PM2.5 concentration).

Independent Variables (X_1, X_2, ..., Xn): Factors that affect pollution (e.g., traffic volume, temperature).

Coefficients (β_1, β_2, ..., βn): These values show how strongly each factor influences the predicted pollution level.

Error Term (ε): Accounts for any inaccuracies in prediction.

Let's predict the PM2.5 level (Y) based on:

X_1 = Traffic volume = 100 vehicles

X_2 = Temperature = 25°C

Given coefficients:

β_0 = 5 (Intercept)

β_1 = 0.2 (Effect of traffic volume)

β_2 = 1.5 (Effect of temperature)

Using the regression formula:

Y=5+(0.2×100) +(1.5×25) Y = 5 + (0.2 \times 100) + (1.5 \times 25)

Calculate the effect of traffic:

0.2×100=200.2 \times 100 = 20

Calculate the effect of temperature:

1.5×25=37.51.5 \times 25 = 37.5

Sum the results:

Y=5+20+37.5=62.5Y = 5 + 20 + 37.5 = 62.5

So, the predicted PM2.5 level is 62.5 μg/m³.

Equation 3: Neural Network Activation Function (ReLU)

The ReLU (Rectified Linear Unit) is an activation function used in deep learning models to add non-linearity (Hinton, 2010).

ReLU Formula (Glorot & Bordes, 2011):

$f(x, \theta) = \text{Wdec} \max(\text{Wenc}x + \text{benc}, 0) + \text{bdec}$.

Where:

x = Input (usually the weighted sum from the previous layer)

f(x) = Output after applying ReLU

If the input x is positive, the output is x itself.

If the input x is negative, the output is 0.

Example 1: If the input is x = 5:

Since x is positive, the output will be:

f(x)=5f(x) = 5

Example 2: If the input is x = -3:

Since x is negative, the output will be:

f(x)=0f(x) = 0

So, ReLU helps in transforming the input values by turning negative inputs to zero while keeping positive values unchanged.

CHALLENGES IN APPLYING BIG DATA AND AI FOR AIR POLLUTION ANALYSIS

Recently, several machine learning approaches have been used by experts, researchers and others, with different parameters combined for air quality prediction. Consequently, it is difficult to understand the reasons for which algorithms are being selected to solve the different ranges of world challenges. This is further made difficult due to the growing number of studies.

Analyzing air pollution trends using Big data, Artificial intelligence and Machine Learning Involves complex problems particularly focuses on data integrity and accessibility. Environmental data frequently comes from various sources like ground sensors, satellites, and weather stations, it could be impacted by missing data, inconsistent formatting, or a low data. Another big problem is real-time data processing. Several pollution monitoring networks need immediate information for quick decisions nut managing constant flow of real-time data need high performance system and dynamic processing systems. Deep learning (DL) models give high precision but due to their black box nature makes them more ambiguous and harder to decode.

This lack of understanding become a major challenge for policymakers who needs simple actions which depends on model outcomes. Connecting data from multiple sources also remains a technical hurdle. Working with sequential data using Deep learning models, particularly like LSTM (Long Short-Term Memory network) can be computationally costly and requires a large amount of time and resources. This could be achieved by edge computing through lowering the latency and bandwidth consumption, by brining processing power closer to the data source (such as IoT sensors).

Major big data technologies and expert's knowledge are required as the huge amount of data improved additionally, the need for processing and storage resources increases. Privacy and ethical problems, especially those pertaining to data anonymization (Bazai et al., 2017; Bazai & Jang-Jaccard, 2019; Bazai et al., 2021; Bazai & Jang-Jaccard, 2020) and legal compliance, surface when pollution data is linked to health or demographic data.

CONCLUSION

The combination of big data analysis, ML, and Artificial intelligence in air pollution is transforming environmental sustainability. Utilizing more understanding in remote sensing has remarkably upgrade the observation of pollution sources and tendency, permit policymakers and researchers to make data-driven decisions. One of the most significant contributions of AI and ML is to allow them to automate air pollution observing through smart sensors and real-time analytics. This method enables cities to embrace accurate estimate, decrease the impact of pollution on public health and the environment. Hybrid Artificial intelligence models that when combined with many machine learning models, the understanding proceed towards have seen promising results in enhancing forecast precision and robustness.

Despite development, challenges like data inconsistencies, absence of standardization, and computational limitation persist. Future research should pivot on improving data integration technology, high transparency in Artificial intelligence models, and optimizing computational efficiency. Further collaboration between data scientists, environmentalists, and policymakers is essential to ensure that Artificial intelligence -driven pollution control measures are effectively implemented. In addition to strong data management, concentrate on AI- driven solutions such as dynamic traffic pricing and AI-driven solutions to support policy suggestions for smart cities. Big data, ML, and Artificial intelligence have the potential to transform air pollution observing and mitigation strategies.

FUTURE RESEARCH DIRECTIONS

The future of air pollution analysis exists in the connection of developed Artificial intelligence, Machine learning and deep learning technologies to increase predicting accuracy. Neural networks, specifically deep learning techniques can learn difficult, non-linear relationships between pollutants and atmospheric factors which allowing more accurate predictions (Rautela & Goyal, 2024). These techniques are expected to deliver better results than conventional techniques which provides early prediction of pollution level in air. By modifying AI Models deployed directly on IoT-integrated sensors without depending on centralized systems, data may be handled locally, which lowers latency and speeds up reaction times in crucial areas. Low-cost portable sensors or smartphone apps allow individuals to participate in air quality monitoring, thereby expanding the amount and quality of environmental data (Zayed, 2024). This kind of involvement can help close gaps in official monitoring networks, particularly in rural or underrepresented areas. Additionally, tying patterns in air pollution to more general studies on climate change will help us better understand how human activity and global warming affect air quality (Sebestyén et al., 2021). Scientists and decision-makers may address both immediate air pollution and long-term environmental sustainability with the aid of such integration.

REFERENCES

Aamir, M., Bazai, S., Bhatti, U. A., Dayo, Z. A., Liu, J., & Zhang, K. (2023). Applications of machine learning in medicine: Current trends and prospects. In *2023 Global Conference on Wireless and Optical Technologies (GCWOT)* (pp. 1-4). IEEE.

Asif, M., Song, H., Zia, M. A., Ali, S., Awwad, E. M., Ghadi, Y. Y., Al-Razgan, M., Sadiq, T., & Bhatti, U. A. (2024). Shedding light on diagnostic precision: GANs for low light endoscopy image enhancements. Research. https://doi.org/DOI: 10.21203/rs.3.rs-421332s1/v1

Baloch, L., Bazai, S. U., Marjan, S., Aftab, F., & Aslam, S. (2023). A Review of Big Data Trends and Challenges in Healthcare. 14(May), 1320–1333. https://doi.org/DOI: 10.14716/ijtech.v14i6.6643

Bazai, S. U., & Jang-Jaccard, J. (2019). SparkDA: RDD-based high-performance data anonymization technique for Spark platform. In International conference on network and system security (pp. 646-662). Cham: Springer International Publishing.

Bazai, S. U., & Jang-Jaccard, J. (2020). In-memory data anonymization using scalable and high performance rdd design. *Electronics (Basel)*, *9*(10), 1732.

Bazai, S. U., Jang-Jaccard, J., & Alavizadeh, H. (2021). A novel hybrid approach for multi-dimensional data anonymization for apache spark. *ACM Transactions on Privacy and Security*, *25*(1), 1–25.

Bazai, S. U., Jang-Jaccard, J., & Wang, R. (2017). Anonymizing k-NN classification on MapReduce. *International Conference on Mobile Networks and Management*, 364–377.

Bhatti, M. A., Zeeshan, Z., Syam, M. S., Bhatti, U. A., Khan, A., Ghadi, Y. Y., & Alsenan, S. (2024d). Advanced plant disease segmentation in precision agriculture using optimal dimensionality reduction with fuzzy C-means clustering and deep learning. IEEE Journal of Selected Topics in Applied Earth Observations and Remote Sensing.

Bhatti, U., Bazai, S., Hussain, S., Fakhar, S., Ku, C., Marjan, S., Por, Y., & Jing, L. (2023a). Deep learning-based trees disease recognition and classification using hyperspectral data. *Computers, Materials & Continua*, *77*, 681–697.

Bhatti, U. A., Bhatti, M. A., Tang, H., Syam, M. S., Awwad, E. M., Sharaf, M., & Ghadi, Y. Y. (2024c). Global production patterns: Understanding the relationship between greenhouse gas emissions, agriculture greening and climate variability. *Environmental Research*. PMID: 38169167

Bhatti, U. A., Hao, T., Khan, A., Ghadi, Y. Y., Bhatti, M. A., & Khan, K. A. (2024b). Investigating the nexus between energy, socio-economic factors and environmental pollution: A geo-spatial multi-regression approach. *Gondwana Research.*

Bhatti, U. A., Masud, M., Bazai, S. U., & Tang, H. (2023). Investigating AI-based smart precision agriculture techniques. [). Frontiers Media SA.]. *Frontiers in Plant Science, 14*, 1237783. PMID: 37496855

Bhatti, U. A., Mengxing, H., Li, J., Bazai, S. U., & Aamir, M. (2024a). *Deep learning for multimedia processing applications: Volume two: Signal processing and pattern recognition.* CRC Press.

Bhatti, U. A., Zeeshan, Z., Nizamani, M. M., Bazai, S., Yu, Z., & Yuan, L. (2022). Assessing the change of ambient air quality patterns in Jiangsu Province of China pre-to post-COVID-19. *Chemosphere, 288*, 132569. PMID: 34655644

Chen, F., Zhang, W., Fawzi, M., Mfarrej, B., & Hamzah, M. (2024). Ecotoxicology and Environmental Safety Breathing in danger : Understanding the multifaceted impact of air pollution on health impacts. 280(December 2023). https://doi.org/ DOI: 10.1016/j.ecoenv.2024.116532

Chen, L., Zhang, Y., Xu, B., Shao, K., Yan, J., & Bhatti, U. A. (2024). A lot-based VMD-CNN-BiGRU indoor mariculture water quality prediction method including attention mechanism. *International Journal of High Speed Electronics and Systems.* Advance online publication. DOI: 10.1142/S0129156425400105

Dubey, S. R., Singh, S. K., & Chaudhuri, B. B. (2022). Activation Functions in Deep Learning: A Comprehensive Survey and Benchmark.

Essamlali, I., Nhaila, H., & El Khaili, M. (2024). Supervised Machine Learning Approaches for Predicting Key Pollutants and for the Sustainable Enhancement of Urban Air Quality. *Systematic Reviews.*

Feng, S., Liu, Q., Patel, A., Bazai, S. U., Jin, C., Kim, J. S., Sarrafzadeh, M., Azzollini, D., Yeoh, J., & Kim, E. (2022). Automated pneumothorax triaging in chest X-rays in the New Zealand population using deep-learning algorithms. *Journal of Medical Imaging and Radiation Oncology, 66*(8), 1035–1043. PMID: 35224858

Glorot, X., & Bordes, A. (2011). Deep Sparse Rectifier. *Neural Networks, 15*, 315–323.

Gómez, N. I. M., Arévalo, J. L. D., & Jiménez, P. A. L. (2021). Air quality and urban sustainable development : The application of machine learning tools. *International Journal of Environmental Science and Technology, 18*(4), 1029–1046. DOI: 10.1007/ s13762-020-02896-6

Guo, Q., Ren, M., Wu, S., & Sun, Y. (2022). Applications of artificial intelligence in the field of air pollution : A bibliometric analysis.

Haider, A., Hussain, L., Tareen, A. W., Bazai, S. U., Aslam, S., Neo, M., & Amphawan, A. (2024). A comparative study of machine learning techniques for accurate disease prediction using symptom-based diagnosis. In AIP Conference Proceedings (Vol. 3153, No. 1). AIP Publishing.

Hamza, M., Bazai, S. U., Ghafoor, M. I., Ullah, S., Akram, S., & Khan, M. S. (2023). Generative adversarial networks (GANS) video framework: A systematic literature review. In *2023 International Conference on Energy, Power, Environment, Control, and Computing (ICEPECC)* (pp. 1-5). IEEE.

Han, H., Bazai, S. U., Bhatti, M. A., Basit, A., Wahid, A., Bhatti, U. A., Ghadi, Y. Y., & Algarni, A. (2024). Hybrid Climate Forecasting: Variational Mode Decomposition and Convolutional Neural Network with Long-Term Short Memory. *Polish Journal of Environmental Studies, 33*(2).

Han, H., Hasnain, A., Bhatti, U. A., Yue, Y., He, Y., & Wei, G., & ur Rahman, W. (2024a). Assessing the spatio-temporal impact of the COVID-19 pandemic lockdown on air quality in Jiangsu Province, China. *Environment, Development and Sustainability*, •••, 1–21.

Han, H., Zeeshan, Z., Assam, M., Khan, F. U., Khan, W., Asif, M., Bhatti, U. A., Hasnain, A., Awwad, E. M., & Sarhan, N. (2024c). *Deep learning techniques for enhanced mangrove land use and land change from remote sensing imagery: A blue carbon perspective*. Big Data Research., DOI: 10.1016/j.bdr.2024.100478

Han, H., Zeeshan, Z., Talpur, B. A., Sadiq, T., Bhatti, U. A., & Awwad, E. M. (2024b). Studying long-term relationship between carbon emissions, soil, and climate change: Insights from a global Earth modeling framework. *International Journal of Applied Earth Observation and Geoinformation, 130*, 118049.

Haq, S. U., Bazai, S. U., Fatima, A., Marjan, S., Yang, J., Yee, L., & Ku, C. S. (2023). Reseek-arrhythmia: Empirical evaluation of ResNet architecture for detection of arrhythmia. *Diagnostics (Basel), 13*(18), 2867. PMID: 37761234

Hasnain, A., Hashmi, M. Z., Khan, S., Bhatti, U. A., Min, X., Yue, Y., He, Y., & Wei, G. (2024). Predicting ambient PM2.5 concentrations via time series models in Anhui Province, China. *Environmental Monitoring and Assessment, 196*, 487. DOI: 10.1007/s10661-024-12644-9 PMID: 38687422

He, B., & Bhatti, U. A. (2024). Smart cities and smart networks: AI applications in urban geography and industrial communication. *International Journal of High Speed Electronics and Systems*. Advance online publication. DOI: 10.1142/S0129156424401220

. Hinton, G. E. (2010). Rectified Linear Units Improve Restricted Boltzmann Machines. 3.

Huang, M., Zhang, X. S., Bhatti, U. A., Wu, Y. Y., Zhang, Y., & Ghadi, Y. Y. (2024). An interpretable approach using hybrid graph networks and explainable AI for intelligent diagnosis recommendations in chronic disease care. *Biomedical Signal Processing and Control*, *91*, 105913.

Kang, G. K., Gao, J. Z., Chiao, S., Lu, S., & Xie, G. (2018). Air Quality Prediction. *Big Data and Machine Learning Approaches.*, *9*(1). Advance online publication. DOI: 10.18178/ijesd.2018.9.1.1066

Kılınç, M. (2024). *Information Systems. December. Kılınç, M. (2024).* Data-Driven Analytical Techniques in Geographic Information Systems.

Kommineni, H. P., & International, M. (2024). Data-Driven Business Intelligence in Energy Distribution : Analytics and Environment- Focused Approaches. December. https://doi.org/DOI: 10.18034/gdeb.v13i1.779

Lee, S., Hyun, C., & Lee, M. (2023). Machine Learning Big Data Analysis of the Impact of Air Pollutants on Rhinitis-Related Hospital Visits. 1–17.

Li, X., Hou, B., Tang, H., Talpur, B. A., Zeeshan, Z., Bhatti, U. A., Liao, J., Liu, J., & Zhang, Y. (2024). Abnormal crops image data acquisition strategy by exploiting edge intelligence and dynamic-static synergy in smart agriculture. *IEEE Journal of Selected Topics in Applied Earth Observations and Remote Sensing*.

Loganathan, A., Sumithra, P., & Deneshkumar, V. (2022)... *Estimation of Air Quality Index Using Multiple Linear Regression.*, (December). Advance online publication. DOI: 10.12691/aees-10-12-3

Mintz, D. (2012). *Technical Assistance Document for the Reporting of Daily Air Quality – the Air Quality Index*. AQI.

Nabeel, S. M., Bazai, S. U., Alasbali, N., Liu, Y., Ghafoor, M. I., Khan, R., & Por, L. Y. (2024). Optimizing lung cancer classification through hyperparameter tuning. *Digital Health*, *10*, 20552076241249661. PMID: 38698834

Nawaz, S. A., Li, J., Bhatti, U. A., Bazai, S. U., Zafar, A., Bhatti, M. A., Mehmood, A., & Ain, Q. ul, & Shoukat, M. U. (2021). A hybrid approach to forecast the COVID-19 epidemic trend. *PLoS One*, *16*(10), e0256971. PMID: 34606503

Noor, R., Wahid, A., Bazai, S. U., Khan, A., Fang, M., Syam, M. S., & Bhatti, U. A. (2024). DLGAN: Undersampled MRI reconstruction using deep learning-based generative adversarial network. *Biomedical Signal Processing and Control*, *93*, 106218.

Noor, S., Bazai, S. U., Ghafoor, M. I., Marjan, S., Akram, S., & Ali, F. (2023). Generative adversarial networks for anomaly detection: a systematic literature review. In 2023 4th International Conference on Computing, Mathematics and Engineering Technologies (iCoMET) (pp. 1-6). IEEE.

Peng, M., Liu, Y., Khan, A., Ahmed, B., Sarker, S. K., Ghadi, Y. Y., Bhatti, U. A., Al-Razgan, M., & Ali, Y. A. (2024b). Crop monitoring using remote sensing land use and land change data: Comparative analysis of deep learning methods using pre-trained CNN models. *Big Data Research*, *36*, 100448. DOI: 10.1016/j.bdr.2024.100448

Peng, M., Liu, Y., Qadri, I. A., Bhatti, U. A., Ahmed, B., Sarhan, N. M., & Awwad, E. M. (2024a). Advanced image segmentation for precision agriculture using CNN-GAT fusion and fuzzy C-means clustering. *Computers and Electronics in Agriculture*, *226*, 109431.

Rautela, K. S., & Goyal, M. K. (2024). Transforming air pollution management in India with AI and machine learning technologies. *Scientific Reports*, •••, 1–19. DOI: 10.1038/s41598-024-71269-7 PMID: 39223178

Sebestyén, V., Czvetko, T., & Abonyi, J. (2021). The Applicability of Big Data in Climate Change Research. *The Importance of System of Systems Thinking.*, *9*(March), 1–26. DOI: 10.3389/fenvs.2021.619092

Subramaniam, S., Raju, N., Ganesan, A., Rajavel, N., & Dixit, S. (2022). Artificial Intelligence Technologies for Forecasting Air Pollution and Human Health : A Narrative Review. 1–36.

Taj, G., & Navi, M. (2020). Air quality monitoring and assessment using. *IoT*. Advance online publication. DOI: 10.1088/1757-899X/955/1/012006

Topbaş, A., Jamil, A., Hameed, A. A., Ali, S. M., Bazai, S., & Shah, S. A. (2021). Sentiment analysis for covid-19 tweets using recurrent neural network (RNN) and bidirectional encoder representations (BERT) models. 2021 International Conference on Computing, Electronic and Electrical Engineering (ICE Cube), 1–6.

Wu, D. D., Lambert, J., Olson, D. L., Ding, S., Wu, D., Zhao, L., & Li, X. (2022). Smart Healthcare Engineering Management and Risk Analytics. https://doi.org/ DOI: 10.1007/978-981-19-2560-3

Wu, G., Bhatti, U. A., Nawaz, S. A., Hasnain, A., Bhatti, M. A., Fahim, A., & Kaleri, A. A. (2022). COVID-19 and Air Pollution : Air Quality Impact in 13 Cities of the Jiangsu Province of China. 31(5), 4907–4916. https://doi.org/DOI: 10.15244/ pjoes/149714

Yu, C., Tan, J., Cheng, Y., & Mi, X. (2024). Data analysis and preprocessing techniques for air quality prediction : A survey. *Stochastic Environmental Research and Risk Assessment*, *38*(6), 2095–2117. DOI: 10.1007/s00477-024-02693-4

Zaland, Z., Bazai, S. U., Marjan, S., & Ashraf, M. (2021). Three-tier password security algorithm for online databases. In 2021 2nd International Informatics and Software Engineering Conference (IISEC) (pp. 1-6). IEEE.

Zayed, R. (2024). Big Data AI System for Air Quality Prediction. January 2022.

Zhang, C., Iqbal, I., Bhatti, U. A., Liu, J., Awwad, E. M., & Sarhan, N. (2024a). ResNet50 in remote sensing and agriculture: Evaluating image captioning performance for high spectral data. *Environmental Earth Sciences*, *83*(23), 1–16.

Zhang, Y., Chen, J., Ma, X., Wang, G., Bhatti, U. A., & Huang, M. (2024b). Interactive medical image annotation using improved Attention U-net with compound geodesic distance. *Expert Systems with Applications*, *237*, 121282. DOI: 10.1016/j. eswa.2023.121282

Zhang, Y., Xu, S., Chen, H., Bhatti, U. A., & Huang, M. (2024c). Context-aware environment online monitoring for safety autonomous vehicle systems: An automata-theoretic approach. *Journal of Cloud Computing (Heidelberg, Germany)*, *13*(6). Advance online publication. DOI: 10.1186/s13677-023-00567-8

Chapter 6
Big Data and Cloud Computing for Climate-Oriented Remote Sensing

R. N. Ravikumar
https://orcid.org/0009-0009-3705-1681
Marwadi University, Rajkot, India

S. Aarthi
https://orcid.org/0009-0006-9064-2091
Marwadi University, Rajkot, India

Sivakumar Nadarajan
https://orcid.org/0000-0001-5039-072X
Marwadi University, Rajkot, India

Gularam Masharipova
Alfraganus University, Tashkent, Uzbekistan

Zokir Mamadiyarov
Termez University of Economics and Service, Termez, Uzbekistan

S. Tamilselvi
Kongu Engineering College, Erode, India

ABSTRACT

Improved climate change surveillance arises from rapid remote sensing via satellites, UAVs, and sensor networks. However, real-time data processing faces storage and format challenges due to growing data volumes. Deep learning enhances remote sensing for climate monitoring when integrated with big data tools in cloud environments. Scientists use multispectral, hyperspectral, and radar systems for analyzing surface temperatures, deforestation, sea levels, and glaciers. Research on Google Earth Engine's structure explores tool performance between AWS and Azure for geolocation-based deep learning. Image analysis benefits from cloud clusters and statistical models to detect climate anomalies. Cloud-powered systems monitor air pollution and floods. To ensure effectiveness, large-scale systems must follow pro-

DOI: 10.4018/979-8-3373-3501-8.ch006

tocols for data processing, optimization, and quality control. The section concludes with data privacy, then shifts to hybrid edge-cloud models and affordable real-time climate analytics. These methods inform policy with critical adaptation data.

INTRODUCTION TO REMOTE SENSING AND CLIMATE CHANGE

Remote measurement has emerged as a transformation tool for monitoring climate change, and delivers multi-resolution, real-time and satellites from satellites, unmanned air runs (UAV) and ground-based sensor networks. These technologies enable tracking of significant environmental variables, such as surface temperature, deforestation, sea level growth, and glacial changes, planetary changes provide unique insights into changes. However, the exponential growth of remote objective data presents important challenges, including storage restrictions, calculation requirements and effective treatment of asymmetrical data sets, (Abdalla, 2024).

To solve these challenges, integration of Big Data Analytics, Cloud Computing and Deep Learning has brought a revolution under climate monitoring. Cloud platforms such as Google Earth Engine, AWS and Microsoft Azure provide scalable solutions for large dataset storage and processing, while deep learning models increase pattern recognition and future abilities. This chapter explains how these advanced technologies coordinate how to improve climate change monitoring, support politics and enable adaptive reactions to global environmental changes, (Chen & Lin, 2025). By taking advantage of these devices, researchers can unlock new opportunities for real -time climate analysis on a planetary scale.

Role of Remote Sensing in Climate Monitoring

Remote measurement has become an indispensable tool for climate scientists and decision makers, and provides a broad, global perspective on environmental changes. Satellite-based sensors, UAVs and observation networks for the ground collect continuous data in several alphabetic bands, such as competent monitoring of large climate indicators such as atmospheric structure, land surface temperature, ice dynamics and botanical changes. These technologies provide unique benefits including global coverage, frequent temporary resolution and ability to reach remote or dangerous areas, (Dewitte, 2020). Multisectoral and hyperspectral imaging enables detailed analysis of changes in land use, while radar systems such as SAR provide all weather capacity to increase sea level and monitor soil moisture. Thermal infrared sensors track urban heating islands and sea temperatures, and the lidar systems measure the forest's canopy structure and ice thickness properly. The long -term data archives for assignments such as Landsat and Modis have created an invaluable time series to detect climate trends, (Gudelė & Visockienė, 2023). Recent progress in sensor technology has improved the spatial resolution of under -measurement levels and temporary resolution for careful monitoring, which enables more accurate drawing of climate phenomena. These possibilities support important applications, including the evaluation of carbon stock, predict the drought and monitoring of extreme weather events, which create foundations related to climate models and moulding strategies, (Li et al., 2021).

Challenges: Data Volume, Heterogeneity, and Real-Time Processing

The exponential growth of remote target data presents important computational and analytical challenges. Modern satellite constellations produce petabytes with annual data, often more than more gigabytes with single-soluble high-resolution images. This data supplies traditional storage systems and processes workflows, and requires innovative effective handling solutions. Asymmetry reduces these challenges, as climate monitoring integrates different data types from optical and radar images to atmospheric sound and in -situ measurement with each different forms, solutions and coordination systems, (Ni et al., 2025). The temperature of data with multiple sources adds another layer of adjustment complexity. The demand for real -time treatment is especially intense for initial warning systems that monitor extreme weather events or rapid environmental changes. Problems with delay in data down, proclamation and analysis can prevent timely decision-making. Data quality problems, including cloud cover, atmospheric intervention and sensor noise, require strong preprocessing pipelines. In addition, the calculation intensity of advanced analysis techniques such as Pixel-Timer chain analysis or 3D modelling is pushing the boundaries of traditional data processing infrastructure, (Novikarany, 2025). These challenges require scalable, automated solutions that can handle a variety of data sets, when providing timely insight, for climate monitoring and response.

Figure 1. Climate data processing challenges

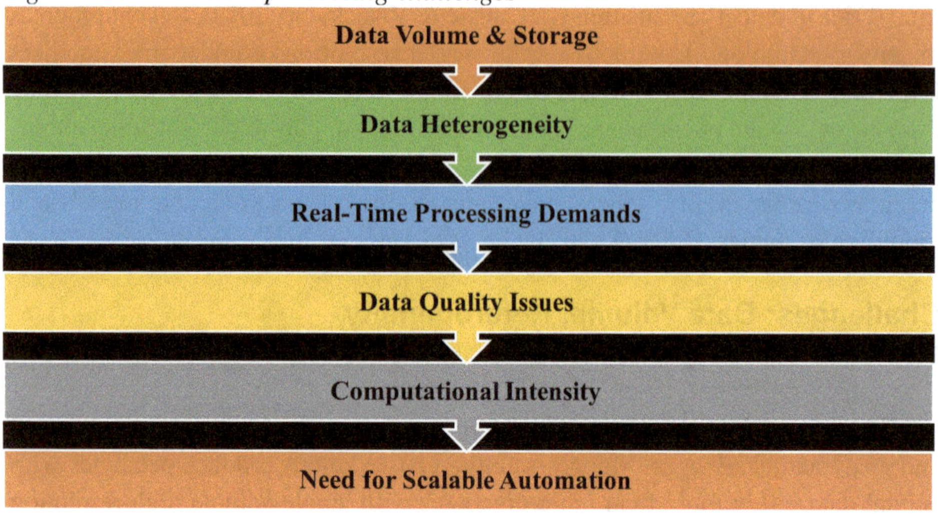

Figure 1 visually represents the key computational and data management challenges in climate remote sensing workflows. It outlines a layered flow from "Data Volume & Storage" to the "Need for Scalable Automation," emphasizing the sequential nature of technical bottlenecks such as heterogeneity, quality, real-time processing, and computational load.

Synergy of Deep Learning, Big Data, and Cloud Computing

The convergence of deep learning, big data technologies and cloud computing has revolutionized climate-oriented remote measurement. Deep learning models, especially CNN and transformer architecture, are distinguished in extracting complex patterns from huge image datasets, enabling automated functional detections in land coverage classification, detection of change and identification of nonconformities, (Park et al., 2021). Large data frames such as Apache Sparks and Hadoops provide the processor power required to handle the Petabyte scale of remote goal archives, while effective data structures such as reshaping tiles and spatial indices adapt to geo-smart questions. Cloud computing platforms provide elastic scalability, which provides access to GPU clusters for model training and altitude calculation resources for large-scale analysis. Services such as Google Earth Engine have a large dataset with calculation resources to democratized access to analysis on the scope of planets, (Pettorelli, 2019). This technical coordination enables end-to-end pipelines for action-rich insights from raw data intake, supports applications ranging from daily forest harvest notification to decade-to-decade-prime analysis. Hybrid architecture enables real-time workflow by combining Edge Computing for data collection with cloud-based models serving, (Wang & Yésou, 2018). The union learning methods allow collaboration model training in distributed datasets while maintaining the data's confidentiality. Together, these technologies convert climate monitoring to an insight rich paradigm from a data limiting, strengthens researchers and decision makers with outstanding analytical abilities.

REMOTE SENSING TECHNOLOGIES AND DATA ACQUISITION

Modern climate monitoring depends on advanced remote measurement technologies that capture high -resolution, multispectral data from rooms, air and ground -based platforms. Satellites such as Landsat, Sentinel and Modis provide global coverage of land surface temperatures, vegetative health and atmospheric structure, while Unmanned Aerial Vehicles (UAV) enables located, high -compatible imaging to study disaster evaluation and ecosystem. Seating networks complement these systems with a real -time measurement of air quality, soil moisture and weather conditions, (Aggarwal & Papay, 2022).

Key remote sensing techniques include:

- Multispectral & Hyperspectral Imaging – Detects subtle changes in vegetation, water quality, and urban heat islands.
- Synthetic Aperture Radar (SAR) – Penetrates clouds to monitor deforestation, flood dynamics, and glacial movements.
- LiDAR (Light Detection and Ranging) – Measures 3D forest structure, ice thickness, and coastal erosion.

These technologies generate vast datasets that feed into AI-driven climate models, helping track carbon emissions, sea-level rise, and extreme weather events. The integration of satellite, aerial, and IoT sensor data creates a comprehensive monitoring framework essential for climate resilience and policy-making, (Cai et al., 2021).

Table 1. Overview of remote sensing technologies and their applications

Remote Sensing Technology	Primary Features	Climate Applications	Advantages
Multispectral Imaging	Captures data in a few broad spectral bands	Vegetation analysis, water body identification	Simplifies land use classification, lower data volume
Hyperspectral Imaging	Collects data in hundreds of narrow spectral bands	Soil composition, crop stress detection, coral reef health	Detailed chemical/material analysis
Synthetic Aperture Radar (SAR)	Active microwave imaging, penetrates cloud cover	Flood monitoring, deforestation tracking, glacier movement	All-weather, day/night operation
LiDAR	Uses laser pulses to generate 3D elevation models	Forest canopy structure, glacier thickness, coastal erosion	High-resolution 3D mapping

Table 1 provides a comparative summary of the most widely used remote sensing technologies Multispectral Imaging, Hyperspectral Imaging, Synthetic Aperture Radar (SAR), and LiDAR highlighting their core features, primary climate-related applications, and operational advantages. These technologies form the foundation of modern environmental monitoring systems and support critical climate research areas such as flood detection, vegetation stress assessment, deforestation tracking, and glacial retreat analysis. By outlining their unique capabilities, this table helps clarify how each modality contributes to a comprehensive climate observation framework.

Satellite, UAV, and Sensor Networks

Satellite, Unmanned Aerial Vehicles (UAV) and sensor networks are integrated parts of modern environmental monitoring and climate analysis systems. The satellites provide observation of macro level, and cover spacious geographical regions with high temporary resolution, which is invaluable to track long -term trends in climate change, vegetation and atmospheric conditions, (Jiao, 2024). UAVs, or drones, provide more local and flexible surveillance solutions, able to capture high-resolution images and data that are difficult to reach hilly areas or disaster areas. These are especially effective for short-term, targeted assignments such as crop health or subsequent storm damage. On the ground, sensors play an important role in collecting real -time data on the variables such as air temperatures, moisture, soil moisture and carbon dioxide levels. When combined, these technologies create an integrated system that is able to provide a broad, multi -level understanding of environmental conditions. Satellite-based microbus and coordination between UAV or Sensor-based micro-Detta allows large-scale trend analysis and the detection of hyperlocal deviations, (Kior, Sukhov, & Sukhova, 2021). In addition, the progress of age calculation and wireless communication increases in real -time treatment and transfer of environmental data. These networks with multiple platforms are important in the processes of decision -related decisions related to climate policy, disaster response, agriculture and sustainable development schemes.

Multispectral/Hyperspectral Imaging, Radar, and LiDAR

Multispectral and hyperspectral offer advanced remote measurement options for environmental monitoring, geological examination and accurate agriculture. Multispectral imaging contains a limited amount of data of specific wavelengths, enabling the basic classification of vegetation, water bodies and land use, (Kior et al., 2024). Hyper -Collects -Connection collects reverse data in hundreds of continuous spectral bands, allowing a detailed analysis of materials and their chemical properties -important to identify environmental toxins, evaluate crop health or monitor coral reef. Radar (Radio Detection and Range) systems, especially synthetic aperture (sar), all-visits, imaging of day-and-night and are useful in supervision of surface formation, deforestation and urban development. Lidar (light detection and range) uses laser pulses to generate high resolution of 3D maps over the region and vegetative structure, making it invaluable in forestry, flood modelling and infrastructure, (Mookambiga & Gomathi, 2016). These technologies complement each other and are often used at the tip of maximum data prose permit. For example, hyperspectral data can detect plant tension while the lidar maps the canopy structure. As the data volumes grow, machine learning and AI apply quickly to analyze these rich data sets, highlight patterns and predicted environmental changes. Together, they represent the state -Art -species of earth observation and environmental diagnosis, (Shu & Sun, 2020).

Key Climate Variables: Surface Temperature, Deforestation, Ocean Height, Glaciers

In order to understand and reduce climate change, it is necessary to monitor the most important climate variables such as surface temperature, deforestation, ocean height and mass of ice. Surface temperature is one of the most basic indicators, which is measured through satellite infrared sensors and ground-based stations, (Yokoya & Iwasaki, 2013). The long-term temperature data reveal changes in global warming trends, heat wave patterns and urban microclimate. Forest harvesting is tracked using high-resolution and radar systems satellite images, which detect changes in land coverage and forest erosion. This is important because forests act as carbon washer and their destruction accelerates climate change. Sea height, or monitoring of sea level is carried out by means of a satellite ultimately system, which measures the height of the sea surface with millimetres level accuracy, (Zhang et al., 2016). The sea level is a direct result of melting rising ice hats and thermal expansion, which threatens coastal communities worldwide. Glacier monitoring includes multisectoral imaging and radar interferometry to track changes in ice mass and flow dynamics. The Retreat of Glaciers is one of the most visible signals of global warming and has a deep impact on the availability of fresh water and the increase in sea level. The exact observation of these variables is important for policy design aimed at climate modelling, risk assessment and optimization and moulding strategies, (Boumrah, Garbaya, & Radgui, 2023).

Table 2. Key climate variables monitored via remote sensing

Climate Variable	Sensing Technology	Primary Data Sources (Satellites/Sensors)	Significance in Climate Modeling
Surface Temperature	Thermal Infrared Sensors	MODIS, Landsat, Sentinel-3 SLSTR	Tracks global warming trends, urban heat islands, and heatwave patterns
Deforestation	Multispectral Imaging, SAR	Landsat, Sentinel-1, Sentinel-2, PlanetScope	Monitors land cover change and forest loss impacting carbon sequestration
Ocean Height	Radar Altimetry	Jason-3, Sentinel-6, TOPEX/Poseidon	Measures sea-level rise due to ice melt and thermal expansion
Glaciers	SAR, LiDAR, Multispectral Imaging	ICESat-2, Sentinel-1, TerraSAR-X, Landsat	Assesses glacier retreat, ice mass loss, and freshwater availability

Table 2 outlines the essential climate variables that are routinely monitored through remote sensing technologies, including surface temperature, deforestation, sea level (ocean height), and glacial mass. For each variable, the table identifies the most suitable sensing technologies (e.g., thermal infrared sensors, SAR, LiDAR), corresponding data sources (such as Landsat, MODIS, or Sentinel satellites), and explains their role in supporting accurate climate change modeling and long-term environmental assessment. By organizing these variables systematically, the table emphasizes their interconnectedness and importance for scientific research, early warning systems, and global climate policy decisions.

BIG DATA CHALLENGES AND SOLUTIONS IN CLIMATE REMOTE SENSING

Climate remote measurement produces huge amounts of data from satellites, UAVs and ground-based sensors, which presents important major data challenges.

These include data storage, real -time processing, integration of heterogeneous data types (eg images, radars and lidar), and secure computer courses and stability over time. In addition, the complexity is in extracting meaningful insights from high-dimensional data set analytical challenges. These problems require scalable cloud-based storage solutions, advanced data compression algorithms and high-performance data processing for real -time processing. Interoperability frames and standardization protocols help to integrate different datasets, (Krämer & Senner, 2015). Machine learning and AI models are used to detect quick computer classification, deviations and automatic modelling. In addition, Edge Computing is used to process data near the source, which can reduce the requirements for delay and bandwidth. By taking advantage of these technological advances, researchers and decision makers can convert raw climate data to action -rich insights, improve environmental monitoring, disaster response and long -term climate strap, (Aarthi & Vinayagasundaram, 2022).

Scalability Issues: Storage, Processing, and Streaming

Climate remote measurement systems face significant scalability challenges due to surrender volumes, speeds and different types of data generated from satellites, UNAs and ground-based sensors. Storage becomes a bottleneck in the form of high-resolution images, multi-spectral data and sensor current, which accumulates on petabyte parameters, (Agbehadji et al., 2023). Traditional databases often struggle with dynamic, unnecessary nature of geophysical data, requires scalable, the sky environment requires distributed storage solutions such as object storage. The treatment is equally demanding-tendency detection, classification or prediction requires high data sets for data processing and parallel. Without effective load balance and resource allocation, treatment can seriously affect the deadline for delayed decision -making. Streaming is another major obstacle, especially for real-time applications such as disaster detection, initial warning system or accurate agriculture. Scalable architecture requires strong, scalable architecture to swallow and analyze continuous data flow and at the same time maintain accuracy and low delay. The solution of these problems includes data processing frames, edge calculation and advanced compression algorithms to reduce data transfer. Development systems are able to scale horizontally by adding more nodes instead of increasing individual engine power it has become necessary, (Anand, 2025). It is important to address these scalability problems to ensure timely and accurate insights from climate remote data, which increases responsibility and stability in environmental monitoring and governance.

Figure 2. Scalability challenges in climate remote sensing

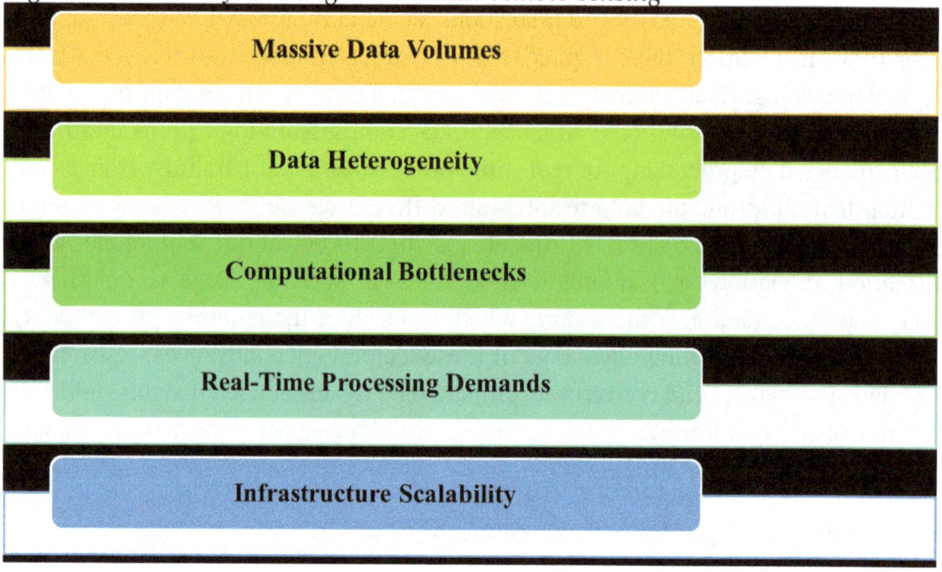

Figure 2 illustrates the core technical challenges in managing climate remote sensing data pipelines. It highlights five key issues massive data volumes, data heterogeneity, computational bottlenecks, real-time processing demands, and infrastructure scalability crucial for designing efficient and responsive climate monitoring systems.

Table 3. Scalability challenges and corresponding technological solutions

Scalability Challenge	Problem Description	Technological Solution	Benefits
Data Storage	Petabyte-scale image and sensor data exceed traditional storage limits	Cloud storage (e.g., AWS S3, Google Cloud Storage, Azure Blob)	Elastic capacity, cost-effective, redundant backups
Data Processing	High-resolution images and real-time analytics strain local infrastructure	Distributed computing (e.g., Hadoop, Apache Spark)	Parallel processing, scalable compute clusters
Data Streaming	Real-time data ingestion from satellites and sensors with low latency needs	Stream processing engines (e.g., Apache Kafka, AWS Kinesis)	Continuous data flow, low-latency alerting for real-time response

Table 3 identifies the major scalability challenges encountered in climate data processing specifically in storage, computation, and streaming. For each challenge, it outlines the nature of the problem and pairs it with an appropriate technological solution, such as cloud storage, distributed computing frameworks, and real-time data streaming tools. The table further explains the benefits of each solution in terms of scalability, efficiency, and responsiveness. It serves as a practical reference for designing robust, high-throughput systems that can manage the volume, velocity, and variety of climate data in real-world monitoring applications.

Distributed Frameworks (Hadoop, Spark) for Geospatial Data

Distributed data processing frame Climate remote measurement such as Hadoop and Apache Spark have become indispensable tools for the management and processing of geopolitical data. Hadoop's ecosystem, built around Hadoop Distributed File System (HDFS), provides a scalable storage infrastructure that supports fault tolerance and high availability. It makes it possible to treat in parallel with large - scale data sets using MapReduce, which is well suited to detect terrain -oriented analysis tasks such as terrain classification, deforestation maps or changes in land use. On the other hand, Apache provides significant performance by enabling sparks, processing and support for recurrent algorithms-the ideal analysis of ideal and climate variables for machine learning models, (Bhagyalakshmi, Harinivas, & Shastry, 2024). Spark Machine also supports a series of libraries such as MLlib for learning and Graphx for graph treatment, which is important for network -based geotastic analysis. Both frameworks support integration with units such as Apache Hive, Pig and localized extensions such as Geomesa and Gospark, allowing for spatial questions and analysis on a scale. They can also be distributed on cloud platforms, making it easier to calculate resources dynamically. By taking advantage of these distributed frameworks, researchers and analysts can effectively handle the complexities of geopolitical large data, achieve real -time insights and develop scalable, reproductive workflows for climate research, policy design and environmental protection, (Huntingon et al., 2017).

Figure 3. Distributed frameworks for scalable climate data processing

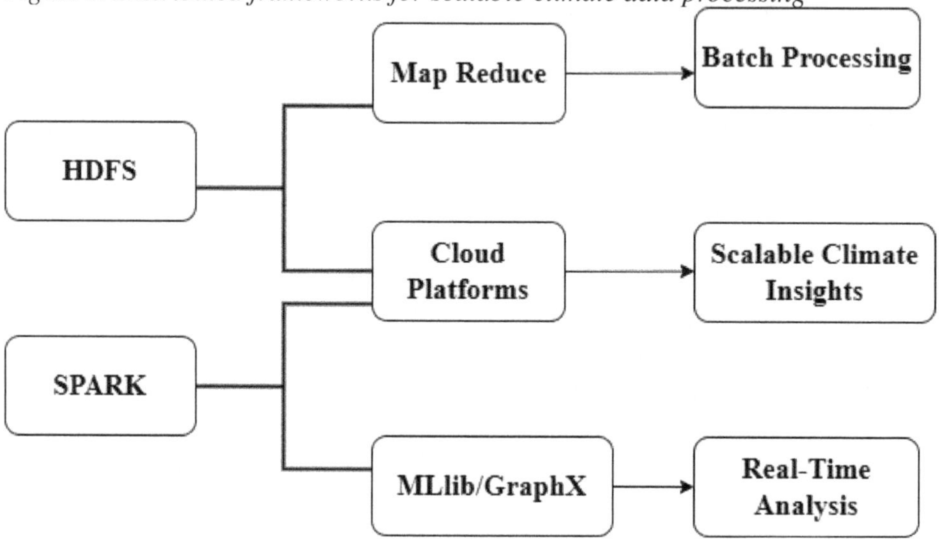

Figure 3 represents a data processing architecture integrating HDFS and Apache Spark for climate analytics. It shows how HDFS supports batch processing via MapReduce, while Spark leverages MLlib/GraphX and cloud platforms to deliver real-time analysis and scalable climate insights.

Real-Time Analytics Pipelines

Pipelines in real -time analyses are required to treat dynamic climate phenomena, such as floods, fires or extreme weather treatment and response. These pipelines enable immediate intake of data, processing and visualization, and ensure that measures are available with minimal delays. Production of such systems includes several components: data sources (eg satellite feed, UAV, IoT sensor), Apache Kafka or AWS Kinesis as intake platforms, real -time treatment engines such as Apache good or kick streaming, and visualization dashboards or warning mechanisms. The pipeline should handle data flow with high wooden reputation, often in different formats, and in real-time changes, filtration and analysis, (Monteleoni, Schmidt, & McQuade, 2013). This may include rejection of nonconformities, incident classification and future modelling. The challenges include maintaining low delays, and ensures data stability and scale to scale the system to accommodate ups and downs. Cloud-Infester reform flexibility and scalability (eg, using Kuber nets) flexibility and scalability. Edge Computing is quickly integrated into real -time pipelines, causing the persecuting data source. This reduces the use of bandwidth and accelerates the decision for timely applications. Real-time analysis in climate remote measurement supports initial warning systems, emergency coordination and adaptive resource management. As climate change increases, the demand for accurate and sharp analysis pipes sometimes becomes more important to protect life and ecosystem, (Sarala, 2025).

CLOUD COMPUTING PLATFORMS FOR CLIMATE DATA

Cloud computing platforms play an important role in management, processing and analysis of large versions of climate data generated by platform satellites, sensors and remote measurement technologies. Platforms such as Amazon Web Services (AWS), Google Cloud Platform (GCP) and Microsoft Azure provide scalable storage solutions such as Amazon S3 or Google Cloud Storage, enabling effective handling of petabyte targets. These platforms also provide powerful data processing resources such as EC2, Big Query or Azure Virtual Machine for parallel processing and advanced analysis, (Chammas et al., 2023). Sky-country equipment supports climate modelling, environmental monitoring and preliminary warning systems (eg Sparks, Tenser flows) and real-time data datapy lines. In addition, integrated machine learning services enable future indication analysis and automated computer classification. The pay-a-u-go model reduces the cost of infrastructure, while access and cooperation are increased through shared cloud work areas. By taking advantage of cloud computing, researchers and organizations can accelerate climate research, support quick decision -making and scale operations in response to global environmental challenges, (Goodfellow et al., 2020.

Comparative Analysis: AWS, Google Earth Engine, Microsoft Azure

Amazon Web Services (AWS), Google Earth Engine (GEE) and Microsoft Azure provide powerful cloud platforms for processing climate data, but each has unique strength. AWS offers highly adaptable services for S3 calculations, EC2 for storage and EMR for Big Data Framework such as Hadoop and Spark. This is ideal for massive, adaptable work flakes. Microsoft Azure Entrepreneurs in Integration offers services such as Azure Synapse Analytics and Compatibility with Windows Systems, making it suitable for institutional distribution, (Coglianese, 2019). The Google Earth engine stands with its pre-developed, ready to use geostatic data sets and especially remote measurement and an intuitive platform designed for earth observation. Unlike AWS and Azure, Gee Abstract's infrastructure management allows researchers to focus on analysis using JavaScript or Python API. While AWS and Azure Complex provide more flexibility for distribution of multiple services, GEE provides fast prototypes and visualization tools that fit environmental monitoring, making it very accessible to climate scientists and geopolitical analysts, (Emami & Marzban, 2023).

Table 4. Comparative features of cloud platforms for climate data analysis

Feature	Amazon Web Services (AWS)	Google Earth Engine (GEE)	Microsoft Azure
Storage Type	Amazon S3 (Object Storage)	Integrated data repository	Azure Blob Storage
Compute Services	EC2, EMR, Lambda	Serverless, browser-based	Azure VMs, Synapse Analytics
AI/ML Integration	SageMaker, TensorFlow, PyTorch support	Built-in classifiers, Random Forest, CART	Azure Machine Learning Studio, PyTorch, ONNX
Geospatial Data Handling	Supports GIS tools, integrates with QGIS	Preloaded with satellite datasets and geoprocessing APIs	Spatial analytics via Azure Maps and Esri ArcGIS
Ease of Use	Requires technical configuration	Highly user-friendly for researchers	Balanced for enterprise and research users
Best Use Case	Flexible infrastructure for custom workflows	Rapid prototyping and environmental monitoring	Institutional deployment and integration

Table 4 presents a comparative overview of three leading cloud platforms Amazon Web Services (AWS), Google Earth Engine (GEE), and Microsoft Azure focusing on their relevance to climate data analysis. It evaluates key features such as storage infrastructure, compute capabilities, AI/ML tool integration, geospatial data handling, and user accessibility. Each platform's strengths and ideal use cases are summarized to assist researchers and institutions in selecting a suitable cloud environment based on their technical needs and project scale. This comparison highlights how cloud ecosystems can support scalable, efficient, and collaborative climate research and decision-making.

Cloud-Based Deep Learning Workflows (GPU/TPU Clusters)

Cloud-based workflows with deep learning used GPU and TPU clusters to intensify training and conclusions of models used in climate remote. AWS (using EC2 with Nvidia GPUs), Google Cloud (TPU and GPU operated AI platform offered), and Microsoft Azure (with Azure ML and Azure ML and NVIDIA GPU VMS) platform scaled, scalable, scalable, room-saving. These resources are crucial for tasks such as land coverage classification, flood exposure and automatic extension of systems from satellite images, (Eshbayev et al., 2022). Frames such as TensorFlow, PyTorch and Kera's are easily distributed in these environments, which enables spontaneous integration with scalable storage and data pipelines. In addition, auto-scaling skills ensure cost-effective resource use by adjusting electricity calculating depending on the demand for workload. The cloud -based Jupiter allows the laptop, model version and distributed exercise researchers to use, repeat and distribute models, (Hino, Benami, & Brooks, 2018). By reducing infrastructure barriers, these platforms democratize access to height demonstration calculation, and intensify the development of AI models for climate monitoring and decision-making of environmental.

Case Study: Google Earth Engine for Geospatial AI

Google Earth Engine (Gee) is a powerful platform that combines a huge list of geophagia data sets with cloud -based environmental analysis processing. A remarkable case study involves the use of Gee to detect forest harvesting in Amazon purely. Researchers used Gee's Global Landsat and Watchdog Image Data to create time series models, using monitored machine learning techniques that automatically detect forest decks, (Singh, 2023). Integrated machine learning tools for the platform, including support for random forests and classification trees, allow fast training and browser perfection directly in the browser. The scalability enabled analysis without downloading or managing a massive dataset in large spatial devices. The study was achieved near Real -Time deforestation notification, which was shared with conservation managers to trigger the intervention. Gees intuitive interfaces and API-based access enabled reproduction and collaboration between global research teams, (Tan et al., 2020). This case illustrates how Gee Empower GE ethnic AI applications, especially remote target functions, require extensive historical data and rapid cloud-based calculations.

DEEP LEARNING MODELS FOR CLIMATE APPLICATIONS

Deep learning models bring revolution in climate science by enabling advanced analysis of complex, high-dimensional remote measurement data. Convolutional Neural Network (CNN) is widely used for image-based tasks such as land covering classification, monitoring of glaciers and deforestation detection from satellite images. The Recurrent Neural Network (RNN) and Long-Card-Term memory (LSTM) The network applies to temporary data to predict climate patterns, rainfall or temperature trends. Autoencoders help to detect dimension reduction and abnormalities in climate data sets, while Generative Outdoor Network (GAN) emerges to follow computer text and future climate, (Alemayehu et al., 2024). These models can learn complex patterns from Multisource data, improve the accuracy and speed of environmental predictions. Cloud-based GPU/TPU acceleration, on a scale of intensive learning work flows in large geostatic data sets. Integration of deep education into climate applications increases early warning systems, ecosystem modelling and policy planning, which provides a powerful tool set to understand and reduce the effect of climate change.

Figure 4. AI models used in climate research

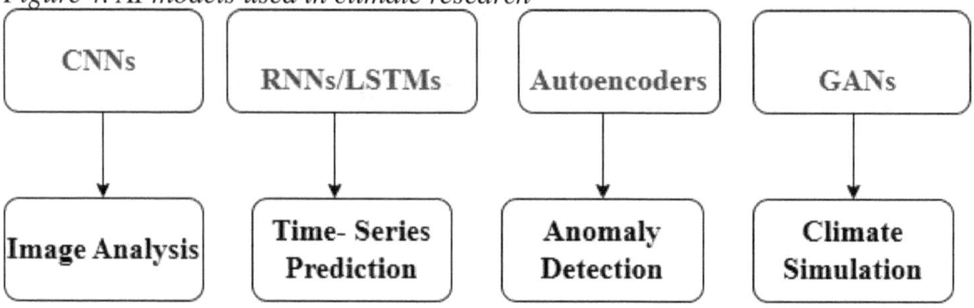

Figure 4 maps deep learning models to their primary climate science applications. It shows how CNNs are used for image analysis, RNNs/LSTMs for time-series prediction, autoencoders for anomaly detection, and GANs for climate simulation, illustrating their roles in environmental monitoring and forecasting.

Table 5. Deep learning models and their climate applications

Model Type	Core Functionality	Climate Application	Advantages
Convolutional Neural Network (CNN)	Extracts spatial features from imagery	Land cover classification, deforestation detection, urban mapping	High accuracy for satellite and aerial image analysis
Recurrent Neural Network (RNN) / LSTM	Captures sequential dependencies in time-series data	Temperature/rainfall forecasting, seasonal pattern prediction	Effective in long-term temporal modeling
Transformer	Parallel processing of time-series with attention mechanisms	High-resolution climate trend forecasting, spatiotemporal modeling	Scalable for long-range dependencies and large datasets
Generative Adversarial Network (GAN)	Learns to generate synthetic data distributions	Climate scenario simulation, image enhancement, anomaly detection	Enhances data availability and realism for model training
Hybrid Models (e.g., CNN-LSTM)	Combines spatial and temporal learning capabilities	Flood prediction, glacier retreat analysis, multisensor fusion	Integrates multi-dimensional data for more comprehensive models

Table 5 categorizes and explains various deep learning models CNNs, RNNs/ LSTMs, Transformers, GANs, and hybrid architectures based on their core functions and suitability for climate-related tasks. It links each model type to specific environmental applications, such as land cover mapping, rainfall prediction, or simulation of climate scenarios. Additionally, it highlights the strengths of each model in processing spatial, temporal, or multimodal data. This structured comparison supports researchers in selecting appropriate AI techniques for solving complex challenges in climate monitoring and predictive modeling.

CNNs for Image Classification (e.g., Land Cover Change)

Convolutional Neural Networks (CNNS) Climate applications have deep learning architecture for classification tasks, especially to analyze satellites and air fantasy. CNN -is distinguished by removing spatial functions and patterns, making them ideal for detecting changes in land coverage such as urban expansion, deforestation and crop health variations, (An, Kim, & Kim, 2021). By exercising CNN on multispectral or hyperspectral images from sources such as Landsat, Sentinel or UAV, researchers can classify different surface types - normal, water, urban, rugged land - with high accuracy. Advanced architecture such as U-Net and ResNet also improves split and the convenience detection of noise or complex areas. CNN -er also supports transmission learning, which enables effective adaptation in new geographical areas with limited labelled data. Combined with GIS tools and cloud platforms, CNN offers scalable solutions to monitor environmental changes over time, (Anderegg et al., 2010). This insight helps insight policy decision makers and protectionists to assess ecosystem's health, monitor climate -driven changes and implement permanent plans for land use plans.

RNNs/Transformers for Time-Series Data (e.g., Temperature Trends)

The Recurrent Neural Network (RNN) and the transformer models are powerful tools for analyzing time series data in climate science, such as temperature trends, precipitation warning and sea swings. RNN -er, including their advanced version LSTM (long -term short -term memory), is designed to capture sequential addiction and long-term patterns, making them ideal for modelling temporary mobility in the environmental dataset, (Carlton et al., 2015). They can predict temperature anomalies, detect seasonal changes and identify long climatic trends by learning from historical records. The transformers developed early for natural language treatment, receive traction in climate applications due to their ability to treat long sequences in parallel and capture global references effectively. These models have been used for high -resolution seasonal assumptions and spatiotemporal analysis of atmospheric variables. Unlike traditional statistical models, RNN and transformers adapt better for non-led, multi-commercial data, (Cobb, 2024). When the climate is trained on computer voltages, these architecture forecasts increase accuracy and support active plans in areas such as agriculture, disaster preparation and management of water resources.

Hybrid Models for Multisensory Data Fusion

Hybrid deep learning models are used quickly to integrate multisensory data-such as optical images, radar, lidar and in-situ measurements for extensive climate analysis. These models add different nervous architecture, such as CNN and RNN or transformer for time modelling for spatial analysis, to extract insight rich from different data sources. For example, CNN-LSTM hybrids are effective in flood prediction, where satellite images capture spatial expansion while time series of rain data provides temporary reference. The meditation mechanism can be included in the sensor to prioritize important properties, (Dee et al., 2011). Multi-sensor data fusion causes microamp environment changes, such as glacier retreat or vegetation voltage, can increase the accuracy of detection, which can be remembered by a single type of data. These models can handle data with different solutions and formation, and use complementary strength for each source. By merging data in domains and dimensions, hybrid deep learning systems enable more general, strong climate modelling-importing intelligent environmental monitoring, influence evaluation and permanent decision-making.

CASE STUDIES: AI-DRIVEN CLIMATE MONITORING

AI-operated climate monitoring has led to transformation results in environmental management. For example, NASA uses deep learning models to detect and predict fires from satellite data, enables initial intervention, (Fer et al., 2020). In Amazon, Google Earth Engine and AI algorithms help trace illegal deforestation near real time. In India, the AI-based crop monitoring system integrates satellite images with weather data to assess the risk of drought. Another case includes IBM's Green Horizon project, which uses AI to predict air pollution patterns in urban areas. The study of this case shows how AI globally increases accurately, speed and scalability in monitoring and responding to climate challenges, (Hanekamp & Bergkamp, 2016).

Table 6. AI-enabled case studies in climate monitoring

Case Study	AI Technique Used	Application Area	Technology Platform	Outcome/Impact
IBM Green Horizon (China)	Machine Learning for pattern recognition	Air pollution prediction and urban planning	Sensor networks, ML pipelines	Improved pollution forecasts and mitigation strategies
Google Flood Forecasting (South Asia)	Deep Learning with hydrological models	Real-time flood prediction	Satellite data, river gauge sensors	Enhanced early warnings for flood-prone areas
Global Forest Watch (Amazon, SE Asia)	CNN-based classification	Deforestation detection	Google Earth Engine, satellite imagery	Near real-time alerts and improved enforcement of forest protection
NASA ICESat-2 (Polar Regions)	Deep Learning + LiDAR integration	Glacier retreat and ice thickness analysis	ICESat-2, cloud computing	Accurate measurement of ice mass loss over time
India Crop Risk Monitoring System	AI with multispectral image processing	Drought and crop failure assessment	Remote sensing, weather data fusion	Enhanced agricultural planning and drought resilience

Table 6 presents a collection of real-world case studies where AI technologies have been applied to address diverse climate monitoring challenges. It highlights the specific AI techniques and technology platforms used in each case, such as machine learning models, deep learning architectures, and multispectral data integration. The applications span critical areas like air quality forecasting, flood prediction, deforestation alerts, glacier analysis, and agricultural risk monitoring. The outcomes demonstrate the tangible benefits of AI in improving decision-making, enabling early interventions, and supporting sustainable environmental policies.

Air Pollution and Emission Tracking

AI brings revolution in supervision of air pollution by enabling trace of real - time and predictions on emission sources. Machine learning models, combined with satellite data (eg these systems integrate meteorological data to estimate pollution spreading and identify industrial hotspots. For example, IBM's green horizontal initiatives AI uses to predict pollution spikes in Chinese cities, and helps decision -making. Mobility patterns and industrial activity to estimate emission shoes with better precision. to fight climate change and public health risk, (Harker-Schuch et al., 2018).

Flood Prediction and Disaster Response

AI plays an important role in the prediction of floods through real -time data analysis and modelling. Machine learning algorithms analyze different data sets including rainfall patterns, river discharge levels, ground satellite and satellite images to identify areas with high risks of floods. The FLOOT prognosis initiative from Google uses trained AI models on hydrological and topological data to provide initial warnings in weak areas such as South Asia, (Haywood et al., 2011). These systems offer granulated predictions to specific societies, which increases the lead time for emergency preparedness. Deep learning models also treat social media and sensor input to track disaster development. AI -run drones and UAV facilitates fast injury evaluation, guidance of rescue operations and resource allocation. By enabling quickly and more accurate prognosis, AI-driven flood monitoring reduces expansion, reduces economic losses and supports climate flexibility in disaster-exposed areas. Integration of AI with IoT networks and cloud computing ensures scalable, real -time flooding for authorities and human agencies.

Deforestation and Glacial Retreat Analysis

Deforestation also supports global commitments to biodiversity conservation and carbon sequestration. The AI-operated equipment has changed how researchers who were automated in the analysis of satellite images over time monitored forest harvesting and ice help. The short neural network (CNN) is used to classify land cover and detect vegetation losses in areas such as Amazon, the Congo pool and Southeast Asia, (Hmielowski et al., 2013). AI and Google Earth offer Global Forest Watch -platforms controlled, near real -time forestry, which allows officials and protectionists to provide a quick response to illegal harvesting. In polar and mountainous regions, the AI models analyze multi-spectral and radar data to monitor the size, flow rates and the retreat patterns of the glacier. For example, NASA's ICESAT-2 data combined with deep learning enables to dilute the exact measurement of ice sheet over time. This insight is important for understanding sea level growth and climate responses. AI enables scalable, continuous and timely environmental monitoring, strengthening data -driven guidelines for forest protection and climate adaptation, (Jaspal, Nerlich, & Koteyko, 2012).

OPTIMIZING PERFORMANCE AND ENSURING DATA INTEGRITY

In climate monitoring systems, adjustment of performance, which ensures data integrity, is necessary for timely insight. Result adaptation involves the streamlining of data pipes using parallel processing, distributing scalable cloud -based architecture to manage giant versions of remote measurement and sensor data, (Lemos & Rood, 2010). Tools such as Apache Sparks, GPU acceleration and edge calculation improve processing speed and efficiency, enabling real -time analysis. In addition, the data integrity ensures that the collected data remains accurate, consistent and tamper -based. It is achieved by using cryptographic methods such as automatic verification routine, strong data cleaning algorithms and blockchain for safe data logging and traceability. Excessive system and fault -tolerant architecture exceeds data continuity during transmission or system failure. Version control, metadata tracking and revision log data ensure openness and responsibility in the life cycle. Together, this practice increases the reliability of the AI-operated climate model, supports scientific fertility and creates confidence in the insights gained from environmental monitoring initiatives, (Lorenz et al., 2016).

Data Preprocessing/Quality Control Protocols

In climate monitoring, data preparation and quality control protocols are important to ensure the reliability and accuracy of AI-operated analysis. Raw data of satellites, UAVs and sensor networks often contain noise, intervals or discrepancies due to atmospheric intervention, error errors or transmission errors. Pre-treatment of step-like normalization, cloud aching, outlier detection and data cleaning data for GAP films standards and downstream modelling. Technology such as statistical threshold, spatial-temporal interpolation and machine learning-based Denoising help maintain data integrity. Quality control protocols often include cross reference, implementation of confirmation rules and allocate confidence in different data segments with ground data, (McCright & Dunlap, 2011). The food data document supports traceability and fertility. Automatic pipelines can flag real -time deviations, which ensure timely corrective functions. These procedures are particularly important when integrating multisensory data sets. Rigid persecuting ensures that the AI model is trained on high quality input, leading to more accurate predictions and strong decisions in climate science and environmental monitoring.

Model Validation and Uncertainty Quantification

It is necessary to validate the AI model and determine uncertainty to ensure the credibility of climate conditions. Model verification involves comparing an output with independent ground data or installing climate-benchmarks to assess accuracy and generality. Techniques such as cross-validation, confusion matrix, RMSE (Root Mein Square error) and ROC curves help to evaluate the performance of models in different data sets and scenarios. In remote measurement, spatial and temporary cross-insert ensures strength in different fields and time periods. Equally significant uncertainty is stagnant the model that understands the level of trust in production, (Monem & Bastawissy, 2016). The Biopiracy methods, enchanted learning and Monte Carlo simulation are usually used to capture the spectrum of potential results and identify sensitive variables. Visual tools such as uncertainty with uncertainty help to convey confidence in the model for decision makers. Verification provides a confidence among the matrix and transparent reporting of uncertainty interests, supports the risk of risk-a convenience and strengthens the scientific basis for AI applications in climate monitoring and environmental forecasting.

Privacy, Security, and Ethical Considerations

AI-driven climate monitoring systems should address privacy, security and moral concerns, especially when handling data with human activities. Privacy conflict occurs when images with high resolution or sensor data accidentally catching identifying information, such as personal characteristics or individuals. The use of data neutralization, spatial masking and moral data insert helps to reduce these risks, (Mutmainah, Hakim, & Syam, 2022). From a security point of view, it is important to protect the integrity of the climate data set from tampering or cyber-attacks - especially when the data indicates disaster response or public policy. Encryption, access control and secure data transfer protocol system ensure flexibility. Ethically, the AI model should be designed to avoid prejudice, ensure openness and prefer environmental justice. Open access to AI tools and data strengthens global equity, but should be balanced against abuse or monitoring of concerns. Following the moral framework, confusing communities and following computer principles can lead the responsible development and distribution of AI in climate applications, (Reimov et al., 2023).

FUTURE TRENDS AND POLICY IMPLICATIONS

The future of the AI-controlled climate monitoring will be shaped by data processing, Federated Learning and Quantum AI, which will enable fast, decentralized and safer analysis of spacious environmental datasets. Integration with digital twins and climate models will increase future abilities for active climate action. Political decision makers should prioritize open data standards, moral AI regime and international cooperation to ensure equal access and openness, (Sain, 2022). Investment in digital infrastructure, capacity production and regulatory structures will be important to maximize AI's ability to reduce the global digital division and support climate flexibility, stability and evidence -based environmental policy.

Edge-Cloud Hybrid Systems for Real-Time Analytics

Edge cloud hybrid systems bring revolution in real-time climate analysis by balancing local processor power with cloud-based scalability. In these architecture, UAV, IoT sensor and smart weather stations make initial data processing at the unit source, reduces the delay and bandwidth requirements, (Santos et al., 2012). This is especially useful for external or high-risk environments where quick decisions are important-as to detect initial flood warnings or forest fire. Processed data can then be sent to cloud platforms for more intensive analysis, model training and long -term storage. Cloud Systems offers high performance data processing resources, centralized data integration dashboards and global monitoring from many sources. The synergy between age and the cloud system allows for real -time response while maintaining analytical depth, (Sarathchandra & Haltinner, 2019). This architecture also increases data security and system flexibility by decentralizing important functions. Since climate monitoring requires both immediate and scalability, the Alder sky hybrid models will play an important role in providing responsible, skilled and adaptive environmentally intelligent systems worldwide.

Low-Cost Solutions for Global Scalability

Cost effects and available solutions are crucial for a broad climate monitoring to be impressive globally, especially in development areas. Innovations such as cheap UAVs, Open-Sources geophagic equipment and mild deep teaching models reduce the obstacle to admissions to resource-limiting nations. Cloud-based platforms such as Google Earth engines provide free or subsidized access to satellite performances and calculation tools, enabling environmental analysis in real time without the need for local infrastructure, (Sarathchandra & Haltinner, 2021). Smartphones and civilian platforms can supplement community -controlled data collection satellite data, especially in areas with limited sensor coverage. Edge Computing Equipment, operated by cheap microcontroller (eg Raspberry Pie), allows local data processing with minimal strength and connection requirements. This innovation supports located climate acts, democratizes access to AI units and promotes inclusive global participation. By focusing on strength, modularity and adaptability, low -cost solutions on the scale of climate monitoring efforts worldwide help and ensure that the weakest areas can also participate in climate flexibility initiatives, (Storch & Bray, 2017).

Policy Frameworks for AI-Enhanced Climate Governance

The estate of effective control is necessary to utilize AI's capacity in climate monitoring. The policy should prioritize moral AI purposes, the data transparency and international cooperation. The standards should make open access, Inter -updated dataset compulsory and ensure that the AI models are explained, audible and free of prejudice, (Uyar & Karamustafaoğlu, 2024). Regulatory structures should protect privacy in especially high -resolution geophagia data, and support the same access to climate technologies. Governments should invest in digital infrastructure, capacity building and public-private partnerships to promote innovation. In addition, the climate rule should encourage global data sharing agreements to increase collective response to cross -border environmental challenges. AI Ethics Board and Climate Councils should be established to review the social implications of new technologies in environmental management. Integration of AI ideas into national climate strategies, SDG guidelines and disaster response frameworks will ensure permanent and inclusive implementation. Such further political structures are important for coordinating technological innovation with climate justice and long -term ecological stability, (Vögt et al., 2023).

CONCLUSION

The rapid speed of climate change requires sophisticated, scalable and probable surveillance solutions. In this context, the convergence of Big Data technologies, cloud computing and deep learning have emerged as a transformative force in climate-oriented remote measurement. This chapter discovered how these digital innovations enable high -resolution climate monitoring and decision -making, how the huge, strange dataset adopted from UAV and Sensor network enables treatment, analysis and interpretation. We began to recognize the exponential growth of climate data, which is driven by the widespread distribution of the rapidly consistent satellite assignments, high -resolution imaging technologies and the Internet of Things (IoT) -designed sensor. This progress has increased our ability to monitor the essential climate variables such as surface temperature, glaciers, deforestation and the height of the sea. However, he has also introduced complex challenges in terms of storage, processing power, data integration and streaming-which are ill to address traditional systems.

To reduce these challenges, we investigated the role of Big Data Framework such as Hadop and Spark in handling large-scale geophagia datasets. These distributed data processing environments enable scalable data processing, which enables effective intake, filtration and changes of real -time climate data. In addition, the development of real -time analytical pipelines have made it possible to perform the nearest environmental assessment for applications such as disaster response and initial warning systems. Cloud Computing Platforms such as Amazon Web Services (AWS), Microsoft Azure and Google Earth Engine (GEE) have emerged as a powerful hosting ecosystem and processing of the climate data on the platform scale. These platforms not only offer storage and calculation infrastructure, but also integrate AI tools, GPU/TPU cluster and application programming interface (APIs), which are in accordance with geophysical analysis. A comparative evaluation of these platforms suggests that Google Earth Engine, in particular, provides a strong environment for fast prototypes and operational practices for Geospatial deep learning models. By taking advantage of gender equality and elasticity in cloud infrastructure, researchers and doctors can use complex workflows with deep learning without the need for expensive resources on the inhabitants.

In addition, this chapter on how advanced a deep learning architecture includes the Convolutional Neural Network (CNN), the Recurrent Neural Network (RNNS), transformers and hybrid fusion models-used on pattern in satellite images and time series data. These models not only increase the accuracy of environmental conditions, but also facilitate the merger of multimodal sensor data, leading to rich, more widespread insight into climate systems. Case studies gave a perspective in the real world on monitoring air pollution, and predicted floods and distribution of these techniques for the observation of forests and glaciers. These examples demonstrated concrete social benefits of AI enhanced remote measurement in handling natural resources and shaping the policy.

Finally, we emphasized the importance of strict data preparation, quality control protocols, model verification technology and moral security measures. Together with the integrity and safety of climate data -present openness and justice in the AI model, the AI model is important for reliable science and inclusive climate rule. Finally, big data and cloud computing provide digital back legs for the next generation of climate monitoring systems. As an edge-sky hybrid model, low-price sensor integration and AI policy frameworks develop, they will unlock the new opportunities for scalable, real-time and democratic climate analysis. This development will strengthen nations, institutions and communities to adapt more efficiently to global environmental challenges and to design active, data -driven reactions ahead of climate change.

REFERENCES

Aarthi, R., & Vinayagasundaram, B. (2022). Effective management of class imbalance problem in climate data analysis using a hybrid of deep learning and data level sampling. *Journal of Intelligent & Fuzzy Systems*, *43*(4), 4187–4199. DOI: 10.3233/jifs-210666

. Abdalla, R. (2024). Framework for assessing the impacts of climate change on urban agglomerations: a gis and remote sensing perspective.. DOI: 10.5772/intechopen.1004284

Agbehadji, I., Mabhaudhi, T., Botai, J., & Masinde, M. (2023). A systematic review of existing early warning systems' challenges and opportunities in cloud computing early warning systems. *Climate (Basel)*, *11*(9), 188. DOI: 10.3390/cli11090188

Aggarwal, L., & Papay, F. (2022). Applications of multispectral and hyperspectral imaging in dermatology. *Experimental Dermatology*. Advance online publication. DOI: 10.1111/exd.14624 PMID: 35671110

Alemayehu, M., Workie, S., Belew, M., Agimas, M., Derseh, N., Abuhay, H., & Bazezew, A. (2024). Knowledge towards the health impacts of climate change and associated factors among undergraduate health sciences students in amhara region: A multi-centered study. *Frontiers in Environmental Health*, *3*. Advance online publication. DOI: 10.3389/fenvh.2024.1363395

An, S., Kim, H., & Kim, S. (2021). Rate-dependent hysteresis of the atlantic meridional overturning circulation system and its asymmetric loop. *Geophysical Research Letters*, *48*(1). Advance online publication. DOI: 10.1029/2020gl090132

Anand, A. (2025). Machine learning and artificial intelligence in environmental change prediction. *Interantional Journal of Scientific Research in Engineering and Management*, *09*(01), 1–9. DOI: 10.55041/ijsrem40801

Anderegg, W., Prall, J., Harold, J., & Schneider, S. (2010). Expert credibility in climate change. *Proceedings of the National Academy of Sciences of the United States of America*, *107*(27), 12107–12109. DOI: 10.1073/pnas.1003187107 PMID: 20566872

. Bhagyalakshmi, V., V., N., Harinivas, S., M., H., S, H., & Shastry, K. (2024). Machine learning for climate data analysis.. DOI: 10.59544/oqxm6494/icrcct24p122

. Boumrah, M., Garbaya, S., & Radgui, A. (2023). Real-time visual analytics for remote monitoring of patient's health., 368-378. DOI: 10.24132/csrn.3301.61

Cai, J., Chatoux, H., Boust, C., & Mansouri, A. (2021). Extending the unmixing methods to multispectral images. *Color and Imaging Conference*, 29(1), 311-316. DOI: 10.2352/issn.2169-2629.2021.29.311

Carlton, J., Perry-Hill, R., Huber, M., & Prokopy, L. (2015). The climate change consensus extends beyond climate scientists. *Environmental Research Letters, 10*(9), 094025. DOI: 10.1088/1748-9326/10/9/094025

Chammas, S., Wang, M., Schneider, T., Ihme, M., Chen, Y., & Anderson, J. (2023). Accelerating large-eddy simulations of clouds with tensor processing units. *Journal of Advances in Modeling Earth Systems, 15*(10). Advance online publication. DOI: 10.1029/2023ms003619

Chen, S., & Lin, Z. (2025). Research on ecological environment monitoring and protection applications based on remote sensing technology. *Academic Journal of Science and Technology, 14*(1), 55–57. DOI: 10.54097/k03eem97

Cobb, C. (2024). The politics of climate denialism and the secondary denialism of economics. *American Journal of Economics and Sociology, 84*(1), 69–88. DOI: 10.1111/ajes.12606

. Coglianese, C. (2019). Deploying machine learning for a sustainable future., 200-208. DOI: 10.2307/j.ctvqc6gcq.26

Dee, D., Uppala, S., Simmons, A., Berrisford, P., Poli, P., Kobayashi, S., & Vitart, F. (2011). The era-interim reanalysis: Configuration and performance of the data assimilation system. *Quarterly Journal of the Royal Meteorological Society, 137*(656), 553–597. DOI: 10.1002/qj.828

Dewitte, S. (2020). Editorial for special issue "earth radiation budget". *Remote Sensing, 12*(20), 3379. DOI: 10.3390/rs12203379

Emami, P., & Marzban, A. (2023). The synergy of artificial intelligence (ai) and geographic information systems (gis) for enhanced disaster management: Opportunities and challenges. *Disaster Medicine and Public Health Preparedness, 17.* Advance online publication. DOI: 10.1017/dmp.2023.174 PMID: 37694385

Eshbayev, O., Mirzaliev, S., Rozikov, R., Kuzikulova, D., & Шакирова, Г. (2022). Nlp and ml based approach of increasing the efficiency of environmental management operations and engineering practices. *IOP Conference Series. Earth and Environmental Science, 1045*(1), 012058. DOI: 10.1088/1755-1315/1045/1/012058

Fer, I., Gardella, A., Shiklomanov, A., Campbell, E., Cowdery, E., Kauwe, M., & Dietze, M. (2020). Beyond ecosystem modeling: A roadmap to community cyberinfrastructure for ecological data-model integration. *Global Change Biology*, *27*(1), 13–26. DOI: 10.1111/gcb.15409 PMID: 33075199

Goodfellow, I., Pouget-Abadie, J., Mirza, M., Xu, B., Warde-Farley, D., Ozair, S., & Bengio, Y. (2020). Generative adversarial networks. *Communications of the ACM*, *63*(11), 139–144. DOI: 10.1145/3422622

. Gudelė, G. and Visockienė, J. (2023). Application of remote sensing for monitoring carbon farming: a review. Mokslas - Lietuvos Ateitis, 15(0), 1-6. DOI: 10.3846/mla.2023.19396

Hanekamp, J., & Bergkamp, L. (2016). The 'best available science' and the paris agreement on climate change. *European Journal of Risk Regulation*, *7*(1), 42–48. DOI: 10.1017/s1867299x00005365

Harker-Schuch, I., Mills, F., Lade, S., & Colvin, R. (2018). Opinions of 12 to 13-year-olds in austria and australia on the worry, cause and imminence of climate change. DOI: 10.1101/333237

. Haywood, A., Ridgwell, A., Lunt, D., Hill, D., Pound, M., Dowsett, H., … & Williams, M. (2011). Are there pre-quaternary geological analogues for a future greenhouse warming?. Philosophical Transactions of the Royal Society a Mathematical Physical and Engineering Sciences, 369(1938), 933-956. DOI: 10.1098/rsta.2010.0317

Hino, M., Benami, E., & Brooks, N. (2018). Machine learning for environmental monitoring. *Nature Sustainability*, *1*(10), 583–588. DOI: 10.1038/s41893-018-0142-9

Hmielowski, J., Feldman, L., Myers, T., Leiserowitz, A., & Maibach, E. (2013). An attack on science? media use, trust in scientists, and perceptions of global warming. *Public Understanding of Science (Bristol, England)*, *23*(7), 866–883. DOI: 10.1177/0963662513480091 PMID: 23825287

Jaspal, R., Nerlich, B., & Koteyko, N. (2012). Contesting science by appealing to its norms. *Science Communication*, *35*(3), 383–410. DOI: 10.1177/1075547012459274

Jiao, Z. (2024). The application of remote sensing techniques in ecological environment monitoring. *Highlights in Science Engineering and Technology*, *81*, 449–455. DOI: 10.54097/7dqegz64

Kior, A., Sukhov, V., & Sukhova, E. (2021). Application of reflectance indices for remote sensing of plants and revealing actions of stressors. *Photonics*, *8*(12), 582. DOI: 10.3390/photonics8120582

Kior, A., Yudina, L., Zolin, Y., Sukhov, V., & Sukhova, E. (2024). Rgb imaging as a tool for remote sensing of characteristics of terrestrial plants: A review. *Plants*, *13*(9), 1262. DOI: 10.3390/plants13091262 PMID: 38732477

Krämer, M., & Senner, J. (2015). A modular software architecture for processing of big geospatial data in the cloud. *Computers & Graphics*, *49*, 69–81. DOI: 10.1016/j.cag.2015.02.005

Lemos, M., & Rood, R. (2010). Climate projections and their impact on policy and practice. *Wiley Interdisciplinary Reviews: Climate Change*, *1*(5), 670–682. DOI: 10.1002/wcc.71

Li, X., Hao, X., Yin, L., Liu, L., Ma, Y., Yang, R., & Song, Q. (2021). Validation of gps-based monitoring and remote sensing of ice-shelf and ice-sheet movement changes. *Sensors (Basel)*, *21*(23), 7822. DOI: 10.3390/s21237822 PMID: 34883824

Lorenz, S., Dessai, S., Forster, P., & Paavola, J. (2016). Adaptation planning and the use of climate change projections in local government in england and germany. *Regional Environmental Change*, *17*(2), 425–435. DOI: 10.1007/s10113-016-1030-3 PMID: 32269501

McCright, A., & Dunlap, R. (2011). Cool dudes: The denial of climate change among conservative white males in the united states. *Global Environmental Change*, *21*(4), 1163–1172. DOI: 10.1016/j.gloenvcha.2011.06.003

Monem, R., & Bastawissy, A. (2016). Dira: A framework of data integration using data quality. *International Journal of Data Mining & Knowledge Management Process*, *6*(2), 37–58. DOI: 10.5121/ijdkp.2016.6204

Monteleoni, C., Schmidt, G., & McQuade, S. (2013). Climate informatics: Accelerating discovering in climate science with machine learning. *Computing in Science & Engineering*, *15*(5), 32–40. DOI: 10.1109/mcse.2013.50

Mookambiga, A., & Gomathi, V. (2016). Comprehensive review on fusion techniques for spatial information enhancement in hyperspectral imagery. *Multidimensional Systems and Signal Processing*, *27*(4), 863–889. DOI: 10.1007/s11045-016-0415-2

Mutmainah, O., Hakim, A., & Syam, M. (2022). Validity of physics teaching materials based on stem to improve climate literacy of high school students. *Jurnal Pendidikan Fisika Dan Teknologi*, *8*(2), 208–216. DOI: 10.29303/jpft.v8i2.4053

Ni, M., Wu, Q., Li, G., & Li, D. (2025). Remote sensing technology for observing tree mortality and its influences on carbon–water dynamics. *Forests*, *16*(2), 194. DOI: 10.3390/f16020194

Novikarany, R. (2025). Benefits of remote sensing for environmental monitoring. *International Journal of Integrative Sciences*, *4*(2), 445–454. DOI: 10.55927/ijis.v4i2.19

Park, S., Jeong, S., Park, Y., Kim, S., Lee, D., Mo, Y., & Park, K. (2021). Phenological analysis of sub-alpine forest on jeju island, south korea, using data fusion of landsat and modis products. *Forests*, *12*(3), 286. DOI: 10.3390/f12030286

. Pettorelli, N. (2019). Satellite remote sensing and anthropogenic pressures on biodiversity., 51-65. DOI: 10.1093/oso/9780198717263.003.0004

. Reimov, M., Mamutov, N., Reymov, P., Statov, V., Khudaybergenov, Y., Sultashova, O., … & Tleumuratova, B. (2023). Some issues of geoinformatics support for agroclimatic zoning in the south aral sea basin. E3s Web of Conferences, 386, 06009. DOI: 10.1051/e3sconf/202338606009

Sain, S. (2022). Data science and climate risk analytics. *Environmetrics*, *34*(2). Advance online publication. DOI: 10.1002/env.2749

. Santos, E., Koop, D., Maxwell, T., Doutriaux, C., Ellqvist, T., Potter, G., … & Silva, C. (2012). Designing a provenance-based climate data analysis application., 214-219. DOI: 10.1007/978-3-642-34222-6_18

Sarala, R. (2025). Machine learning and spatio-temporal patterns in climate change. *Journal of Information Systems Engineering & Management*, *10*(15s), 77–85. DOI: 10.52783/jisem.v10i15s.2431

Sarathchandra, D., & Haltinner, K. (2019). Trust/distrust judgments and perceptions of climate science: A research note on skeptics' rationalizations. *Public Understanding of Science (Bristol, England)*, *29*(1), 53–60. DOI: 10.1177/0963662519886089 PMID: 31691642

Sarathchandra, D., & Haltinner, K. (2021). A survey instrument to measure skeptics' (dis)trust in climate science. *Climate (Basel)*, *9*(2), 18. DOI: 10.3390/cli9020018

Shu, C., & Sun, L. (2020). Automatic target recognition method for multitemporal remote sensing image. *Open Physics*, *18*(1), 170–181. DOI: 10.1515/phys-2020-0015

Singh, G. (2023). Environmental monitoring with machine learning. *Epra International Journal of Multidisciplinary Research (Ijmr)*, *208-212*. Advance online publication. DOI: 10.36713/epra13330

Storch, H., & Bray, D. (2017). Models, manifestation and attribution of climate change. *Meteorology Hydrology and Water Management*, *5*(1), 47–52. DOI: 10.26491/mhwm/67388

Tan, L., Guo, J., Mohanarajah, S., & Zhou, K. (2020). Can we detect trends in natural disaster management with artificial intelligence? a review of modeling practices. *Natural Hazards*, *107*(3), 2389–2417. DOI: 10.1007/s11069-020-04429-3

Uyar, F., & Karamustafaoğlu, O. (2024). Polar science awareness of science teachers. *International E-Journal of Educational Studies*, *8*(16), 32–44. DOI: 10.31458/iejes.1386430

Vögt, V., Harrs, J., Reinhart, V., Hollenbach, P., Bühler, M., & Tewes, T. (2023). Implementing agile data workflows to unlock climate-resilient urban planning. *Climate (Basel)*, *11*(9), 174. DOI: 10.3390/cli11090174

Wang, Y., & Yésou, H. (2018). Remote sensing of floodpath lakes and wetlands: A challenging frontier in the monitoring of changing environments. *Remote Sensing*, *10*(12), 1955. DOI: 10.3390/rs10121955

. Yokoya, N. and Iwasaki, A. (2013). Hyperspectral and multispectral data fusion mission on hyperspectral imager suite (hisui)., 4086-4089. DOI: 10.1109/igarss.2013.6723731

Zhang, C., Filella, I., Garbulsky, M., & Peñuelas, J. (2016). Affecting factors and recent improvements of the photochemical reflectance index (pri) for remotely sensing foliar, canopy and ecosystemic radiation-use efficiencies. *Remote Sensing*, *8*(9), 677. DOI: 10.3390/rs8090677

Chapter 7
Deep Learning in Remote Sensing for Disaster Management and Humanitarian Aid

Qazi Mudassar Ilyas
https://orcid.org/0000-0003-4238-8093
Sultan Sharif Ali Islamic University, Brunei Darussalam

Abid Mehmood
https://orcid.org/0000-0001-9974-9537
Dakota State University, USA

ABSTRACT

Disaster management refers to the preparation, response, and recovery activities to mitigate or minimize the losses caused by a disaster. Remote sensing plays a crucial role in disaster management by gathering information about the affected areas through satellite and other aerial imagery platforms. This chapter explores the role of deep learning techniques in processing remote sensing data and supporting the decision-making process in disaster management. We explain how various deep learning architectures, such as convolutional neural networks, sequential models, transformers, and autoencoders, can be used in real-life disaster situations. The chapter details how multi-modal and multi-temporal data fusion integrates optical, synthetic-aperture radar, and light detection and ranging imagery into a single analytic pipeline. A discussion on datasets and benchmarking protocols is followed by model deployment and real-time considerations. The chapter concludes with limitations of the discussed techniques, open challenges, and future research directions.

DOI: 10.4018/979-8-3373-3501-8.ch007

INTRODUCTION

The frequency and severity of disasters have increased in recent years due to several natural and human-caused factors. Natural factors such as climate change, tectonic activity, extreme weather events, sea-level rise, soil erosion, land degradation, solar and cosmic events, and hydrological variability have contributed to a sharp increase in natural disasters, such as floods, hurricanes, landslides, wildfires, and droughts. Large-scale uncontrolled human activity has further deteriorated the situation, (Huang et al., 2024; Zhang et al., 2024a; Peng et al., 2024a;). The human-caused factors include armed conflicts, greenhouse gas emissions, deforestation, urbanization, poor land use and agricultural practices, mining and industrial activity, dam and reservoir construction, and consumption of fossil fuels. The Centre for Research on the Epidemiology of Disasters (CRED) estimates suggest that 387 natural disasters caused economic losses of over USD 123 billion and affected 185 million people in 2022 (CRED, n.d.). Table 1 summarizes major natural and human-caused disasters worldwide and their causes, locations, years, and verified damages, (Bhatti et al., 2024; Li et al., 2024; Han et al., 2024a; Noor et al., 2024; Han et al., 2024b; Bhatti et al., 2023; Noor et al., 2023). The losses of this magnitude call for immediate actions and measures before, during, and after a disaster. Disaster management has four main phases: preparedness, early warning, response, and recovery, (Bhatti et al., 2022; Bhatti et al., 2024a; Peng et al., 2024b; Han et al., 2024c; Zhang et al., 2024b; He & Bhatti 2024). Disaster preparedness deals with preparation activities. i.e., planning and training to prepare for any potential disasters. Early warning systems predict any disasters and inform the stakeholders in advance. The response phase comprises all activities to minimize health impacts and save lives and infrastructure. Figure 1 gives an overview of this disaster management process (Zhang et al., 2024c; Asif et al., 2024; Chen et al., 2024; Hasnain et al., 2024).

Researchers are actively working on the latest technologies to achieve these goals, such as artificial intelligence Hamza et al. 2023; Nawaz et al. 2021), geographic information systems, the Internet of Things, big data analytics along with privacy concerns (Bazai et al., 2017; Bazai & Jang-Jaccard, 2019; Bazai et al., 2021; Bazai & Jang-Jaccard, 2020), and early warning systems. Remote sensing is a cornerstone technology for disaster preparedness and mitigation, which aims to obtain information about various objects or areas from a distance through satellite and other aerial platforms (Fernández-Montes De Oca et al., 2025).

Table 1. Major natural and human-caused disasters worldwide, including causes, locations, years, and verified damages

Type	Disaster Name	Cause	Location	Year	Damages
Natural	Indian Ocean Tsunami	Undersea 9.1–9.3 magnitude earthquake	Southeast Asia (esp. Indonesia, Sri Lanka, India, Thailand)	2004	~230,000 deaths, 1.7 million displaced, ~$15 billion in damages
	Hurricane Katrina	Category 5 hurricane	USA (Louisiana, Mississippi)	2005	1,833 deaths, over $125 billion in damages, massive infrastructure loss
	Haiti Earthquake	Magnitude 7.0 earthquake	Haiti	2010	220,000+ deaths (est.), 300,000 injured, $8–14 billion in damages
	Japan Tōhoku Earthquake & Tsunami	Magnitude 9.0 earthquake and tsunami	Japan	2011	~20,000 deaths/missing, $360 billion in damages (most costly natural disaster ever)
	Nepal Gorkha Earthquake	Magnitude 7.8 earthquake	Nepal	2015	~9,000 deaths, 22,000+ injured, $7 billion in damages
	Typhoon Haiyan (Yolanda)	Super typhoon (Category 5 equivalent)	Philippines	2013	6,300+ deaths, 1.1 million homes damaged, $5.8 billion in damages
	European Heatwave	Extreme heat (climate-linked event)	Europe	2003	70,000 excess deaths across Europe, agriculture loss, no precise cost figure
	Pakistan Floods	Monsoon rainfall intensified by climate change	Pakistan	2022	1,739 deaths, $30+ billion in damages, 33 million affected
	Cyclone Nargis	Category 4 tropical cyclone	Myanmar (Burma)	2008	138,000 deaths, $10 billion in damages, 2.4 million severely affected

continued on following page

Table 1. Continued

Type	Disaster Name	Cause	Location	Year	Damages
Human-caused	Chernobyl Nuclear Disaster	Reactor explosion due to human/operator error	Ukraine (then USSR)	1986	31 immediate deaths, 4,000+ cancer deaths estimated, long-term exclusion zone
	Bhopal Gas Tragedy	Toxic gas leak from pesticide plant	India	1984	3,800+ deaths, over 500,000 injured, long-term birth defects and diseases
	Deepwater Horizon Oil Spill	Offshore drilling explosion	Gulf of Mexico	2010	11 deaths, 4.9 million barrels of oil spilled, $65+ billion in damages
	Fukushima Nuclear Disaster	Reactor meltdown (design flaws + tsunami)	Japan	2011	154,000 evacuated, $200+ billion in cleanup and compensation costs
	Rana Plaza Collapse	Poor construction and safety negligence	Bangladesh	2013	1,134 deaths, 2,500+ injured, global scrutiny of garment industry
	Amazon Rainforest Fires	Illegal deforestation and slash-burning	Brazil	Recurring	Millions of hectares burned, loss of biodiversity, global carbon emissions
	Minamata Disease	Industrial mercury poisoning	Japan	1950s–1960s	900+ deaths, thousands with lifelong neurological damage
	Syrian Refugee Crisis	Civil war and armed conflict	Syria & neighboring states	Ongoing	Over 500,000 dead, 13+ million displaced, humanitarian crises
	Love Canal Waste Disaster	Hazardous waste buried in residential area	USA	1970s	Hundreds relocated, increased cancer and birth defect rates, $250M cleanup
	Texas City Refinery Explosion	Industrial accident due to safety violations	USA	2005	15 deaths, 170+ injured, $1.5 billion in fines and damages

Figure 1. Disaster management process phases

Advanced remote sensing satellites, such as Sentinel-1, Landsat-8, and PlanetScope, generate humongous quantities of high-resolution and multi-spectral imagery at high speed, which cannot be processed using conventional remote sensing methods relying on manual interpretation and analysis. This poses a significant challenge and bottleneck in the real-time processing of such data in an emergency and hampers the decision-making for rescue activities. A promising technique for addressing such a problem is deep learning – a subset of the machine learning domain modeled on the working of the human brain – which is being used successfully in several domains involving vast amounts of imagery data (Aamir et al., 2023; Feng et al. 2022 ; Haider et al. 2024) . Several deep learning techniques, such as recurrent neural networks, convolutional neural networks, transformers, and generative models, possess immense potential to extract meaningful patterns from high-quality remote sensing data.

Integrating deep learning with remote sensing technology possesses immense potential in disaster management. Some potential areas include object detection (e.g., damaged infrastructure and road blockages), semantic segmentation (e.g., flood and wildfire coverages), and predictive modeling (e.g., landslide and fire spread probabilities). For example, Li et al. (W. Li et al., 2022) used deep ConvNets to extract buildings from high-resolution satellite imagery. Similarly, Doan and Le-Thi proposed a deep learning model for flood detection using remote sensing images (Doan & Le-Thi, 2025).

A unique strength of deep learning techniques in disaster management is their ability to generalize and transfer learning across different domains and contexts, as it is impossible to provide labeled data in a crisis situation. In such cases, transfer and few-shot learning techniques are especially useful and powerful. For instance, Hasnaoui et al. used transfer learning for flood mapping in the coastal area of Algeria (Hasnaoui et al., 2025).

Most disasters require response times in terms of minutes or hours. Hence, the operational deep learning systems must be scalable, interpretable, and efficient. Although deep learning models are generally computationally expensive, the recent developments in cloud computing, edge AI, and distributed systems enable real-time deep learning applications. Open source frameworks, such as TensorFlow, Keras, and PyTorch, coupled with remote sensing libraries like Rasterio and GDAL, enable public access and low-cost development, thus providing access to these expensive technologies to more developers worldwide.

Another challenge in deep learning models is interpretability. Deep learning models generally suffer from poor interpretability, making them less than ideal solutions for sensitive application areas, such as disaster management. A stakeholder would like to know the clear reasoning process and rationale behind a recommendation produced by a deep learning system, for example, a road blockage or flooded area.

Recent innovations in explainable AI, such as Layer-wise Relevance Propagation (LRP), Grad-CAM, and SHAP values, are trying to bridge this gap and make the deep learning solution more explainable and trustworthy.

Other challenges in the adoption of deep learning solutions for disaster management include bias in the training data, challenges in domain adaptation, lack of standardization frameworks, and ethical challenges. Addressing these challenges is crucial to the success of deep learning solutions in disaster management to realize the full potential of these powerful technologies and frameworks.

RELATED WORK

Many researchers have explored the application of deep learning solutions in remote sensing in the last decade. We highlight major contributions in the key aspects of disaster management, including building damage detection, flood mapping, earthquake assessment, wildfire prediction, and oil spill detection (Figure 2). Table 2 gives a summary of these techniques, along with their strengths and limitations.

Table 2. Deep learning applications in disaster management

Reference	Application	Techniques	Strengths	Limitations
(Al Shafian & Hu, 2024)	Building damage assessment	Bibliometric review of 370 articles	Comprehensive overview; identifies trends	No experimental results
(J. Z. Xu et al., 2019)	Earthquake building damage (Haiti 2010)	CNN architectures comparison	Benchmarks multiple CNNs; generalization tested	Uses only arXiv preprint; not peer-reviewed
(Adriano et al., 2020)	Multi-disaster damage mapping	Multi-modal CNN on SAR + optical	Global dataset; supports various disasters	Preprint; lacks journal validation
(Wu et al., 2023)	Flood mapping	CNN on Sentinel-1 SAR	Near-real-time; robust to speckle	Validated only on Yangtze basin
(Jenifer et al., 2022)	Agricultural flood detection	Dual-patch CNN	High accuracy in crop zones	Untested in urban/ rural diversity
(Singh et al., 2025)	Earthquake building damage	Transformer + imagery + metadata	SOTA multi-class accuracy; metadata boosts performance	Preprint; no field deployment yet
(Kizilay et al., 2024)	Drone disaster damage detection	YOLOv8, Detectron2, VGG16 with transfer learning	Real-time capability; clear architecture comparison	Limited to drone imagery; small area tested
(Rao et al., 2023)	Earthquake damage mapping	CNN + metadata + InSAR + building footprints	Modular, wide applicability, tested in multiple events	Only ~50% detection; needs more spatial diversity

continued on following page

Table 2. Continued

Reference	Application	Techniques	Strengths	Limitations
(Z. Xu et al., 2024)	Wildfire risk assessment	Systematic review of DL models (CNN, RNN, ConvLSTM, Vision Transformers	Comprehensive taxonomy; focus on interpretability and data fusion	Does not propose a new model; limited empirical benchmarking
(Huot et al., 2021)	Wildfire risk assessment	CNN on multivariate EO data	Strong AUC (~0.83); multi-data fusion	Retrospective; lacks real-time evaluation
(Prapas et al., 2021)	Daily fire danger mapping	ConvLSTM time-series	High AUC (0.926); surpasses RF baseline	Requires large-scale EO cube; heavy compute
(Basit et al., 2024)	Oil spill detection	Encoder-decoder CNN on SAR	Discovered new spills; large temporal coverage	Performance metrics not fully disclosed
(Mahmoud et al., 2023)	Oil spill segmentation	Dual-Attention U-Net	Enhanced contextual capture; robust segmentation	Tested on limited dataset; scalability unknown
(Bianchi et al., 2020)	Operational oil spill pipeline	Segmentation + classification neural net	Human-level accuracy; operational-ready	Preprint; needs peer review

Figure 2. Deep learning applications in disaster management

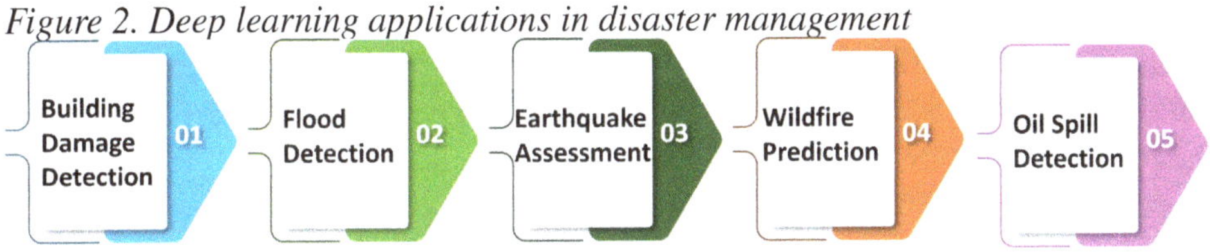

Post-Disaster Building Damage Detection

Al Shafian and Hu present a thorough review of deep learning and remote sensing techniques for post-disaster building damage assessment based on 370 journal articles published from 2014-2024 and indexed by the Scopus database (Al Shafian & Hu, 2024). The study identifies key trends in the area, including UAV imagery, data fusion, and CNN architecture. The major challenges include generalization and real-time processing.

Xu et al. (2019) systematically benchmarked CNN-based building damage detection techniques after the 2010 Haiti earthquake. The study proposed a CNN-based model after comparing several deep learning architectures and evaluating their generalization capabilities. However, the model is based on a single event, making its generalization to other events challenging.

Adriano et al. (2020) developed an advanced deep learning model for building damage assessment by combining optical and synthetic aperture radar (SAR) imagery. The study proposed a multi-modal and multi-event model, thus achieving high generalization across various disasters, such as earthquakes, tsunamis, and typhoons.

Flood Detection and Mapping

Wu et al. (2023) prepared a dataset of 16 flood events in the Yangtze River basin area using SAR images and analyzed various deep learning techniques for flood prediction efficiency. The study concluded that vertical transmit-horizontal receive (VH) polarization of SAR images achieved better flood detection compared to vertical transmit-vertical receive (VV) polarization.

Jenifer et al. (2022) developed a dual patch CNN model to detect floods from Sentinel-1 imagery. The proposed model produced excellent results in crop areas but lacks validation in urban areas.

Earthquake Disaster Assessment

Singh et al. (2025) proposed a transformer-based architecture by integrating ground motion data, SAR images, and soil metadata to predict building damage in the aftermath of the 2023 Turkey-Syria earthquakes. The model achieved an excellent multi-class performance. However, the study is limited by a lack of deployment of the proposed solution.

Kizilay et al. (2024) compared several deep learning models, including YOLOv8, Detectron2, and adapted VGG16 for real-time earthquake damage assessment using drone images. The study found that YOLOv8 achieved the best combination of accuracy and speed.

Rao et al. (2023) proposed a semi-automated machine learning framework for earthquake building damage assessment using building information, earthquake ground shaking intensity maps, and interferometric synthetic aperture radar (InSAR) data The study achieved an accuracy of 50%, highlighting the need for more spatial diversity and improved model architecture.

Wildfire Risk Prediction

Xu et al. (2024) provide an excellent exposé of the recent deep learning techniques used for wildfire risk prediction and assessment. The study emphasizes the use of model outputs in real-life scenarios. The study also highlights the limitations of the current wildfire risk prediction techniques and discusses the potential to exploit multi-modal data in future models.

Huot et al. (2021) trained a deep learning model to predict wildfires using vegetation, weather, and topography data features. Although the proposed method achieved a relatively low AUC score of 83%, it shows the potential of using deep learning techniques for wildfire prediction.

Prapas et al. (2021) explored several deep learning models to predict daily wildfires. The study achieved the best results, with a 92.6% AUC for convolutional long short-term memory (ConvLSTM) network.

Oil Spill Detection

Basit et al. (2024) developed an encoder-decoder CNN to detect oil spills in Pakistan's exclusive economic zone. The study used SAR images of the area from 2017-2023 and successfully detected 92 unreported oil spills.

Mahmoud et al. (2023) employed a dual attention UNet for oil spill detection using SAR images. A comparison of the proposed method with a baseline of four models, including a fully convolutional network, PSPNet, LinkNet, and traditional UNet, shows an increase of 3.2% in the overall prediction accuracy, reaching 94.2%.

Bianchi et al. (2020) proposed oil fully ConvNet, a variation of UNet specifically designed to detect oil spill detection using image segmentation. The proposed method achieved impressive results, with the highest oil spill prediction of 93.8% and an F1 score of 0.91.

DEEP LEARNING FOUNDATIONS FOR GEOSPATIAL ANALYSIS

Recent advancements in deep learning and the resulting application in geospatial analysis have enabled several opportunities in the domain that were not possible or inefficient with conventional remote sensing methods. Advanced pattern recognition, semantic segmentation, and spatiotemporal forecasting are a few applications enabled by highly sophisticated deep learning models. This section explores the basic concepts and architecture of deep learning models at the core of remote sensing and geospatial analysis. Table 3 gives a summary of deep learning models used for geospatial analysis, highlighting their key features, strengths, and limitations.

Figure 3. Deep learning models for disaster management

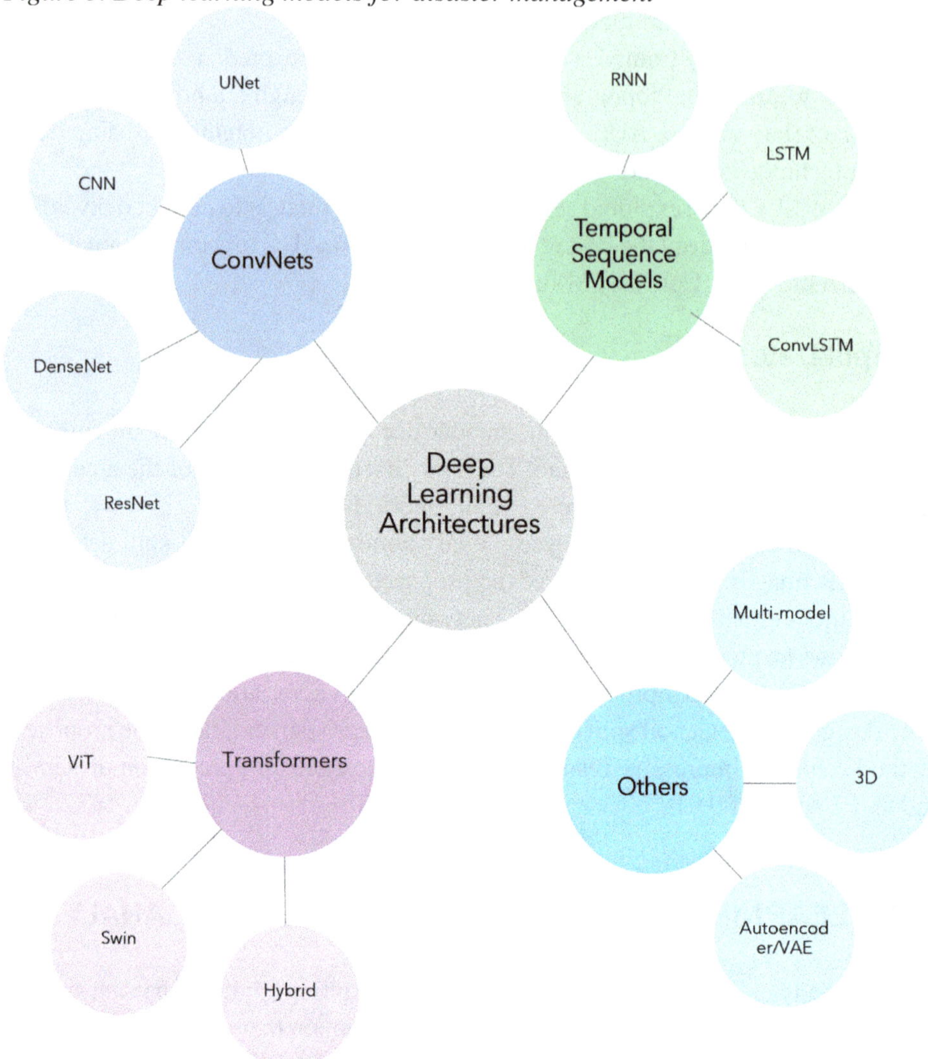

Convolutional Neural Networks (CNNs) and Spatial Representation

CNNs are the basic building block of geospatial analysis because of their ability to extract interesting features from images using convolutional and pooling layers. They can be used in various applications, such as image classification, object detection, and semantic segmentation of satellite and drone images. The main limitations of

CNNs are their inability to capture long-range dependencies and limited temporal understanding of the image features. In recent years, encoder-decoder variants of CNNs, like UNet, have become popular in performing pixel-wise tasks, such as semantic segmentation. However, there is an increase in performance at the cost of high computational requirements. They are also more sensitive to scale variations. Another variation of CNNs is deep residual CNNs. Their deep hierarchical feature extraction capabilities make them ideal for deep scene classification. However, they require a long training time and are susceptible to overfitting.

Recurrent Neural Networks (RNNs) and Temporal Dynamics

While the basic spatial feature learning capabilities of CNNs suffice for several disaster management applications, others call for temporal dynamics to perform advanced geospatial analysis. To this end, recurrent neural networks (RNNs), particularly long short-term memory (LSTM) networks, perform excellently in capturing temporal dependencies. The temporal understanding of remote sensing data is crucial for dynamic geospatial applications, such as wildfire spread and flood progression. The temporal sequence modeling capabilities of RNNs and LSTMs can capture long-term dependencies in time series data. However, they are poor at spatial representation and hard to parallelize.

Transformers and Attention-Based Models

Transformer architecture was initially developed for natural language processing to capture long-term dependencies, which are essential for natural language processing. However, remote sensing applications that require long-term dependencies, such as disaster progression and change detection, also exploit this strength of transformers. Two variations of transformers stand out in this regard: vision transformers (ViT) and Swin transformers. Their strength lies in their ability to model global context and the scalability of large datasets. On the other hand, their major limitations are large data requirements and lack of inductive bias for spatial locality.

Researchers have also proposed a combination of CNNs and transformers architecture. For instance, Xu et al. developed a hybrid Swin UNet model to capture local and global context for building damage detection (S. Xu et al., 2022).

Multi-Modal Architectures

Remote sensing applications and geospatial analysis usually require integrating diverse data sets to perform unified analysis and reasoning. For instance, in a disaster situation, data might be generated from various sensors (e.g., optical and SAR) with varying resolutions and temporal spans. Spatiotemporal fusion and multi-modal architectures are designed to address these integration challenges. Multi-modal CNNs can handle heterogeneous inputs and are robust enough to handle missing modalities. However, they are relatively difficult to train, and data misalignment might affect their performance.

Table 3. Deep learning models used for geospatial analysis

Model / Architecture	Type	Primary Applications	Strengths	Limitations
CNN (Convolutional Neural Network)	Spatial feature extractor	Image classification, object detection, land cover mapping	Captures local spatial patterns; translation invariance	Struggles with long-range dependencies; limited temporal understanding
U-Net / Encoder–Decoder CNN	Semantic segmentation	Flood mapping, building footprint extraction	High-resolution segmentation; effective with small datasets	Computationally expensive; sensitive to scale variation
ResNet / DenseNet	Deep residual CNNs	Scene classification, feature extraction	Deep hierarchical features; avoids vanishing gradients	High training time; risk of overfitting with limited data
RNN / LSTM	Temporal sequence model	Vegetation change, weather prediction	Captures long-term dependencies in time series	Poor spatial representation; hard to parallelize
ConvLSTM	Spatiotemporal model	Flood prediction, wildfire forecasting	Integrates spatial and temporal context; good for EO sequences	Limited in capturing long-range spatial features
Vision Transformer (ViT)	Attention-based spatial model	Land use classification, damage assessment	Models global context; scalable to large datasets	Requires large data; lacks inductive bias for spatial locality
Swin Transformer	Hierarchical transformer	Land cover mapping, change detection	Combines local and global context; efficient on high-res images	Architecture complexity; training requires careful tuning
Hybrid CNN + Transformer	Multi-modal spatial model	Building damage mapping, disaster assessment	Leverages strengths of CNNs and transformers	Integration complexity; model interpretability
Multi-modal CNN	Data fusion model	SAR-optical fusion, disaster detection	Handles heterogeneous inputs; robust to missing modalities	Training difficulty; data alignment challenges

continued on following page

Table 3. Continued

Model / Architecture	Type	Primary Applications	Strengths	Limitations
3D CNNs	Spatiotemporal volume model	Volumetric EO data analysis	Joint modeling of spatial and temporal patterns	High memory usage; fewer pre-trained models available
Autoencoders / Variational Autoencoders (VAEs)	Feature compression / generation	Data denoising, anomaly detection, pretraining	Learns latent representations; unsupervised	Less interpretable; may discard fine-grained spatial details

MULTI-MODAL AND MULTI-TEMPORAL DATA FUSION

Recent advances in deep learning have pushed multimodal (multi-sensor) and multitemporal (multi-snapshot) fusion from proof-of-concept demos to operational tools. Architectures such as M^{3}Fusion for land-cover mapping (Benedetti et al., 2018) and Multi3Net for rapid damage assessment (Rudner et al., n.d.) illustrate how optical, Synthetic-aperture radar (SAR) and Light Detection and Ranging (LiDAR) imagery can be melded into a single analytic pipeline. Success stories now span change-detection (Ebel et al., 2021), rapid disaster response (Algiriyage et al., 2021) and hybrid optical-SAR-LiDAR classification models (Hong et al., 2021). Two open challenges remain: i) heterogeneity, which refers to efficiently integrating data that differ in geometry, radiometry and revisit rate; and ii) data scarcity, i.e., the limited number of large, well-annotated multimodal benchmarks (J. Li et al., 2022).

Against this backdrop, in this section we explain why fusion matters, what kinds of sensors and time-scales can be combined, how modern architectures implement fusion, and what we can learn from flagship studies.

Why Fuse? From Blind Spots to Situational Awareness

Disasters rarely occur under ideal viewing conditions. Optical satellites stop "seeing" when night falls, smoke fills the air, or thick clouds cover the scene. SAR can operate day or night and through clouds, but it records only the strength of microwave echoes and cannot show colors or thermal patterns. Social-media photos and text provide ground-level reports, yet their locations are often imprecise and their images lack metric geometry. Each source therefore captures only part of the picture. Modal fusion stitches these sources together, combining the all-weather reach of SAR, the spectral detail of optical and hyperspectral data, the height information from LiDAR, and the contextual cues from crowd-sourced content to build a fuller description of the scene (Benedetti et al., 2018; Ebel et al., 2021). Temporal fusion then adds the time dimension, aligning pre-event baselines, near-real-time snapshots, and post-event recovery images so that both sudden damage and gradual rebuilding become visible (J. Li et al., 2022). When used together, these two forms of fusion transform a set of partial observations into a robust, information-rich view that can guide rapid, evidence-based disaster response.

Fusion offers three practical gains for disaster mapping: greater observability, higher robustness, and faster timeliness. Table 4 organizes these benefits and illustrates how they manifest at three levels: sensor modality, time, and spatial resolution.

Table 4. Benefits of multi-sensor fusion illustrated across modality, time, and spatial resolution

Benefit	Modal illustration	Temporal illustration	Resolution illustration
Observability	SAR pierces clouds that block optical imagery	Pre- and post-event pairs expose change	Sub-meter UAV data fills the 10 m gaps in Sentinel-2 mosaics
Robustness	Independent noise sources (e.g., SAR speckle vs. optical haze) add redundancy	Voting across multi-snapshot stacks cuts false alarms	Down-/up-scaling mitigates errors when resolutions differ
Timeliness	Medium-resolution Sentinel-2 gives the first look; later VHR images refine detail	Hourly geostationary IR frames track wildfire fronts	A lightweight CNN handles the first pass; a heavier transformer refines it as bandwidth allows

Note that each row links a benefit to concrete design choices. For observability, SAR complements optical sensors by operating through cloud or smoke, while fine-grained UAV imagery closes the spatial gaps left by 10 m multispectral tiles. Robustness emerges when uncorrelated noise sources (or different revisit times) allow the system to cross-validate predictions. Timeliness combines fast-downlink sensors and lightweight models for an early "good-enough" map, followed by high-resolution data and heavier networks for final detail.

Empirical results back these patterns. On the ONERA-CD benchmark, fusing SAR with optical imagery raises change-detection IoU by more than ten percentage points (Ebel et al., 2021). In rapid-flood mapping, the three-sensor Multi3Net pipeline cuts product latency from days to mere hours (Rudner et al., n.d.). Taken together, these studies confirm that the benefits listed in Table 4 are not only theoretical but also measurable in real-world deployments.

Dimensions of Modality

A modality is defined by the physical principle that produces a signal (i.e., solar reflectance, microwave back-scatter, laser ranging, or human reporting) and by the spatial and temporal scales at which that signal is collected. In practice, disaster analysts must keep several of these data streams in play at once, because each illuminates the emergency from a different angle. Table 5 summarizes the six modalities most often used in disaster response, spelling out what each one does well and where each falls short.

Table 5. Core sensing modalities used in disaster response

Modality	Physical cue & typical resolution	Main value for DM/HA	Key limitation
Multispectral (MS)	Solar reflectance; 10–30 m; 5-day revisit	Land-cover maps, flood-water indices	Blocked by clouds and darkness
SAR (C/X-band)	Microwave back-scatter & coherence; $\approx$10 m; 6–12 h revisit	All-weather, day-night change detection	Speckle noise; layover in mountains
LiDAR / DSM	Laser elevation & roughness; 1–5 m; campaign-based	Debris volume, building-height loss	Collected infrequently; costly
Hyperspectral (HS)	> 200 bands; $\approx$30 m; 16-day revisit	Material chemistry, soil moisture, oil spills	Low signal-to-noise; large files
VHR optical / UAV	0.3–1 m; on-demand	Fine-scale damage, blocked roads	Narrow swath; weather-limited
Crowd / social media	Text & images; real-time	On-the-ground confirmation, population needs	Noisy content; imprecise geo-tags

The table pairs each sensor family with three pieces of context: the underlying physics and its usual spatial-temporal footprint, the main advantage that physics provides for disaster response, and the operational caveat that limits routine use. Thus, SAR's microwave backscatter appears alongside its all-weather, day-night value for change detection, while the same row reminds readers of speckle noise and layover effects in rugged terrain (Ebel et al., 2021). LiDAR's entry emphasizes its strength in measuring debris height yet flags the infrequency and cost of airborne campaigns; the social-media row highlights real-time ground truth but cautions that geotagging is often imprecise (Alam et al., 2020). In each row, a clear pattern emerges: every modality offers a unique benefit but also carries a weakness that another sensor can offset; underscoring why multi-modal fusion is a necessity rather than a luxury. Thus, a quick scan shows that no single row covers every requirement: optical sensors supply rich color but fail under smoke; SAR works all weather but lacks color; social media adds context but little metric accuracy. Fusion is therefore indispensable. With the diversity of sensors clarified, we can now turn to the temporal dimension and examine how revisit rates and multi-snapshot sequences further shape disaster-response workflows.

Temporal Granularity

The moment an image is collected can be as important as the sensor that records it. Disaster-response workflows usually rely on three time-scales, each suited to a different question. Table 6 lists these time arcs, explains how they are used, and notes the main technical hurdles.

Table 6. Temporal scales in disaster response workflows

Time-scale	Typical span	Main purpose	Example workflow
Bi-temporal	One image before the event (T_0) and one after (T_1)	Rapid damage mapping	A Sentinel-1/2 before-and-after pair highlights collapsed buildings within hours of an earthquake and raises IoU by > 10 pp on the ONERA-CD benchmark (Ebel et al., 2021).
Short-arc multi-temporal	Hours to a few days	Tracking fast-moving hazards—flood water, wildfire fronts	High-revisit SAR or geostationary IR provides near-continuous coverage; occasional very-high-resolution frames confirm detail and guide field teams (Rudner et al., n.d.).
Long-arc recovery	Weeks to years	Monitoring rebuilding and ecosystem rebound	Monthly Sentinel-2 mosaics trace roof repairs and vegetation regrowth across seasons, provided radiometric calibration is kept stable (J. Li et al., 2022).

Each row in Table 6 links a time window to a typical task. The first row shows how a simple before/after set can reveal sudden change. The second extends the window to days so that frequent but coarse images can be blended with rare high-detail shots, which is essential when a flood rises hour by hour. The third row looks months or years ahead to measure recovery, a task that relies on consistent imaging over long periods.

One key challenge in this context is that operational archives are often irregular: some dates are missing, revisit intervals vary, and illumination conditions shift. Models must therefore handle sequences of unequal length and quality without a drop in performance. Recent transformer and recurrent designs that allow optional or masked time steps are a step toward this goal. With the time dimension clear, the next section shows how modern networks combine these temporal cues with the spatial and modal information discussed earlier.

Fusion Strategies & Deep-Learning Architectures

In a neural network, fusion can happen early, midway, or late. Selecting the right depth is a trade-off between how much raw detail is kept, how robust the model is to mis-alignment, and how much compute the hardware can spare. Table 7 outlines the three common tiers, the deep-learning patterns that implement them, and where each tier performs best and worst.

Table 7. Fusion tiers, their deep-learning patterns, and typical trade-offs

Fusion tier	Typical DL pattern	Main strength	Main limit	Typical use-case
Pixel-level (early)	Stack all bands or time steps into one tensor; 2-D or 3-D CNN	Keeps full physical detail	Needs precise co-registration	Estimating flood depth from SAR plus altimetry grids
Feature-level (hybrid)	Separate encoders; concatenate or apply attention to latent codes	Balances detail and abstraction	Heavier models, more GPU memory	Building-damage mapping with optical imagery fused to a DEM (Hong et al., 2021)
Decision-level (late)	Independent models; ensemble or Mixture-of-Experts	Sensor-agnostic and easy to extend	May lose subtle cross-modal cues	Rapid triage by stacking SAR, optical, and social-media classifiers (Rudner et al., n.d.)

Each row in Table 7 shows where the merger happens inside the network, the building block that makes it work, and the trade-off that comes with that choice. Early fusion keeps every raw pixel value but fails if the inputs are not perfectly aligned. Late fusion is lenient to mis-alignment and lets new sensors be attached, yet it can miss fine interactions across modalities. Hybrid, feature-level designs sit in the middle, mixing separate encoders with attention or concatenation, but they demand more compute and memory. The key point here is that no single tier wins in every setting. Practical systems often combine them, e.g., an early optical-plus-SAR block to capture physics, followed by a late ensemble that adds social-media cues once text streams arrive.

Case-Study Gallery

Real deployments, not just benchmark scores, show what fusion can and cannot do. Table 8 lists four well-cited systems, the data they fuse, and the gains they report. The studies were chosen to span different fusion tiers, sensor mixes, and time-scales.

Table 8. Representative fusion studies and their reported gains

Study	Modalities / Temporal scope	Reported gains
Multi3Net (Rudner et al., n.d.)	Sentinel-1, Sentinel-2, VHR; pre-/post event	Late fusion raised flooded-building mIoU from 56% (VHR-only) to 75% and delivered maps within ~3 h of first down-link.
Siam-U-Net (Ebel et al., 2021)	Sentinel-1 & 2 bi-temporal pairs	Dual encoders plus a shared decoder lifted change-detection F1 by 0.8 pp over optical-only baselines and introduced the ONERA-CD-M dataset.
M³Fusion (Benedetti et al., 2018)	34-image Sentinel-2 series + SPOT-6/7	A GRU-CNN hybrid reached 90.7% overall accuracy; F-measure for "greenhouse crop" jumped from 0.25 to 0.58 compared with Random Forest.
Cross-Modal Transformer (Algiriyage et al., 2021)	Sentinel-1, Sentinel-2, PlanetScope, crowd labels	Hierarchical attention chose sensor weights on the fly, tripling throughput with < 1% mIoU loss—promising for low-power deployments.

In Table 8, the first band shows a late-fusion design that waits for all sensors, then merges logits. The second uses feature-level fusion via shared decoding. The third confirms that multi-temporal signals pair well with multi-scale imagery (Haq et al. 2023; Nabeel et al. 2024; Topbaş et al. 2021) The fourth points to attention mechanisms that adapt when a sensor drops out. These key studies highlight practical design patterns for fusing multimodal remote sensing data as described below.

Case 1 — Multi3Net: Each sensor stream produces a quick segmentation as soon as its data arrive; a late-stage combiner refines the map when higher-resolution imagery downloads. This "progressive refinement" sliced product latency from days to hours.

Case 2 — Siam-U-Net: Parallel encoders tame SAR speckle and optical radiance separately, then align latent spaces through shared skip connections—an elegant fix for modality mismatch.

Case 3 — M³Fusion: Time-series (GRU) and patch-texture (CNN) branches each carry an auxiliary loss, forcing them to be useful on their own; the ensemble gains appear consistently across five random splits.

Case 4 — Cross-Modal Transformer: Attention scores rise or fall with sensor quality, so the model degrades gracefully when clouds hide optical data. This is an attractive property for field work where outages are common.

Hene, across flood mapping, change detection, and land-cover classification, well-designed fusion consistently delivers double-digit gains in IoU or F-measure and, in several cases, cuts map-delivery time by an order of magnitude.

Practical Insights & Pitfalls

After reviewing successful fusion systems, several operational lessons stand out. Table 9 groups these lessons into four themes: alignment, labelling, computation, and ethics (Figure 4). The table also notes both the risk each issue poses and the technique most often used to reduce that risk.

Figure 4, Issues in multi-sensor AI

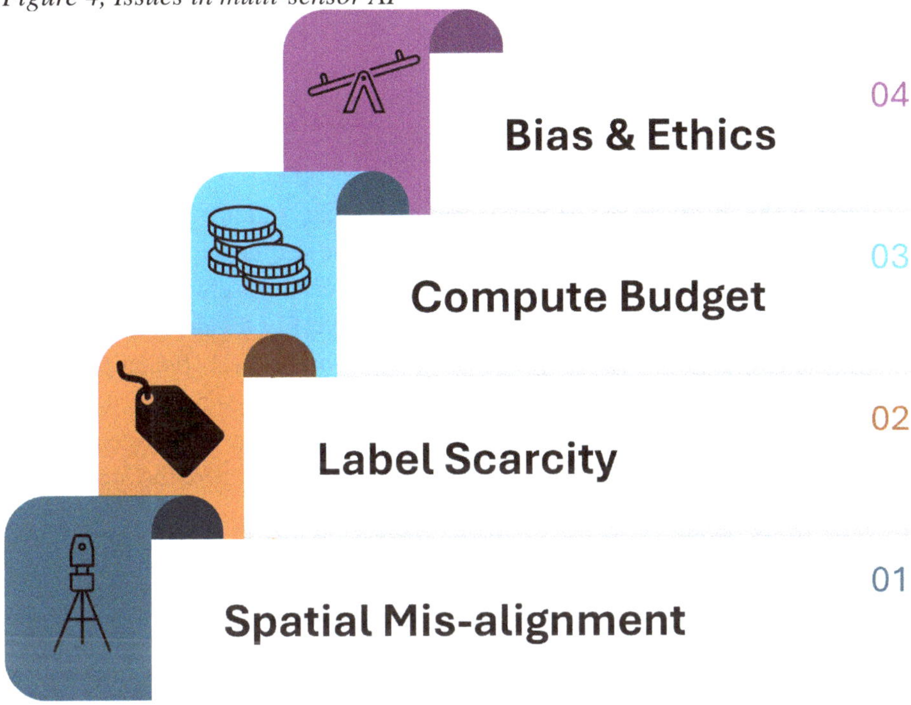

Table 9: Representative risks in multi-sensor AI and common mitigation techniques

Issue	Why it matters	Common mitigation
Spatial mis-alignment	Sub-pixel shifts between sensors turn true change into false change, inflating error rates	Deformable alignment layers or coarse-to-fine image warping before fusion (Ebel et al., 2021)
Label scarcity	Pixel-accurate annotations for multiple sensors are costly and slow to obtain	Self-supervised or contrastive pre-training, which can cut labelled data needs by ~60% (J. Li et al., 2022)
Compute budget	Transformer attention grows quadratically with image size and time steps	Windowed or sparse attention plus mixed-precision inference to stay within GPU or edge limits (Ijaz et al., 2023)
Bias and ethics	Well-imaged regions dominate training data, leaving vulnerable areas under-served	Uncertainty maps, bias audits, and balanced dataset design; visual tools to show confidence to end-users (Alam et al., 2020)

In Table 9, the first band explains why careful co-registration is non-negotiable; even a one-pixel shift can masquerade as damage. The second shows that modern self-supervised methods can offset scarce labels, an especially useful trick for SAR or hyperspectral data. The third highlights the need to budget FLOPs and memory early. The final band highlights a social dimension: without bias checks, fusion pipelines may amplify data gaps and steer aid away from the very regions that need it most.

These practical notes lead directly into the next section, where the focus turns to building fair, well-documented datasets and benchmarks, which are the foundation on which reliable fusion pipelines stand.

DATASETS AND BENCHMARKING PROTOCOLS

Reliable models rest on reliable data. For disaster management, "reliable" means more than a large image count: the training set must span many hazards, many sensors, and many time-scales, and the community must agree on how those data are split, scored, and shared. This section explains why such breadth is essential, surveys the most widely used public datasets, and outlines the protocol rules that keep results fair and reproducible.

Why Dedicated Benchmarks Matter

Deep-learning models thrive only when their training data mirror the visual, temporal, and contextual variety they will face once deployed. In disaster management, that variety spans three main axes, summarized in Table 10.

Table 10. Axes of diversity in disaster imagery and their impact on model robustness

Diversity axis	What varies	Why it matters	Typical failure without it
Hazard	Floods, fires, earthquakes, hurricanes, slow-onset crises	Each hazard leaves a different spectral or geometric trace	A model tuned on floods mislabels burn scars as water (Wiguna et al., 2024).
Sensor	VHR optical, SAR, multispectral, LiDAR, crowd photos	Fusion pipelines rely on cross-modal cues	Training on optical alone collapses when clouds force a switch to SAR (Ebel et al., 2021)
Temporal	Single snapshots, paired images, long time series	Change-detection and recovery tracking need multi-date data	A single-date classifier cannot spot new debris or regrowth (J. Li et al., 2022)

The first band reminds us that fire burn scars are radiometrically distinct from floodwater; covering both hazards in one benchmark stops a model from over-specializing. The second band shows that cross-modal systems must be trained and scored on all sensors they will fuse in the field. The third band points to the need for paired or multi-date images whenever the task is "what changed?" rather than "what is there?".

Because of these requirements, a benchmark plays a dual role: it is training fuel and it is the yard-stick by which the community measures progress. A recent survey of more than 500 Earth-observation datasets calls for "large-scale, high-richness" collections that are chosen with explicit coverage criteria rather than convenience sampling (J. Li et al., 2022). The next subsection reviews how well current public datasets meet those criteria and where the gaps remain.

Landscape of Public Datasets

Two trends often guide the choice of training and test sets. First, early remote-sensing scene benchmarks, most notably AID and the Meta-Dataset compiled by Hurt et al.(Hurt et al., 2018), established fixed train/validation/test splits and class-balance rules that later crisis-specific corpora copied. Second, newer disaster datasets now stress multi-task labels and strict duplicate filtering. Social-media collections such as HumAID (Xia et al., 2017a) and CrisisMMD (Alam et al., 2021) drop near-identical tweets and images and supply multi-label annotations, a practice shown to prevent inflated scores when duplicates leak from train into test folds.

Table 11 groups widely used datasets by sensing modality and principal task, noting scale and design choices that shape downstream model behavior.

Table 11. Widely used disaster datasets by modality, task, scale, and design focus

Dataset (year)	Modality	Main task	Scale	Notable design choice
AID (2017) (Xia et al., 2017b)	VHR RGB	Aerial scene classification	> 10 000 images, 30 classes	Fixed 50/50 train-test split plus 20% validation encourages reproducibility
HumAID (2021) (Alam et al., 2021)	Twitter text	Humanitarian information triage	77 000 labelled tweets, 11 categories	Covers 19 disasters (2016–2019) and balances event types
CrisisMMD (2018) (Alam et al., 2018)	Twitter text + image	Multi-task (informative, humanitarian, damage)	18 082 images with three label sets	Early multimodal baseline; duplicate-aware split
HRSCD (2019) (Daudt et al., 2019)	RGB image pairs	Semantic change detection	291 pre/post pairs, 5 land-cover classes	Pixel-wise labels for both dates enable dense CD training
S2Looking (2021) (Shen et al., 2021)	Sentinel-2 RGB time series	Binary change	5 000 city-scale pairs	Tailored to building-level CD at 10 m resolution
fMoW-RGB (2018) (Christie et al., 2018)	Multi-temporal RGB	Land-use classification	≈ 1 M images, 63 classes	Ships with metadata and bounding boxes for multi-task studies
EarthNets Unified Set (2024) (Xiong et al., 2024)	RGB, MS, SAR	Classify / Detect / Segment	Five high-coverage subsets	Common loaders and metric scripts span tasks and sensors

In the table above, each entry states the sensor mix, the task granularity (scene-level, pixel-level, or social-media text), and the quantitative scale. The "Notable design choice" field highlights factors such as class balance, duplicate filtering, paired imagery that influence a model's real-world reliability beyond simple sample count. Below are the key observations.

- **Bigger is not always richer:** AID's 30 well-separated classes still anchor many baselines despite its modest size because its categories are visually distinctive and evenly distributed.
- **Duplicate control is critical in social media:** Alam et al. (Alam et al., 2021) report F1 drops of more than five percentage points when near-identical images slip into both train and test sets; HumAID and CrisisMMD therefore apply perceptual-hash screening before they generate splits .
- **Change-detection corpora lack geographic spread:** HRSCD is European; S2Looking is Chinese. Work is under way to build globally balanced alternatives, for example the Sentinel-1/2–based DynamicEarthNet (Toker et al., 2022).

These observations set the stage for the protocol discussion that follows, where we ask how to split, score, and report results so that gains on paper translate into gains in the field.

Benchmarking Protocols

Good datasets are only half the story; without clear protocols the scores they produce cannot be trusted or compared. Four points, i.e., splits, duplicate control, shared tooling, and metric choice, now form the accepted checklist for disaster-oriented benchmarks (Figure 5). Table 12 summarizes each point and cites the datasets that set the current standard.

Figure 5. Accepted checklist for disaster-oriented benchmarks

Table 12. Best practices in dataset protocols and the benchmarks that define them

Protocol element	What it enforces	Typical practice	Reference datasets
Split strategy	Keeps training and testing independent	50/50 random split for scene sets (AID); 70/10/20 train/dev/test for social-media corpora	AID; CrisisMMD; HumAID
Duplicate filtering	Prevents the same or near-identical image leaking into train and test folds	Perceptual-hash screening; duplicates moved to training only	HumAID removes 5 593 near-dups across 12 sources, avoiding a > 5 pp F1 inflation (Alam et al., 2020)
Unified evaluation stack	Eliminates drift in pre-processing and metric code	Shared loaders, metric scripts, and baseline models	EarthNets bundles fMoW, DIOR, GeoNRW, giving "plug-and-play" reproducibility
Metric choice	Aligns scores with task needs	Pixel-wise IoU for segmentation, weighted-F1 for multi-label social media, mAP for object detection; report all three when feasible	EarthNets recommends multi-metric reports for fair comparison

In the table above, the first row shows why even a simple random split must be task-aware: social-media data need a reserved dev fold to tune threshold-based classifiers. The second row emphasizes that duplicate control is not a cosmetic detail. Alam et al. (Alam et al., 2020) document performance drops of more than five percentage points when duplicates slip across folds. The third row highlights tooling: common loaders and metric scripts, such as those shipped with EarthNets, prevent silent changes in preprocessing from skewing results. The last row suggests to choose metrics that match the annotation granularity; pixel-wise IoU, for instance, is meaningless for tweet-level labels.

Together these protocol rules ensure that improvements reported in the literature reflect genuine model gains rather than artefacts of data leakage or metric drift. They also make results portable: a team deploying to the field can have confidence that a model's quoted F1 or IoU will hold when the same code is run on new, but similarly curated, data.

To conclude, the last five years have transformed the benchmark landscape from a handful of scene-classification corpora into a rich ecosystem spanning multimodal social media streams, multi-temporal change pairs and foundation-scale remote-sensing compilations. While no single dataset can cover all disaster scenarios, thoughtful fusion of *task-aligned* datasets—guided by rigorous split and evaluation protocols—offers a clear path to robust deep-learning models for humanitarian response.

The next section will translate these benchmarking insights into *deployment considerations*, showing how dataset characteristics (e.g., resolution, noise, class imbalance) influence on-device model design and real-time inference pipelines.

MODEL DEPLOYMENT & REAL-TIME CONSIDERATIONS

In disaster recovery, field teams need maps and analytics while roads are still open and batteries still hold a charge. That requirement shifts the design emphasis from pure accuracy to a three-way balance of latency, reliability, and power budget. Four levers dominate current practice:

- **Architecture slimming:** Depth-wise, grouped, or dilated kernels widen the receptive field without inflating parameter count.
- **Hardware-aware compilation:** Pruning, quantization, and TensorRT fusion squeeze networks onto Jetson-class GPUs and phone-grade NPUs.
- **Edge–cloud hand-off:** An on-board "first-look" model returns a coarse mask; a heavier cloud model refines it when bandwidth allows.
- **On-the-fly adaptation:** Few-shot fine-tuning or confidence-gating handles sudden shifts in sensor, atmosphere, or geography.

Case Studies

Table 13 collects six deployments that make the aforementioned ideas concrete, spanning watt-scale drones, ARM edge clusters, and cloud back-ends.

Table 13. Representative real-world deployments and their key operational takeaways

Study (Year)	Platform & task	Key numbers	Principal lesson
EmergencyNet (2020) (Kyrkou & Theocharides, 2020)	Jetson TX2; UAV scene classifier	25 fps @ < 6 W, F1 ≈ 0.88	Dilated kernels plus feature fusion rival deeper nets on a micro-GPU
UAV-Assisted Edge (2023) (Ijaz et al., 2023)	Swarm UAVs + ARM nodes; multi-hazard mapping	≤ 82 ms per 512^2 tile, 97% recall	A tiling → YOLOv5-Lite → MQTT stack keeps latency and bandwidth low
Post-Disaster UNet (2022) (Gholami et al., 2022)	Cloud UNet; Sentinel-2 + WorldView	Precision drops when test tiles = 2 × training tiles	Deployment scripts must auto-adjust tile size and color space
Drones4Good (2023) (Merkle et al., 2023)	Delivery drone; road/ building/person detection	2 s for 16 MP frame; P 54%, R 66% for people	"Good-enough" accuracy can still gate safety-critical rerouting
Refugee-Camp Mapping (2018) (Quinn et al., 2018)	Mask-RCNN; 12 camps	PR ≈ 0.9 after 10% analyst corrections	Human-in-the-loop beats static models during fast iterations
Global-Model Audit (2024) (Wiguna et al., 2024)	Field test of global land-cover net	IoU ↓ 18% under haze & oblique view	Quick local fine-tunes or confidence gates are essential

Row 1 in the table above shows that architectural thrift yields real-time speed on a 6 W board. Row 2 demonstrates that end-to-end optimization (e.g., tiling, lightweight inference, message queuing, and GIS display) can match model tuning in cutting latency. Row 3 notes that tiling rules and spectral drift can undo a cloud model unless deployment scripts adapt. Rows 4-6 highlight that accuracy targets are mission-specific, analyst feedback accelerates adaptation, and geographic drift remains costly.

Key Findings from the Case Studies

A close reading of the six deployments yields several lessons.

First, context matters more than depth on battery-limited platforms. EmergencyNet (Kyrkou & Theocharides, 2020) widens its receptive field with atrous (dilated) kernels rather than stacking additional layers, allowing the network to run at 25 fps on a 6 W Jetson-TX2 while matching the accuracy of far deeper backbones. The result shows that careful kernel design, not raw parameter count, is the decisive factor when energy is the budget.

Second, optimizing the entire data path can rival model pruning for latency gains. In the UAV-Assisted Edge framework (Ijaz et al., 2023), a tiling scheduler, a lightweight YOLOv5-Lite detector, and an MQTT message queue together keep per-tile inference below 82 ms and sustain 97% recall during live wildfire drills. Those milliseconds shaved from I/O and messaging are as valuable as the FLOPs removed by quantization.

Third, deployment scripts must guard against tile and sensor drift. The Post-Disaster UNet (Gholami et al., 2022) lost double-digit precision when inference switched from the 256^2-pixel tiles used in training to 512^2 tiles streamed in the field; cross-sensor spectral shifts compounded the error. A simple pre-inference resize and a shallow color-space fine-tune would have preserved accuracy.

Fourth, "good-enough" accuracy is mission-specific. The Drones4Good system (Merkle et al., 2023) achieves 54% precision and 66% recall in person detection, yet those numbers are sufficient to abort an airdrop if crowds appear below the drone's delivery route. The case reminds us that operational thresholds, not leaderboard ranks, define success.

Fifth, human-machine teaming accelerates adaptation. In refugee-camp mapping (Quinn et al., 2018), analysts corrected roughly one prediction in ten and allowed the Mask-RCNN to retrain during the same session; precision–recall curves climbed toward 0.9 after that modest intervention. Interactive feedback loops therefore offer a pragmatic bridge between "perfect now" and "perfect later."

Finally, domain drift remains the silent performance killer. A field audit of a globally trained land-cover model reported an 18% IoU drop when coastal haze and oblique viewing angles differed from the training set (Wiguna et al., 2024). Small on-device fine-tunes or confidence-based masks are inexpensive safeguards against such unexpected shifts.

Cross-Cutting Recommendations

Experience from the six deployments points to four design rules that hold across hardware platforms and hazard types.

- **Budget energy, not only FLOPs:** Edge devices care about joules per frame. Networks that use grouped or dilated kernels such as EmergencyNet's atrous design deliver real-time speed on a 6 W Jetson while matching the accuracy of much deeper backbones (Kyrkou & Theocharides, 2020). Counting energy per inference during architecture search therefore pays greater dividends than chasing marginal accuracy gains with heavier layers.
- **Process in stages:** A tiered workflow cuts both latency and bandwidth. Lightweight edge models generate coarse segmentation masks that are small enough to transmit over a shaky link; once a stable connection exists, a cloud model refines those masks to pixel precision. The UAV-Assisted Edge framework shows that this "first-look then refine" pattern keeps per-tile latency below 100 ms while still reaching 97% recall in flood and wildfire drills (Ijaz et al., 2023).
- **Carry a mini-finetune kit:** Domain drift (e.g., new haze levels, different viewing angles, unfamiliar building styles) can wipe out double-digit accuracy. A compact head (≈1 MB) plus a few-shot script stored on the device can recover 10–20% IoU in minutes, as seen in the Global-Model field audit where a quick local update offset an 18% drop caused by coastal haze (Wiguna et al., 2024).
- **Show what the model is thinking:** Simple saliency or confidence maps let field analysts validate predictions and correct errors on the spot. In refugee-camp mapping, exposing class-activation overlays helped annotators lift precision–recall curves toward 0.9 after editing only one in ten masks (Quinn et al., 2018). Visual explanations thus speed human-in-the-loop adaptation and build trust in automated outputs.

LIMITATIONS, OPEN CHALLENGES & FUTURE DIRECTIONS

The case studies in the previous sections prove that multi-sensor deep learning can work in the field; yet they also expose recurring weak points that still derail deployments. This section groups those weak points into data, model, and operational/ethical tiers, then sketches research paths that could close each gap.

Data-Centric Limitations

Most real-world failures trace back to the dataset, not the network. Table 14 organizes four common data bottlenecks and shows the risk each one poses.

Table 14. Common data bottlenecks in crisis AI and the risks they introduce

Bottleneck	Evidence	Operational risk
Geographic bias: training images cluster in Europe & North America	A model that topped xBD lost 53 pp F1 on the 2023 Türkiye-Syria quake (Wiguna et al., 2024)	Wrong priorities when the first crisis occurs outside the "seen" domain
Instance-scale mismatch: object size vs. ground-sample distance varies by platform	Person detection fell when drones flew higher; small Mozambican houses were missed in counts (Merkle et al., 2023)	Systematic under-counting in dense informal settlements
Label sparsity & noise: few pixel masks, many duplicate tweets	Cloud-covered flood scenes remain hard; F1 inflates by > 5 pp when duplicates leak across folds (Alam et al., 2021)	Training looks good; live performance collapses
Missing modalities: SAR, LiDAR, crowdsourced photos rarely co-exist in one set	Drone studies plead for cross-hazard, cross-platform corpora (Ijaz et al., 2023)	Fusion pipelines cannot be tested end-to-end

All four problems stem from poor representativeness. A network trained on only one region, scale, or sensor learns brittle shortcuts that break elsewhere. True robustness demands broader coverage *and* clean, balanced labels across all modalities.

Model-Centric Limitations

Even perfect data cannot rescue a model that is breakable by design.

- **Generalization collapse:** Wiguna et al. (Wiguna et al., 2024) report ≥ 50% F1 drops when a network meets an unseen hazard or sensor shift, which is classic over-fitting to benchmark quirks.
- **Latency-accuracy tension:** Field partners chose a 3× speed-up over a two-point IoU gain when selecting building-damage tools (Gholami et al., 2022); raw accuracy is meaningless if the map arrives too late.
- **View-angle sensitivity:** In Drones4Good (Merkle et al., 2023), oblique shots or dense vegetation cut recall despite unchanged network weights.
- **Limited interpretability:** Saliency maps seldom meet humanitarian audit standards; mis-labelled pixels can misdirect scarce aid.

Thus, optimization must navigate the full speed–accuracy–robustness cube, and reliability diagnostics must be visible to non-experts.

Operational and Ethical Challenges

A working model must also survive real-world constraints:

- **Bandwidth and power ceilings:** Edge devices cap networks at ≈ 20 M parameters; satellite links fail after quakes.
- **Privacy and consent:** On-board inference keeps pixels local but complicates oversight; low-altitude people detection raises legal concerns (Drones4Good blurs faces by design).
- **Human-AI teaming:** Analyst edits raised PR from 0.7 → 0.9 in refugee-camp mapping with only 10% extra labels, yet few pipelines embed such feedback loops (Quinn et al., 2018).
- **Bias amplification:** Urban-trained models lag in rural and informal areas, risking inequitable aid.

To conclude, technical tweaks alone will not fix these issues; deployment plans must interweave legal, social, and human-factor safeguards into the engineering pipeline.

Research Gaps and Future Directions

Practitioners often ask where the next research should land. The answers fall on a timeline.

In the near term (one to two years) the quickest wins lie in making today's networks more robust to distribution shift. Simple spectral augmentations, adversarial style transfer, and few-shot fine-tuning have already shown they can recover much of the 53-percentage-point F1 drop that occurs when a benchmark-trained model meets an unfamiliar geography or sensor mix (Wiguna et al., 2024). A second near-term target is active and semi-supervised learning. Self-supervised pre-training paired with a handful of human labels has cut annotation needs by roughly 60% in multimodal flood datasets (J. Li et al., 2022); wider adoption would ease the chronic label-scarcity problem flagged in Section 7.1.

Over the medium horizon (three to five years) the field is likely to be reshaped by foundation-scale models trained on petabyte streams from many sensors. Early cross-modal transformers already outperform single-sensor CNNs for flood mapping but remain too heavy for edge devices; lighter backbones and smarter token pruning are active topics. Equally important is uncertainty-aware decision support—methods that pass pixel-level confidence maps to logistics dashboards so responders can weigh automated output against ground reports. Transparent uncertainty would also feed directly into the bias audits called for in Section 7.3.

Looking five years and beyond, two lines of work promise structural change. First is edge–cloud federated learning, where drones and field laptops adapt models locally and share only weight updates, not raw imagery. Such schemes address privacy, bandwidth, and geographic bias in one stroke, while keeping models fresh. Second is a suite of ethics-by-design toolkits—built-in bias checks, consent logs for on-board person detection, and interpreter dashboards that let non-experts trace a prediction back to the pixels and sensors that produced it. Aligning technical progress with Red Cross and UN guidelines will be essential as AI moves deeper into humanitarian workflows.

Each of these research threads maps directly onto the bottlenecks catalogued earlier: robust training and semi-supervision tackle data gaps; foundation models and uncertainty layers harden the network itself; federated learning and ethics dashboards guard against operational and social failure. Solved together, they would move the field from promising prototypes to dependable, equitable tools for crisis response.

CONCLUSION

Deep learning has carried remote sensing well beyond single-sensor, after-the-fact mapping into an era of fast, multi-source situation awareness. Multi-modal and multi-temporal fusion now turns scattered views—optical, SAR, LiDAR, hyperspectral, and crowd photographs—into a single, resilient picture of a crisis. In practice this blend delivers double-digit accuracy gains and, more importantly, trims hours from map-delivery time, as illustrated by systems such as Multi3Net, Siam-U-Net, and M^{3}Fusion.

Those gains depend on the second pillar: carefully curated benchmarks. Early scene sets like AID established reproducible splits; later corpora such as HumAID, CrisisMMD, and HRSCD added multi-task labels, duplicate filtering, and change-detection masks. The most useful benchmarks today pair task-aligned data with clear protocols and report not only precision and recall but also seconds-per-square-kilometer—a metric that speaks directly to field needs.

A third pillar makes these models practical: edge-aware engineering. Slim networks such as EmergencyNet run at real-time speeds on six-watt drones. TensorRT-compressed models can process half a million tiles an hour on a single Jetson Xavier. Interactive pipelines let analysts correct errors on the fly, raising precision–recall with only modest extra labels.

The remaining barrier is robustness. Cross-event tests show that a model trained on North-Atlantic imagery can lose half its F-score when haze, oblique angles, or unfamiliar building styles appear. Data gaps—geographic bias, scale mismatch, sparse labels, missing modalities—combine with model brittleness—latency-accuracy tension, limited interpretability—to produce this fragility. Encouragingly, active research already targets each weakness: domain-robust training, semi-supervised annotation, foundation-scale multi-sensor models, uncertainty layers, federated adaptation, and ethics-by-design dashboards.

The path to dependable, equitable disaster response therefore lies in integrated pipelines that co-optimize data, models, hardware, and ethics. Short-term work should focus on robust training and active learning; medium-term progress will come from multi-sensor foundation models and confidence-aware interfaces; the long-term points toward federated edge–cloud learning and built-in bias audits. By embracing these principles—designing for fusion, benchmarking honestly, matching models to mission envelopes, planning for drift, and embedding guard-rails—deep-learning-driven remote sensing can deliver the right map at the right moment to the people who need it most.

FUNDING ACKNOWLEDGEMENT

This work was supported by the Deanship of Scientific Research, Vice Presidency for Graduate Studies and Scientific Research, King Faisal University, Saudi Arabia [Grant No. KFU252308].

REFERENCES

Aamir, M., Bazai, S., Bhatti, U. A., Dayo, Z. A., Liu, J., & Zhang, K. (2023). Applications of machine learning in medicine: Current trends and prospects. In *2023 Global Conference on Wireless and Optical Technologies (GCWOT)* (pp. 1-4). IEEE.

Adriano, B., Yokoya, N., Xia, J., Miura, H., Liu, W., Matsuoka, M., & Koshimura, S. (2020). Learning from Multimodal and Multitemporal Earth Observation Data for Building Damage Mapping (No. arXiv:2009.06200). arXiv. https://doi.org//arXiv.2009.06200DOI: 10.48550

Al Shafian, S., & Hu, D. (2024). Integrating Machine Learning and Remote Sensing in Disaster Management: A Decadal Review of Post-Disaster Building Damage Assessment. *Buildings (Basel, Switzerland)*, *14*(8), 8. Advance online publication. DOI: 10.3390/buildings14082344

Alam, F., Ofli, F., & Imran, M. (2018). Crisismmd: Multimodal twitter datasets from natural disasters. Proceedings of the International AAAI Conference on Web and Social Media, 12(1). https://ojs.aaai.org/index.php/ICWSM/article/view/14983

Alam, F., Ofli, F., Imran, M., Alam, T., & Qazi, U. (2020). Deep Learning Benchmarks and Datasets for Social Media Image Classification for Disaster Response. *2020 IEEE/ACM International Conference on Advances in Social Networks Analysis and Mining (ASONAM)*, 151–158. https://doi.org/DOI: 10.1109/ASONAM49781.2020.9381294

Alam, F., Qazi, U., Imran, M., & Ofli, F. (2021). HumAID: Human-Annotated Disaster Incidents Data from Twitter with Deep Learning Benchmarks. *Proceedings of the International AAAI Conference on Web and Social Media*, 15, 933–942. https://doi.org/DOI: 10.1609/icwsm.v15i1.18116

Algiriyage, N., Prasanna, R., Stock, K., Doyle, E. E. H., & Johnston, D. (2021). Multi-source Multimodal Data and Deep Learning for Disaster Response: A Systematic Review. *SN Computer Science*, *3*(1), 92. DOI: 10.1007/s42979-021-00971-4 PMID: 34870241

Asif, M., Song, H., Zia, M. A., Ali, S., Awwad, E. M., Ghadi, Y. Y., Al-Razgan, M., Sadiq, T., & Bhatti, U. A. (2024). Shedding light on diagnostic precision: GANs for low light endoscopy image enhancements. Research. https://doi.org/DOI: 10.21203/rs.3.rs-4213321/v1

Basit, A., Siddique, M. A., Bashir, S., Naseer, E., & Sarfraz, M. S. (2024). Deep Learning-Based Detection of Oil Spills in Pakistan's Exclusive Economic Zone from January 2017 to December 2023. *Remote Sensing*, *16*(13), 13. Advance online publication. DOI: 10.3390/rs16132432

Bazai, S. U., & Jang-Jaccard, J. (2019). SparkDA: RDD-based high-performance data anonymization technique for Spark platform. In International conference on network and system security (pp. 646-662). Cham: Springer International Publishing.

Bazai, S. U., & Jang-Jaccard, J. (2020). In-memory data anonymization using scalable and high performance rdd design. *Electronics (Basel)*, *9*(10), 1732.

Bazai, S. U., Jang-Jaccard, J., & Alavizadeh, H. (2021). A novel hybrid approach for multi-dimensional data anonymization for apache spark. *ACM Transactions on Privacy and Security*, *25*(1), 1–25.

Bazai, S. U., Jang-Jaccard, J., & Wang, R. (2017). Anonymizing k-NN classification on MapReduce. *International Conference on Mobile Networks and Management*, 364–377.

Benedetti, P., Ienco, D., Gaetano, R., Ose, K., Pensa, R. G., & Dupuy, S. (2018). M^3\textFusion: A Deep Learning Architecture for Multiscale Multimodal Multitemporal Satellite Data Fusion. *IEEE Journal of Selected Topics in Applied Earth Observations and Remote Sensing*, *11*(12), 4939–4949. DOI. 10.1109/JSTARS.2018.2876357

Bhatti, M. A., Zeeshan, Z., Syam, M. S., Bhatti, U. A., Khan, A., Ghadi, Y. Y., & Alsenan, S. (2024a). Advanced plant disease segmentation in precision agriculture using optimal dimensionality reduction with fuzzy C-means clustering and deep learning. IEEE Journal of Selected Topics in Applied Earth Observations and Remote Sensing.

Bhatti, U., Bazai, S., Hussain, S., Fakhar, S., Ku, C., Marjan, S., Por, Y., & Jing, L. (2023). Deep learning-based trees disease recognition and classification using hyperspectral data. *Computers, Materials & Continua*, *77*, 681–697.

Bhatti, U. A., Masud, M., Bazai, S. U., & Tang, H. (2023). Investigating AI-based smart precision agriculture techniques. [). Frontiers Media SA.]. *Frontiers in Plant Science*, *14*, 1237783. PMID: 37496855

Bhatti, U. A., Mengxing, H., Li, J., Bazai, S. U., & Aamir, M. (2024). *Deep learning for multimedia processing applications: Volume two: Signal processing and pattern recognition*. CRC Press.

Bhatti, U. A., Zeeshan, Z., Nizamani, M. M., Bazai, S., Yu, Z., & Yuan, L. (2022). Assessing the change of ambient air quality patterns in Jiangsu Province of China pre-to post-COVID-19. *Chemosphere*, *288*, 132569. PMID: 34655644

Bianchi, F. M., Espeseth, M. M., & Borch, N. (2020). Large-scale detection and categorization of oil spills from SAR images with deep learning (No. arXiv:2006.13575). arXiv. https://doi.org//arXiv.2006.13575DOI: 10.48550

Chen, L., Zhang, Y., Xu, B., Shao, K., Yan, J., & Bhatti, U. A. (2024). A lot-based VMD-CNN-BiGRU indoor mariculture water quality prediction method including attention mechanism. *International Journal of High Speed Electronics and Systems*. Advance online publication. DOI: 10.1142/S0129156425400105

Christie, G., Fendley, N., Wilson, J., & Mukherjee, R. (2018). Functional map of the world. Proceedings of the IEEE Conference on Computer Vision and Pattern Recognition, 6172–6180. https://openaccess.thecvf.com/content_cvpr_2018/html/Christie_Functional_Map_of_CVPR_2018_paper.html

CRED. (n.d.). EM-DAT - The international disaster database. Retrieved May 31, 2025, from https://www.emdat.be/

Daudt, R. C., Le Saux, B., Boulch, A., & Gousseau, Y. (2019). Multitask learning for large-scale semantic change detection. *Computer Vision and Image Understanding*, *187*, 102783.

Doan, T.-N., & Le-Thi, D.-N. (2025). A Novel Deep Learning Model for Flood Detection from Synthetic Aperture Radar Images. *Journal of Advances in Information Technology*, *16*(1), 57–70. DOI: 10.12720/jait.16.1.57-70

Ebel, P., Saha, S., & Zhu, X. X. (2021). FUSING MULTI-MODAL DATA FOR SUPERVISED CHANGE DETECTION. The International Archives of the Photogrammetry, Remote Sensing and Spatial Information Sciences, XLIII-B3-2021, 243–249. XXIV ISPRS Congress <q>Imaging today, foreseeing tomorrow</q>, Commission III - 2021 edition, 5–9 July 2021. https://doi.org/DOI: 10.5194/isprs-archives-XLIII-B3-2021-243-2021

Feng, S., Liu, Q., Patel, A., Bazai, S. U., Jin, C., Kim, J. S., Sarrafzadeh, M., Azzollini, D., Yeoh, J., & Kim, E. (2022). Automated pneumothorax triaging in chest X-rays in the New Zealand population using deep-learning algorithms. *Journal of Medical Imaging and Radiation Oncology*, *66*(8), 1035–1043. PMID: 35224858

Fernández-Montes De Oca, A. I., Ghilardi, A., Kauffer, E., Mas, J. F., Sánchez-Cordero, V., & Gallardo-Cruz, J. A. (2025). Building a Methodological Reference Framework for Quantifying Tropical Deforestation with Remote Sensing. *Sustainability, 17*(4), 1394. DOI: 10.3390/su17041394

Gholami, S., Robinson, C., Ortiz, A., Yang, S., Margutti, J., Birge, C., Dodhia, R., & Ferres, J. L. (2022). On the Deployment of Post-Disaster Building Damage Assessment Tools using Satellite Imagery: A Deep Learning Approach. *2022 IEEE International Conference on Data Mining Workshops (ICDMW)*, 1029–1036. https://doi.org/DOI: 10.1109/ICDMW58026.2022.00134

Haider, A., Hussain, L., Tareen, A. W., Bazai, S. U., Aslam, S., Neo, M., & Amphawan, A. (2024). A comparative study of machine learning techniques for accurate disease prediction using symptom-based diagnosis. In AIP Conference Proceedings (Vol. 3153, No. 1). AIP Publishing.

Hamza, M., Bazai, S. U., Ghafoor, M. I., Ullah, S., Akram, S., & Khan, M. S. (2023). Generative adversarial networks (GANs) video framework: A systematic literature review. In *2023 International Conference on Energy, Power, Environment, Control, and Computing (ICEPECC)* (pp. 1-5). IEEE.

Han, H., Bazai, S. U., Bhatti, M. A., Basit, A., Wahid, A., Bhatti, U. A., Ghadi, Y. Y., & Algarni, A. (2024). Hybrid Climate Forecasting: Variational Mode Decomposition and Convolutional Neural Network with Long-Term Short Memory. *Polish Journal of Environmental Studies, 33*(2).

Han, H., Hasnain, A., Bhatti, U. A., Yue, Y., He, Y., & Wei, G., & ur Rahman, W. (2024). Assessing the spatio-temporal impact of the COVID-19 pandemic lockdown on air quality in Jiangsu Province, China. *Environment, Development and Sustainability*, •••, 1–21.

Han, H., Zeeshan, Z., Assam, M., Khan, F. U., Khan, W., Asif, M., Bhatti, U. A., Hasnain, A., Awwad, E. M., & Sarhan, N. (2024). *Deep learning techniques for enhanced mangrove land use and land change from remote sensing imagery: A blue carbon perspective*. Big Data Research., DOI: 10.1016/j.bdr.2024.100478

Haq, S. U., Bazai, S. U., Fatima, A., Marjan, S., Yang, J., Yee, L., & Ku, C. S. (2023). Reseek-arrhythmia: Empirical evaluation of ResNet architecture for detection of arrhythmia. *Diagnostics (Basel), 13*(18), 2867. PMID: 37761234

Hasnain, A., Hashmi, M. Z., Khan, S., Bhatti, U. A., Min, X., Yue, Y., He, Y., & Wei, G. (2024). Predicting ambient PM2.5 concentrations via time series models in Anhui Province, China. *Environmental Monitoring and Assessment, 196*, 487. DOI: 10.1007/s10661-024-12644-9 PMID: 38687422

Hasnaoui, Y., Tachi, S. E., Bouguerra, H., & Yaseen, Z. M. (2025). Transfer learning-based deep learning models for flood and erosion detection in coastal area of Algeria. *Earth Science Informatics*, *18*(2), 380. DOI: 10.1007/s12145-025-01866-1

He, B., & Bhatti, U. A. (2024). Smart cities and smart networks: AI applications in urban geography and industrial communication. *International Journal of High Speed Electronics and Systems*. Advance online publication. DOI: 10.1142/S0129156424401220

Hong, D., Gao, L., Yokoya, N., Yao, J., Chanussot, J., Du, Q., & Zhang, B. (2021). More Diverse Means Better: Multimodal Deep Learning Meets Remote-Sensing Imagery Classification. *IEEE Transactions on Geoscience and Remote Sensing*, *59*(5), 4340–4354. DOI: 10.1109/TGRS.2020.3016820

Huang, M., Zhang, X. S., Bhatti, U. A., Wu, Y. Y., Zhang, Y., & Ghadi, Y. Y. (2024). An interpretable approach using hybrid graph networks and explainable AI for intelligent diagnosis recommendations in chronic disease care. *Biomedical Signal Processing and Control*, *91*, 105913.

Huot, F., Hu, R. L., Ihme, M., Wang, Q., Burge, J., Lu, T., Hickey, J., Chen, Y.-F., & Anderson, J. (2021). Deep Learning Models for Predicting Wildfires from Historical Remote-Sensing Data (No. arXiv:2010.07445). arXiv. https://doi.org//arXiv.2010.07445DOI: 10.48550

Hurt, J. A., Scott, G. J., Anderson, D. T., & Davis, C. H. (2018). Benchmark Meta-Dataset of High-Resolution Remote Sensing Imagery for Training Robust Deep Learning Models in Machine-Assisted Visual Analytics. *2018 IEEE Applied Imagery Pattern Recognition Workshop (AIPR)*, 1–9. https://doi.org/DOI: 10.1109/AIPR.2018.8707433

Ijaz, H., Ahmad, R., Ahmed, R., Ahmed, W., Kai, Y., & Jun, W. (2023). A UAV-Assisted Edge Framework for Real-Time Disaster Management. *IEEE Transactions on Geoscience and Remote Sensing*, *61*, 1–13. DOI: 10.1109/TGRS.2023.3306151

Jenifer, A. E., Aparna, A., Sudha, N., & Kumar, A. (2022). AgriFloodNet: A dual patch CNN architecture for mapping flooded agricultural lands via bi-temporal multi-sensor images. *Geocarto International*, *37*(26), 13618–13637. DOI: 10.1080/10106049.2022.2082549

Kizilay, F., Narman, M. R., Song, H., Narman, H. S., Cosgun, C., & Alzarrad, A. (2024). Evaluating fine tuned deep learning models for real-time earthquake damage assessment with drone-based images. *AI in Civil Engineering*, *3*(1), 15. DOI: 10.1007/s43503-024-00034-6

Kyrkou, C., & Theocharides, T. (2020). EmergencyNet: Efficient Aerial Image Classification for Drone-Based Emergency Monitoring Using Atrous Convolutional Feature Fusion. *IEEE Journal of Selected Topics in Applied Earth Observations and Remote Sensing*, *13*, 1687–1699. DOI: 10.1109/JSTARS.2020.2969809

Li, J., Hong, D., Gao, L., Yao, J., Zheng, K., Zhang, B., & Chanussot, J. (2022). Deep learning in multimodal remote sensing data fusion: A comprehensive review. *International Journal of Applied Earth Observation and Geoinformation*, *112*, 102926. DOI: 10.1016/j.jag.2022.102926

Li, W., Sun, K., Zhao, H., Li, W., Wei, J., & Gao, S. (2022). Extracting buildings from high-resolution remote sensing images by deep ConvNets equipped with structural-cue-guided feature alignment. *International Journal of Applied Earth Observation and Geoinformation*, *113*, 102970. DOI: 10.1016/j.jag.2022.102970

Li, X., Hou, B., Tang, H., Talpur, B. A., Zeeshan, Z., Bhatti, U. A., Liao, J., Liu, J., & Zhang, Y. (2024). Abnormal crops image data acquisition strategy by exploiting edge intelligence and dynamic-static synergy in smart agriculture. *IEEE Journal of Selected Topics in Applied Earth Observations and Remote Sensing*.

Mahmoud, A. S., Mohamed, S. A., El-Khoriby, R. A., AbdelSalam, H. M., & El-Khodary, I. A. (2023). Oil Spill Identification based on Dual Attention UNet Model Using Synthetic Aperture Radar Images. *Photonirvachak (Dehra Dun)*, *51*(1), 121–133. DOI: 10.1007/s12524-022-01624-6

Merkle, N., Bahmanyar, R., Henry, C., Azimi, S. M., Yuan, X., Schopferer, S., Gstaiger, V., Auer, S., Schneibel, A., Wieland, M., & Kraft, T. (2023). Drones4Good: Supporting Disaster Relief Through Remote Sensing and AI. *2023 IEEE/CVF International Conference on Computer Vision Workshops (ICCVW)*, 3772–3776. https://doi.org/DOI: 10.1109/ICCVW60793.2023.00407

Nabeel, S. M., Bazai, S. U., Alasbali, N., Liu, Y., Ghafoor, M. I., Khan, R., & Por, L. Y. (2024). Optimizing lung cancer classification through hyperparameter tuning. *Digital Health*, *10*, 20552076241249661. PMID: 38698834

Nawaz, S. A., Li, J., Bhatti, U. A., Bazai, S. U., Zafar, A., Bhatti, M. A., Mehmood, A., & Ain, Q. ul, & Shoukat, M. U. (2021). A hybrid approach to forecast the COVID-19 epidemic trend. *PLoS One*, *16*(10), e0256971. PMID: 34606503

Noor, R., Wahid, A., Bazai, S. U., Khan, A., Fang, M., Bhatti, U. A., & Ghadi, Y. Y. (2024). DLGAN: Undersampled MRI reconstruction using deep learning based generative adversarial network. *Biomedical Signal Processing and Control*, *93*, 106218.

Noor, S., Bazai, S. U., Ghafoor, M. I., Marjan, S., Akram, S., & Ali, F. (2023). Generative adversarial networks for anomaly detection: a systematic literature review. In 2023 4th International Conference on Computing, Mathematics and Engineering Technologies (iCoMET) (pp. 1-6). IEEE.

Peng, M., Liu, Y., Khan, A., Ahmed, B., Sarker, S. K., Ghadi, Y. Y., Bhatti, U. A., Al-Razgan, M., & Ali, Y. A. (2024). Crop monitoring using remote sensing land use and land change data: Comparative analysis of deep learning methods using pre-trained CNN models. *Big Data Research, 36*, 100448. DOI: 10.1016/j.bdr.2024.100448

Peng, M., Liu, Y., Qadri, I. A., Bhatti, U. A., Ahmed, B., Sarhan, N. M., & Awwad, E. M. (2024). Advanced image segmentation for precision agriculture using CNN-GAT fusion and fuzzy C-means clustering. *Computers and Electronics in Agriculture, 226*, 109431.

Prapas, I., Kondylatos, S., Papoutsis, I., Camps-Valls, G., Ronco, M., Fernández-Torres, M.-Á., Guillem, M. P., & Carvalhais, N. (2021). Deep Learning Methods for Daily Wildfire Danger Forecasting (No. arXiv:2111.02736). arXiv. https://doi .org//arXiv.2111.02736DOI: 10.48550

Quinn, J. A., Nyhan, M. M., Navarro, C., Coluccia, D., Bromley, L., & Luengo-Oroz, M. (2018). Humanitarian applications of machine learning with remote-sensing data: Review and case study in refugee settlement mapping. *Philosophical Transactions. Series A, Mathematical, Physical, and Engineering Sciences, 376*(2128), 20170363. DOI: 10.1098/rsta.2017.0363 PMID: 30082308

Rao, A., Jung, J., Silva, V., Molinario, G., & Yun, S.-H. (2023). Earthquake building damage detection based on synthetic-aperture-radar imagery and machine learning. *Natural Hazards and Earth System Sciences, 23*(2), 789–807. DOI: 10.5194/nhess-23-789-2023

Rudner, T. G. J., Rußwurm, M., Fil, J., Pelich, R., & Bischke, B. (n.d.). *Rapid Computer Vision-Aided Disaster Response via Fusion of Multiresolution*. Multisensor, and Multitemporal Satellite Imagery.

Shen, L., Lu, Y., Chen, H., Wei, H., Xie, D., Yue, J., Chen, R., Lv, S., & Jiang, B. (2021). S2Looking: A satellite side-looking dataset for building change detection. *Remote Sensing, 13*(24), 5094.

Singh, D., Hoskere, V., & Milillo, P. (2025). Multiclass Post-Earthquake Building Assessment Integrating Optical and SAR Satellite Imagery, Ground Motion, and Soil Data with Transformers (No. arXiv:2412.04664). arXiv. https://doi.org//arXiv .2412.04664DOI: 10.48550

Toker, A., Kondmann, L., Weber, M., Eisenberger, M., Camero, A., Hu, J., Hoderlein, A. P., Şenaras, Ç., Davis, T., & Cremers, D. (2022). Dynamicearthnet: Daily multi-spectral satellite dataset for semantic change segmentation. Proceedings of the IEEE/CVF Conference on Computer Vision and Pattern Recognition, 21158–21167. https://openaccess.thecvf.com/content/CVPR2022/html/Toker_DynamicEarthNet_Daily_Multi-Spectral_Satellite_Dataset_for_Semantic_Change_Segmentation_CVPR_2022_paper.html

Topbaş, A., Jamil, A., Hameed, A. A., Ali, S. M., Bazai, S., & Shah, S. A. (2021). Sentiment analysis for covid-19 tweets using recurrent neural network (rnn) and bidirectional encoder representations (bert) models. 2021 International Conference on Computing, Electronic and Electrical Engineering (ICE Cube), 1–6.

Wiguna, S., Adriano, B., Mas, E., & Koshimura, S. (2024). Assessment of Deep Learning Models Trained Using Global Remote Sensing Imagery in Real-Context Emergency Response. IGARSS 2024 - 2024 IEEE International Geoscience and Remote Sensing Symposium, 1736–1740. https://doi.org/DOI: 10.1109/IGARSS53475.2024.10641821

Wu, X., Zhang, Z., Xiong, S., Zhang, W., Tang, J., Li, Z., An, B., & Li, R. (2023). A Near-Real-Time Flood Detection Method Based on Deep Learning and SAR Images. *Remote Sensing*, *15*(8), 2046. DOI: 10.3390/rs15082046

Xia, G.-S., Hu, J., Hu, F., Shi, B., Bai, X., Zhong, Y., Zhang, L., & Lu, X. (2017a). AID: A Benchmark Data Set for Performance Evaluation of Aerial Scene Classification. *IEEE Transactions on Geoscience and Remote Sensing*, *55*(7), 3965–3981. DOI: 10.1109/TGRS.2017.2685945

Xia, G.-S., Hu, J., Hu, F., Shi, B., Bai, X., Zhong, Y., Zhang, L., & Lu, X. (2017b). AID: A Benchmark Data Set for Performance Evaluation of Aerial Scene Classification. *IEEE Transactions on Geoscience and Remote Sensing*, *55*(7), 3965–3981. DOI: 10.1109/TGRS.2017.2685945

Xiong, Z., Wang, Y., Zhang, F., & Zhu, X. X. (2024). One for all: Toward unified foundation models for earth vision. IGARSS 2024-2024 IEEE International Geoscience and Remote Sensing Symposium, 2734–2738. https://ieeexplore.ieee.org/abstract/document/10641637/?casa_token=PKATjEH_jZQAAAAA:UdZ0Jci0u8hIJuB5PXAGg_9XTbDtpfWsUS8jAyQwTgXAnHRyYuhucbJE_ZeNwHUqW2chDw

Xu, J. Z., Lu, W., Li, Z., Khaitan, P., & Zaytseva, V. (2019). Building Damage Detection in Satellite Imagery Using Convolutional Neural Networks (No. arXiv:1910.06444). arXiv. https://doi.org//arXiv.1910.06444DOI: 10.48550

Xu, S., He, X., Cao, X., & Hu, J. (2022). Damaged Building Detection with Improved Swin-Unet Model. *Wireless Communications and Mobile Computing, 2022*(1), 2124949. DOI: 10.1155/2022/2124949

Xu, Z., Li, J., Cheng, S., Rui, X., Zhao, Y., He, H., & Xu, L. (2024). Wildfire Risk Prediction: A Review (No. arXiv:2405.01607). arXiv. https://doi.org//arXiv.2405 .01607DOI: 10.48550

Zhang, C., Iqbal, I., Bhatti, U. A., Liu, J., Awwad, E. M., & Sarhan, N. (2024). ResNet50 in remote sensing and agriculture: Evaluating image captioning performance for high spectral data. *Environmental Earth Sciences, 83*(23), 1–16.

Zhang, Y., Chen, J., Ma, X., Wang, G., Bhatti, U. A., & Huang, M. (2024). Interactive medical image annotation using improved Attention U-net with compound geodesic distance. *Expert Systems with Applications, 237*, 121282. DOI: 10.1016/j. eswa.2023.121282

Zhang, Y., Xu, S., Chen, H., Bhatti, U. A., & Huang, M. (2024). Context-aware environment online monitoring for safety autonomous vehicle systems: An automata-theoretic approach. *Journal of Cloud Computing (Heidelberg, Germany), 13*(6). Advance online publication. DOI: 10.1186/s13677-023-00567-

Chapter 8
Unsupervised, Semi-Supervised, and Self-Supervised Learning for Climate Applications

Gaurav Kumar
https://orcid.org/0000-0002-8294-582X
Lovely Professional University, India

Saran Kumar
Coimbatore Institute of Technology, India

M. Sankar
https://orcid.org/0000-0002-8614-7956
Acharya Institute of Graduate Studies, India

Fateh Bahadur Kunwar
https://orcid.org/0009-0007-5749-9892
GITS, Rajasthan Technical University, India

A. R. Arunarani
SRM Institute of Science and Technology, India

V. Bhoopathy
https://orcid.org/0000-0003-2175-6328
Sree Rama Engineering College, India

C. S. Sandeep
Jawaharlal College of Engineering and Technology, India

ABSTRACT

AI is rapidly becoming used for predictive modelling in all sectors of life. It is slowly replacing physical model-based climate projections in weather prediction. This poll gathers our knowledge on ML's weather and climate prediction capabilities. This intriguing sector is becoming increasingly significant in agriculture and catastrophe management. Eight strategies were identified as promising for improving short-term environmental circumstances and long-term climate forecasts. It evaluated over 20

DOI: 10.4018/979-8-3373-3501-8.ch008

academic methodologies. The survey indicated that ML is good at short-term climate prediction but not mid- to permanent climate prediction because to numerous environmental elements and data limits. Short-term or moderate-to-long-term weather estimation is usually the focus of the literature. It rarely considers the relationship between the two or modelling structure and subsequent advancements. This survey fills gaps by examining frameworks across length scales. It will then set the stage for future joint research on this rapidly changing topic.

INTRODUCTION

Generating investments that will have an effect on the decades to come requires some foresight. Nevertheless, our capacity to foresee and make choices is hindered by uncertainty regarding foreseeable demands, circumstances, and advancements like warming temperatures, as well as societal, economic, and technological improvements. To successfully deal address unpredictable futures and reduce regretting, responding approaches continue to be promoted. The components of an adaptability plan include both a short-term measures—usually ones with little to no regret—that allow future choices to remain accessible and, if required, a long-term possibilities that can adjust to unpredictable and dynamic circumstances (Y. Wang et al., 2022). There are various methods for creating responsive contingencies. Some of these include the ABP, APM, RDM, Multiplied the objective Reliable Decision Making, EAO and the DAPP. Doing what you've been necessary right away and keeping an eye for potential modifications that signal the demand for new choices to handle altered circumstances is the ideal method to cope with apprehension according to all of these strategies. In the scientific community, there are two distinct kinds of responsive designs unchanging responsiveness, and these seeks to safeguard an initial plan from failure through contingencies preparation and surveillance evolving adaptation, which seeks to track how the plan is progressing towards goals and create alternate schedules of action that can be shifted to when needed (Myakala, 2025). The domains of transportation, military, and groundwater and construction management provide solid foundations for adaptation plan concepts (Akhare, 2023). The concept of adaptive administration is derived from managing ecological systems. Improving the mechanism's ability to respond through experiment-based learning serves the primary goal of flexible administration. The following points to a crucial function for keeping tabs on the system's operational efficiency. In adaptation preparing, the decision-making process on the plan's subsequent activities is based on surveillance (Haasnoot et al., 2018).

WHAT IS ML?

One of the main goals in a number of sciences is to create a structure that can explain the connection involving some quantifiable qualities and some connected characteristics. Through observing what is visible and using the calculated theoretical framework, one can forecast the probable values of the necessary attributes. It is not possible to effectively simulate many actual-life events as closed-form inputs to outputs relationships due to their complexity. By analyzing the data at hand and optimizing for a problem-specific achievement metric, neural network algorithms can simply construct a digital representation featuring these intricate interactions. This algorithm's learning and what information is required to create it automatically are two terms that go hand in hand. By using the model that was established to anticipate unexpected input elements that weren't in the initial validation information we can gain novel information on the mapping of input information to product. Algorithmic methods for frequently necessitate substantial quantities of data to be trained in order to acquire the ability to produce a realistic model(Souza, 2024). Accordingly, gathering a big collection of typical training instances and storing them in a computationally acceptable format is a crucial initial step in utilizing neural network training methodology. The use of machine learning algorithms has become feasible in numerous fields due to current improvements regarding online gathering, archiving, and analyzed power. These fields include robotics, biochemical technology, and analyses of social networks, health care diagnostics, and biochemistry. Regarding every particular artificial intelligence task, there is typically at least one trainable statistical paradigm. Picking a specific model or approach does not adhere to any hard and fast rules (Hur & Roth, 2020). The quantity and quality of data used to train, the makeup and complexity of the connections amongst the variables that are inputs and outputs, and technological restrictions like memory availability and experimental time are among the several aspects that determine the efficiency of a model (Pup & Atzori, 2023). Finding the best framework or engine typically requires trying a few various versions, considering the situation. Thankfully, conventional computer programs exist that integrate hundreds of algorithms into a single the structure for example (Hoyer et al., 2023). Assuming the information is already in the right structure, this set of files can make testing out the various possibilities much easier. Think about the challenge of classifying a potential arrangement of strands as micron or otherwise. Finding a collection of brief amino acid sequences which comprise components that are identified miRNA and non-miRNA structures and building a set of principles around the presence of these genome words would constitute a straightforward method. A region of DNA that contains the characters AGCACU is more likely to constitute a micron versus not, according to any of such criteria. Following this, alternative nucleotides might be easily labeled according to those guidelines (Nasteski, 2017). Due to the complexity of the connecting as well as the large number of potential polynucleotide words, building a system utilizing rules is challenging in practice. Methods used by machine learning can mechanically construct a numerical model from these nucleotides words rather than requiring an extensive set of parameters to be specified by hand. Although the entire training procedure is automated, these kinds of models may be developed by utilizing enormous sampling of biological data. In artificial intelligence, features that are established for the input information are used for establishing such rules, in this case a nucleic acid hexane (Allmer, 2014). Example for supervised learning shown in Figure 1.

Figure 1. Example for supervised learning

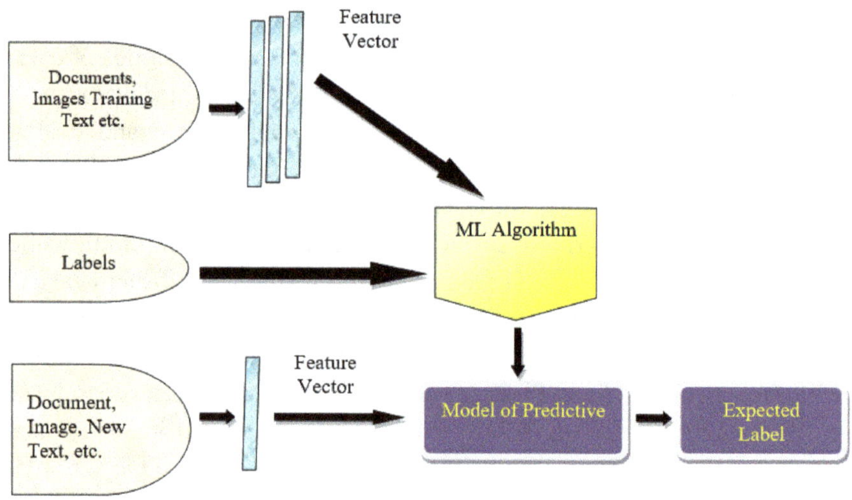

AREAS OF MACHINE LEARNING APPLICATIONS

The conceptual foundations of the algorithm for machine learning are presented in the preceding section. Here we shall provide details regarding the fields and research that currently make use of neural networks. The application of computational learning has grown substantially in recent times. Despite the belief that neural networks are solely applicable to large-scale research, many individuals encounter it on everyday basis (Hashmi et al., 2025). This is a list of research investigations and applications.

Education

Instruction is a major area of implementation, and there seems to be numerous recent researches aimed at identifying and improving results in this subject. Not much progress is being realized regarding schooling following several initiatives in the past decade or so. The malfunction is caused by many different things (Bahlali & Bachir, 2025). Nevertheless, the relative importance of the factors contributing to this failure remains unclear. In this situation, artificial intelligent illustrations were able to successfully forecast the success of the pupils in the classes using an examination administered to pupils attending high school. Representations using algorithmic learning, which yielded successful outcomes. In a comparable manner there are examinations that aim to ascertain the proficiency levels of college students. Research at the resort town of Pam The institution in 2007 found that how well pupils did in mathematics classes indicated whether they were at danger of dropping out of school. The results of 434 applicants' admission to college's exams—including those in a science, the sciences. Turkish, and their highest school graduation—were determined to be significant predictors of mathematical achievement (Agastya et al., 2021). The research investigation utilized data from the 289 pupils for instructional purposes and data from 145 those for evaluating found that 86% among pupils who succeeded the mathematical class were accurately predicted. The courses were accurately predicted (Vera et al., 2020). Learning by machine has also found useful uses in additional areas that include learning, such as

Image Processing

The goal of this approach is to enhance captured images through processing. Outlined are a few examples of situations in which you might see the graphic processing unit in action (ÇELİK, 2018).

- Protection measures
- Facing recognition
- Medical care
- Defense agencies
- Detecting movement
- Recognizing objects

Computational Biology

- Creation of drugs
- Tumors detection

- Genome sequence

Natural Language Processing

Studying and analyzing the grammatical arrangements of native tongues is the goal. Various tasks can be accomplished with the use of conversational treatment

- The process of translating textual materials automatically
- Robots that have responses to questions
- Summing up text technology
- Acquiring the ability of comprehending and employing

Automotive, Aviation and Production

- Identifying potential problems in advance
- The manufacturing of driverless cars

Retail

- Individualized product evaluations for consumers
- Machines that make suggestions
- Predictions of materials and inventory
- Purchase patterns - demanded patterns

Finance

- Evaluations of risk and safety measures for credit
- Purchasing with algorithms

Agriculture

- Using aerial pictures to forecast harvests or shortages

Human Resources

- Choose the most eligible contender from a large pool

Energy

- Determining the temperature and thereby warming requirements for construction plans
- Analyzing electrical consumption
- Streamlined managing networks

Meteorology

- Analyzed weather information gathered from microphones

Health

- Assessing information regarding patients to provide warnings and diagnoses
- Classifying diseases
- Assessment of health services

Cyber Security

- Avoiding malicious network communications detection
- Identifying home scamming

ML GENERIC MODELS

Automatic data mining is applied to a wide range of situations that necessitate algorithms to learn. There are three characteristics of a classroom issue, which include

- Classes of tasks
- Needs improvement on the performance metric
- Gaining the Experience Process

One way to describe the instructional challenge in a pastime like squares is as follows:

- Activity T: Engaging in participating in games
- Measure of Performance: Defeating someone else in plays is an accomplishment criterion.
- Experiences E: honing one's skills by competing with oneself in digital media and steadily getting better.

Six parts make up the general theory of ML, and they are all algorithm-agnostic. You can see three main parts in Figure 2.

Figure 2. Generic ML models components

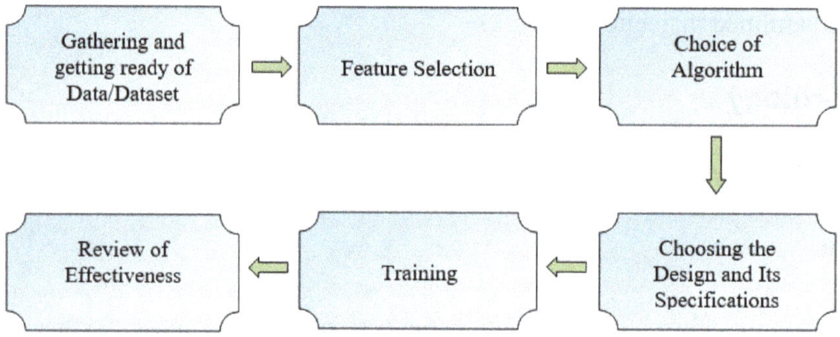

The following section details the individual responsibilities held by the system's constituent parts.

1) Gathering and Data Preparation: Collecting and preparing information to be stored in an arrangement that can be supplied to the engine is the main work in machine learning. There might be a lot of money for any issue. An abundance of noise, or unnecessary and superfluous data, characterizes most internet data, which is often uncontrolled. Therefore, pre-processing information to a standardized shape and cleaning it up are necessary.
2) Selection of Features: Knowledge collected in the previous stage could have several characteristics, some of those might not be useful for education (LIU, 2022). It is necessary to extract only the absolute most crucial aspects after removing them wherever they go.
3) Algorithm Choice: When it comes about machine gaining knowledge, not every algorithm will work. The preceding section outlined the reasons why numerous approaches excel at solving different types of class problems. For optimal outcomes, it is crucial to choose the most appropriate artificial intelligence a method for the given task. In the sixth chapter, we cover the different ML strategies (Alzubi et al., 2018).
4) Selection of Parameters and Models: The majority of techniques for machine learning necessitate human involvement in the initialization process to determine the best settings for the attributes.
5) Training: Instruction the mathematical framework using some data from the database is the next step after choosing the right algorithms and parameters.

6) Evaluation of Performance: It is necessary to test the predictive algorithm on untested data in order to measure the amount of learning using several measures of performance such as recollection, perfection, and specificity before implementing the technology in immediate effect.

FOUNDATIONS OF UNSUPERVISED, SEMI-SUPERVISED, AND SELF-SUPERVISED LEARNING

Unsupervised Learning

The supervised learning techniques, including regression and differentiation, occupy the bulk of this work.

Principal Components Analysis

The term the head office element processing describes the steps employed to determine the primary elements and then apply them to the analysis of information. Since PCA comprises just a collection of parameters $V1, V2, \ldots, V_N$ and no corresponding reaction Z it is an unregulated method. Data visualization is another useful application of PCA in addition to the production of extracted characteristics with use in controlled learning issues. In continuation of that chapter's theme, we are going to go into more depth on PCA with an emphasis on PCA's utility as instruments for exploring unattended datasets (Ashok et al., 2022). This can be exactly what PCA offers as a resource for (James et al., 2000). It seeks out the data collection's low-dimensional representations that capture the most amount of volatility. There are n deductions, and each one resides in p-dimensional dimensions; however, every one of such dimensions holds identical relevance (T. D. Le et al., 2024). The goal of PCA is to find the smallest possible amount of remarkable dimensions of where fascinating is defined as the degree to how data points vary across a given direction. All of the PCA-discovered dimensionality is linear combinations of the p characteristics (Jin et al., 2024). The location of this measurement, sometimes called primary elements, will be described immediately. The normalized straight combination of characteristics $V1, V2, \ldots, V_N$ is the primary aspect of the location shown in Equation 1.

$$Y_1 = \delta_{11} V_1 + \delta_{21} V_2 + \cdots + \delta_{n1} V_{2n} \tag{1}$$

Which exhibits the most variation. This means that $\sum_{l=1}^{n} \delta_{l1}^2 = 1$.The parts of the equation $\delta_{11}, \ldots, \delta_{n1}$ are called the forces of the fundamental primary components. The primary components unloading vector $\delta_1 = (\delta_{11}\,\delta_{21}\ldots\delta_{n1})^R$ is constructed from all of these deliveries. Because an extremely big variation might occur as the consequence of setting all of these variables to extremely large value units, we confine them to loadings such that the product of rectangles is approximately one.

Methods for Clustering

Clustering is an umbrella term for a variety of methods that can be used to divide information set into smaller, more manageable pieces. The goal of clustering is to divide a data set's information into smaller, more manageable groups, with the goal that results beneath every collection are very similar to one another and that accounts across categories are significantly divergent. Obviously, in order to put this into practice, we need to establish what it signifies too many discoveries to be comparable or dissimilar (Qi & Luo, 2022). In fact, understanding the information that is at hand is usually necessary for making this domain-appropriate choice. While PCA and clumping both seek to reduce data complexity by using only a couple of explanations, the ways in which they do this are distinct

- A low-dimensional approximation of the information that could clarify an effective amount of the variation is sought after using PCA.
- Aggregation attempts to classify the evidence into similar subsets.

Clustering K-Means. An attractive and straightforward method for dividing a dataset into I separate, not overlapping, groupings is K-means classification. The K-means method will organize all observations into only one of the predetermined *I* groups when we tell it what number of them to use. An excellent groupings according to the clustering method known as K-means, is one with little within-cluster variance. An estimate of the degree to which observations inside an association contrast with one another is the inside-cluster deviation for colony k, denoted as $W(A_l)$ It follows that we wish to address the issue (Hashmi et al., 2025).

$$minimize\ A_1,\ldots,A_K \left\{ \sum_{l=1}^{I} W(A_l) \right\} \tag{2}$$

To put it another way, that equation states that we aim to divide these findings into multiple clusters of I so that the total amount of within-cluster shifts, when added up across all those clusters, is as modest as feasible in equation 2. K-Means classification is shown in Algorithm 1.

Algorithm 1: Classification with K-Means
1. Pluck a number between one and I and put it onto every measurement. These constitute the starting points for grouping your findings into clusters. 2. Repeat until there is no longer any change in the grouping paperwork ✓ Finding the exact centre of each cluster is your task. The representation of the probability of the p features for the data obtained in the Ith clustered is the Ith clustered centre point. ✓ Put all of the observations in the collection with the centre that is nearest in geography to it, by employing the distance from each point as a metric.

During every stage, Algorithm 1 will unavoidably diminish the objective's significance.

Semi Supervised Learning

Semi-supervised algorithmic learners are a way to use the best parts of the two methods of learning. It may prove useful in data mines and machine learning when there is already unlabelled information because acquiring labeled data is an exhausting procedure. There are quite a lot of different types of learners who are semi-supervised. Some of these are talked about the one that follows

Model Generatives

One of the most elderly semi-supervised methods of learning is computational modeling. They use something like $N(v,z) = N(z)\ N(v/z)$ *where* $N(v/z)$ is an incorporated distribution of wealth, such as Gaussian models a combination models. You can tell which mixed components are in the unlabelled data. One labeled example for every single part is all you need to check the combination transportation (Nasteski, 2017).

Self-Training

In training yourself, a machine learning algorithm learns from a small set of labeled data. Next, the machine learning algorithm gets data that doesn't have labels. In the instruction set, the unlabelled viewpoints and the label predictions are put to one another. After that, same process is done again. The moniker "self-training" comes from the fact that the classification algorithm learns on its own (Ghanbari et al., 2024).

Transductive SVM

The transductive SVM, also known as or TSVM is an improved version of the SVM. TSVM takes into account both labeled and unlabelled data. It can be utilized to put labels on data that doesn't have any labels yet, in order to make the space in between the labeled and unlabelled data is as big as possible (Shou et al., 2024). Determining the exact answer with TSVM is an NP-hard task.

Self Supervised Learning

SML is really usual when it comes to characterization problems because the goal is to help the system learn the conceptual structure that was made. In SML, information that has been documented is called a set for learning, and the data that hasn't been classified is called a testing set for short. Whenever commentary is separate from each other, they are termed class labels. Whenever they are perpetual, they are designated as continues target values. Regression and identification of classes are two types of SML tasks. The categorization is a sort of SML from which the outcome has an exact value and the goal is to guess exactly what values that fit a certain class (Seneviratne et al., 2022). A trend is a form of SML that uses datasets marked with labels to make predictions about new data. The predictions are based on values that are constant. Some of the well-known SML methods are the methods of LR, logistic regression, logistical regression the DT, a SVM and others.

Linear Regression (LR)

LR is the easiest way to do regression analyses. It uses a linear structure to show how a response that is scalar is related to one or more informative elements. Some instances involving the LR the method anticipate the retail value of a stock, test scores, and so on. In another word, it is a regression-based procedure that is generally utilized for solving the regression equation problem in ML and is used for proactive monitoring. Let's say that the amount of input (x) and the consequence value (y) have a connection that is linear. The fact y can be calculated exactly by adding together several inputs (x). When there is a single quantity in the input box, it is called simple LR. When there are a number of numbers in the input into the system, it frequently goes by the term doubles LR. As an illustration, think of a linear formula that takes a group of (x) input variables and gives the anticipated result (y) for that group of inputs.

Figure 3. Linear regression

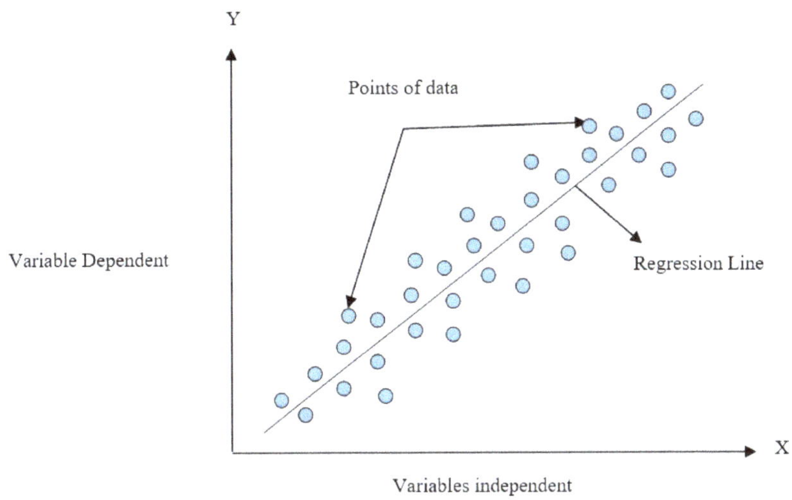

So, both the number that was entered (x) and the resulting value are numbers. There is one increasing factor for every value that is put in the horizontal equation. This is termed a coefficient, or coefficient for short. There's additionally an additional supplementary coefficient that is typically called the angle of incidence. To understand the model used by the LR, you need to figure out the coefficient ratios that were utilized to show the information. There are many ways for modeling the data, but the most frequent is OLS. Figure 3 shows how the information in the highlights alongside the LR line collaborates together.

Logistic Regression

Logistic regression is a widely recognized machine learning an algorithm that falls within the SML strategy. It is used to guess a dependent variable based on a set of autonomous factors, to solve cases of classification, and it is based on the idea of probabilities. The process of LR finds the return on investment of a factor that is dependent. So, the result is a single value. There are three possibilities: yes, no, zero, or one, and valid or invalid. But compared to presenting the exact value like 0 and 1, it gives the approximate probabilities that fall between 0 and 1. For example, imagine that you are offered a lot of challenges or evaluate to figure out which topic you are good at. This investigative services outcome might be seen as a geometry-related problem that is 70% likely to come apart (Mahmoud et al., 2022)V. The following inquiry is about history, and there's only a thirty percent likelihood of discovering the answer. It about the time you found the junk mail. This occurrence uses LR is shown in equation 3, although there is a limit on how many can be set based on the categorization If the group of emails is unwanted and the prediction is always 0.4, then this data is not unwanted messages mail, and that can lead to the result happening more quickly (Shetty et al., 2022).

$$\omega(r) = \frac{g^r}{g^r + 1} = \frac{1}{1 + g^{-r}} \tag{3}$$

There are three types of logistic regression, a binary, multinomial, and ordinary. Currency can only have one or two potential values: yes or no, or true or incorrect. Multinomial can have as many as three potential values, and ordinal supervises target components with categorizations For example, you could set up an overall rating as very poor, poor, great, or outstanding. Gives a clear definition of logistical regression. Regressive behavior with logistics shown in Figure 4.

Figure 4. Logistic regression

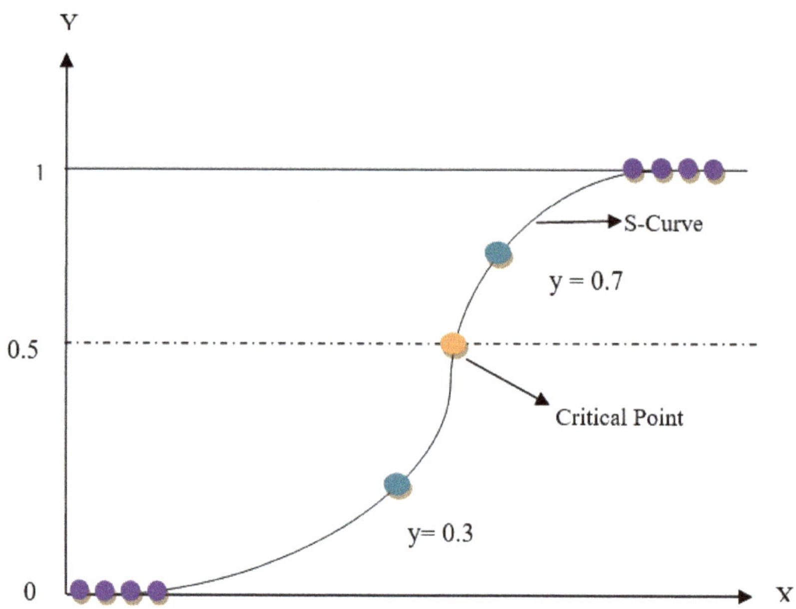

Support Vector Machine (SVM)

SVMs are a powerful but flexible sort of SML that may be used for analysis as well as regression. They are mostly used to solve categorization challenges. The kernel's the seating capacity is an important idea for most of the academic operation. These mathematical techniques create the hyper plane that is used to put every category into groups. The SVM makes the hyper plane over and over again, trying to make the error as little as possible. SVM's goal is to find a MMH by dividing each data set into separate categories. You can find MMH through performing these procedures.

- SVM makes the hyper planes over and over again until they divide classes in the most favorable way possible.
- After that, it chooses the hyper plane that divides every class correctly.

For instance, let's look at three tags that are blue and black and include information characteristics N and *S*. You give the machine learning model a pair of organizes *N*, *S* and it tells you if what it produces is blue or black. SVM looks at the information in points that make up the hyper plane that divides the labels into sections. This line is called an administrative boundary. Anything that is that falls to the side of the line will be blue and everything that falls to the next spot will be black. SVMs are used in ML because they can find complicated links between data despite needing to make a lot of modifications. It is a great choice particularly if you collaborate with smaller databases that have between dozens and hundreds of features. Because they can handle small, complicated historical data, they usually get more accurate results whenever they compare their computations to others. The narrow strip that separates the two separate educations is shown in Figure 5.

Figure 5. Support vector machine

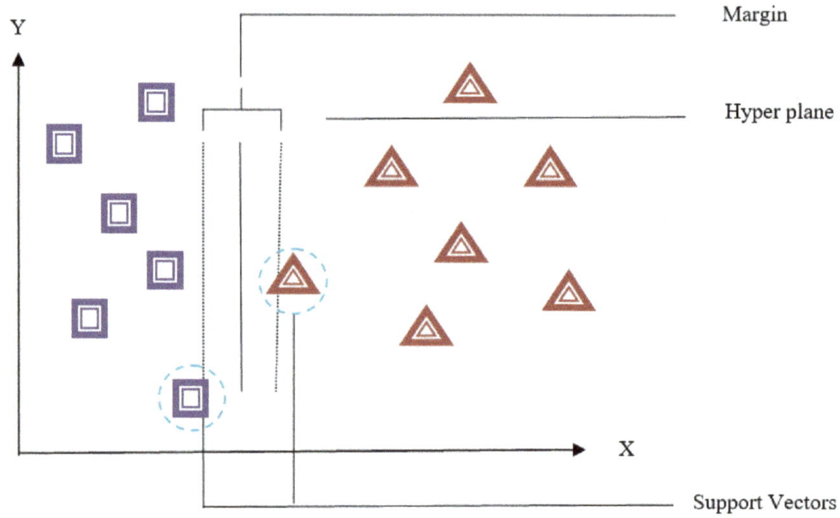

Decision Tree (DT)

The numerical parameters of the elements affect the collections in a DT. They use the information gained technique to find the best element in the dataset, which makes it a node with the root, and so on, until they can sort all of the cases in the database. There is a branch in the DT for each part of the information being considered. They are considered to be the most common ways to figure out how to classify things. The DT is used to show the choice being made and the process of making it. It uses a tree-like picture of choices, as the name suggests. Tree frameworks are based on objective variables that can take an independent settlement of values dubbed as filtering trees; in this kind of tree framework, leaves symbolize the class labels, and the combination of characteristics related to labels for the classes become represented by the branches. Think about making an overview of learners who are eligible for the placement campaign. Now, the question is if the student can go to the drive or not. There are n various things that need to be looked at before a decision can be made for a good choice. The variables that go into the choice are things like how much the learner has passed the test, that which the threshold for passing is, and so on. So, the option tree emulate is made up of the subsequent parts

- Root additionally Node: In the present instance, the fundamental node is the "eighth grade."
- From within Node: the structures in the middle that have one incoming amounts edge as well as more compared with two exiting edges.
- A node known as a leaf is a node that doesn't have a connecting edge. It's sometimes called an intermediate node.

In this example, the DT that is currently being worked on first tests the test condition from the root hub and sends the control to one of the active edges. This means that the condition is tried again and a hub is given. When all the test conditions lead to a leaf hub, the tree should finish. The leaf hub has class names that vote for or regarding the choice.

APPLICATIONS IN CLIMATE SCIENCE

Combining without supervision, semi-supervised, and self-directed instruction in research on climate has led to new ways of comprehending and forecasting complicated weather phenomena. These methods are very useful when there isn't much labeled data or it's not full. They make it easier to analyze large, high-dimensional weather files.

Climate Pattern Discovery

Unsupervised learning approaches like clustering with k-means, hierarchical clustering, and Gaussian mixed-model models have proved very helpful in finding and studying climate trends that happen over and over again. Using clustering methods on SST anomalies, scientists have found phenomena like this phenomenon and the phenomenon known as La which are important parts of the ENSO system. These computerized models immediately aggregate SST structure without any labels, which helps scientists tell the difference from the ocean's normal, warming, and cooling phases. Also, scientists studying climate can use dimensionality reduction methods like PCA and autonomous encoders to find the main patterns of variance in multiple variables historical data, like temperatures, humidity, and pressures. This helps them find connections and patterns that are hard to see something else.

Anomaly Detection in Satellite Data

Warning systems for emergencies and disaster prevention depend on being able to find strange things like extreme weather events, sudden temperature shifts, or unique air components. Being isolated Forests, One-Class SVMs, and deep autonomous encoders are examples of without supervision algorithms for detecting anomalies that have been used efficiently to find these occurrences in satellite images and data collected over time. Those techniques build a model of normal climatic behavior and find things that are different from it. These things could be wildfires, volcanic activity, or storms that happen at the wrong time of year. Self-supervised frameworks make recognizing anomalies in noisy or uncompleted satellite databases even more reliable by learning strong pattern representations from data that doesn't have labels.

Downscaling Climate Models

GCM don't work well for regional or regional climate impact assessments since they don't have very high proposals. Self-directed training can help by creating spatially-aware illustrations that can be used to improve the elimination of climate simulated events. This is called statistical down scaling. When playing like a convolution automated encoders and transforms of vision that are trained on pretext-related tasks like concealed region prediction or interpolation of frames understand little details from high-resolution images. These attributes are then utilized to make not high-quality outputs from GCMs bigger, which makes predictions of local climatic variables like high and low temperatures, velocity of winds, or the quantity of rainfall more exact.

Semi-Supervised Classification of Extreme Events

Extreme weather-related events like tropical cyclones, heat waves, and flash floods don't happen very often, thus it's hard and expensive to put together labeled datasets for them. Semi-supervised methods of learning fix this by using a few labeled samples to help them learn from a much larger pool of unlabelled data. Graph-based algorithms and reliability regularization techniques make it easier to find and categories extreme incidents in different places and times. For example, a semi-supervised convolution neural network, also called a CNN that has been trained on a few hundred labeled cyclone photos can learn to find new storms that haven't been seen before in real-time satellite communication feeds. This makes disaster preparedness and response plans much better (Souto & Oliveira, 2022).

Sea Ice and Glacier Monitoring

To gain insight into how the changing climate affects both polar and alpine areas, it is important to keep an eye on things like the amount of sea ice and how glaciers move. The conventional method supervised methods generally fail since there aren't many categorized satellite photos available in these remote places. Self-supervised instructional approaches, especially comparative learning regulations like Similar or Mock, have done an amazing job at learning how to discern the difference between satellite images without labels. These drawings make it easier to conduct things like classifying sea ice, segmenting glaciers, and keeping track of calving occurrences with a lot of accuracy. Having the capacity to learn from DNA sequences of photos over time also makes it possible to forecast longer-term patterns in the retreat of glaciers or the thinning of sea ice, which are important signs of climate change (Voigt et al., 2023).

RESULT AND ANALYSIS

Figure 6, a sample of a ROC curve, shows the connection across FP Rate or TP Rate, represented on both the vertical and horizontal axes correspondingly. When compared using the ROC, the RF framework clearly comes out on top.

Figure 6. ROC curve

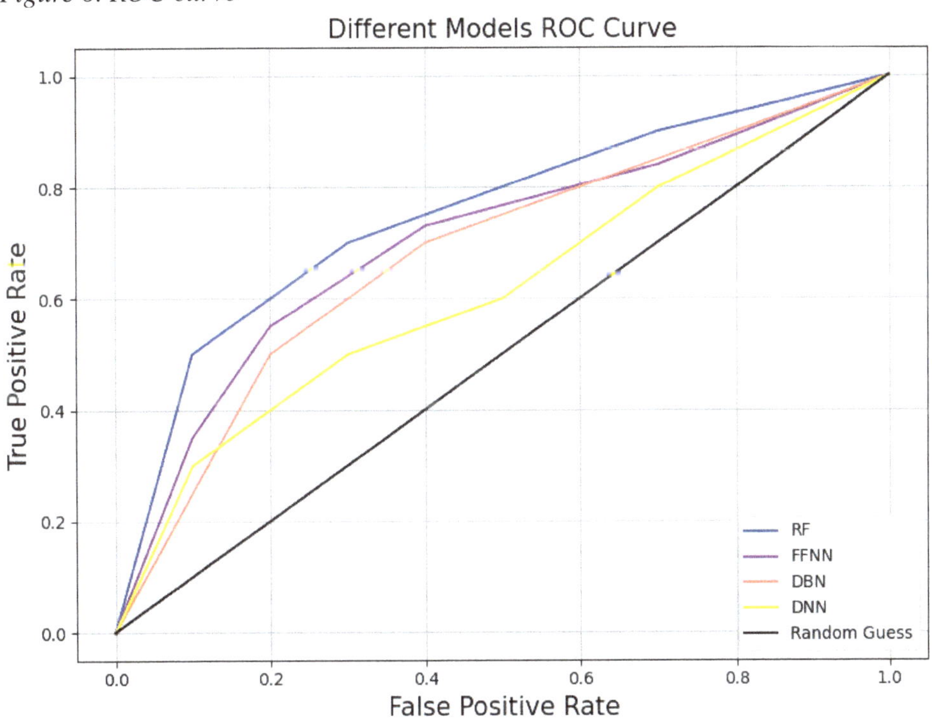

The hybrid of the model's outcomes curve, as shown in Figure 7, has the input data quantity on the bottom axis as well as performance on the axes that are horizontal. Using the variations in Example Accuracy as a benchmark, the RF technique emerges as the clear winner.

Figure 7. Different model accuracy

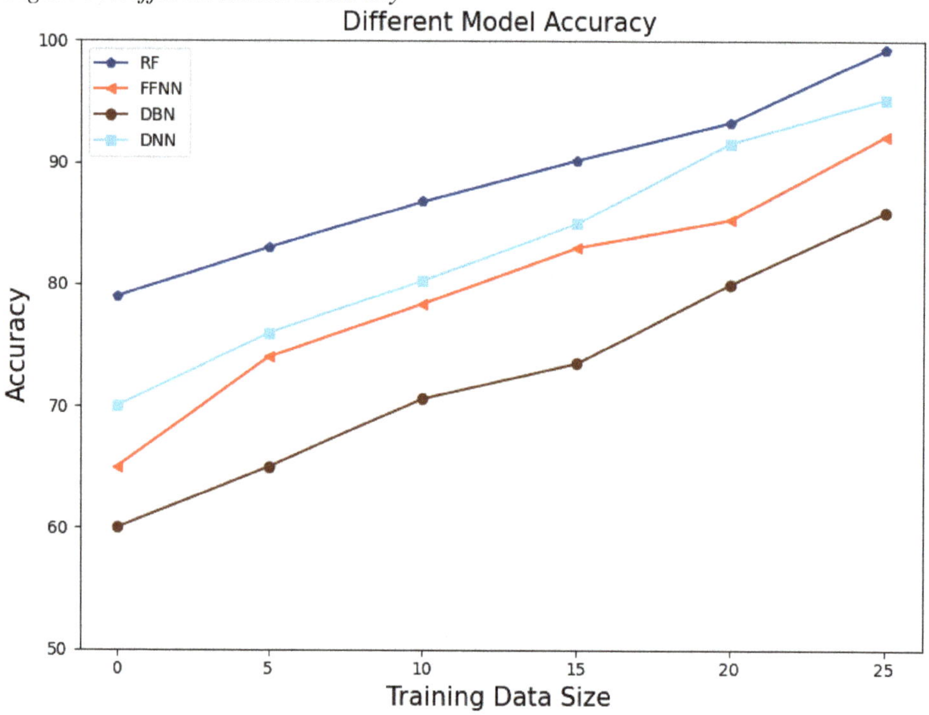

All four settings were appropriately selected, as shown in Table 1, since the predicted results were obtained from each. Accuracy, Precision, Recall, and the bonding-measurement F1-Score were among the measures used to assess the frameworks' performance. Compared to the other models, this RF model is obviously the best.

Table 1. Performance of comparison

Models	Accuracy	Precision	Recall	F1-Score
RF	**99.28**	**97.26**	**95.24**	**99.69**
FFNN	92.18	90.16	88.14	92.58
DBN	85.92	83.90	81.88	85.98
DNN	95.24	93.22	91.20	95.74

Accuracy, Precision, Recall, and F1=score metrics were used to evaluate the models from deep learning for weather prediction. Figure 8 is an indicator that shows the differences between the five supervised learning scenarios. The RF model achieves better results than the other approach, as demonstrated in the visual.

Figure 8. Performance comparison

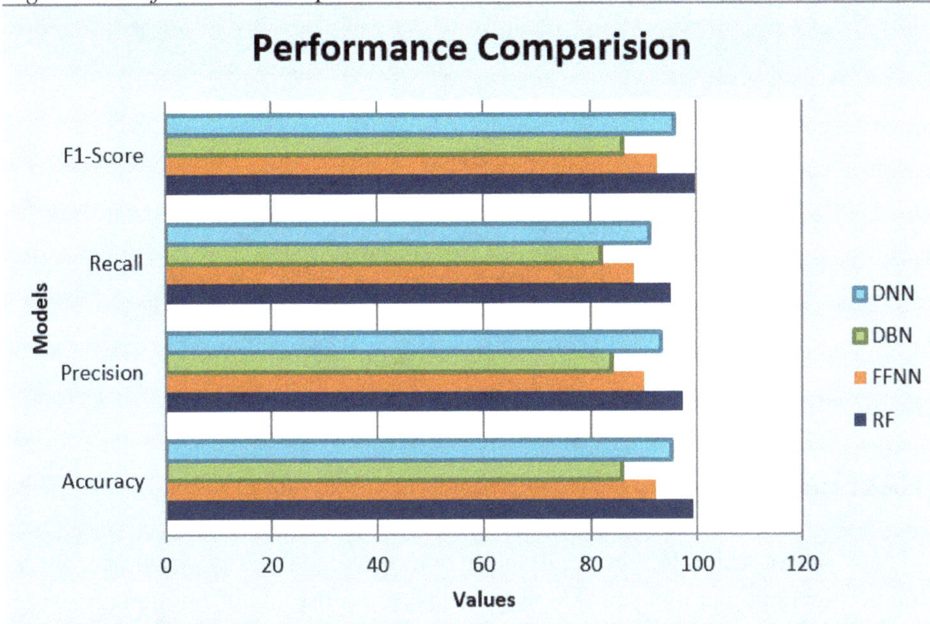

The computational time needed for the RF model presented so far is shown in Figure 9. In comparison to more typical situations, the one that is offered is superior.

Figure 9. Computational time

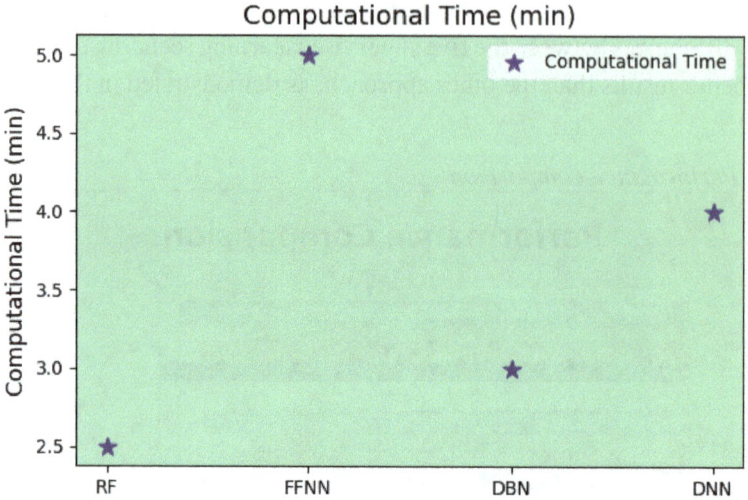

CHALLENGES

Using sensor technology for cultivation could change the way growing crops is done in confronting of many problems, giving farmers useful information about the health of their crops across various measures throughout one's the period of cultivation. Ranchers can use advanced methods like geography, IOT, Big Data investigation, and AI to make smart choices about their agriculture (Astruc et al., 2024). PA uses this new technology in smart ways to get the most out from agricultural inputs, make making more efficient, and cut down on suggestions decreases (Tang et al., 2023). Throughout the past several decades, the use of sensor technology in accurate farming has grown a lot. But to get the most out of it, we need to create a process for real-time application of satellite imagery in precise agriculture that is both easy to use and dependable. This job is very important because imaging processing is very complicated and requires a lot of skills and expertise in technology. The development of systems that are both precise and easy to use is expected to lead to more widespread usage of sensor technology in both business-related and unrelated to and not farming use cases.

Monitoring Depends on a Small Number of Indicators

The sustainable governance strategy focusses on fewer things by only allowing only a few conditions. This makes it easier to comprehend how things relate to each other in space and time. The reduction in complexity often leads to incorrect decisions by managers (Khaertdinov & Asteriadis, 2022). To keep an eye on the many levels of challenges in and a natural system, it is best to use things from different levels of the environmentally sound the organizational structure. So, one of the main problems is finding a set of examinations that give straightforward indications, can be used to keep an eye on ecological situations at an accessible price, and cover the complete spectrum of environmental variations.

Ecological Variable Choices

If your desired outcomes and objectives aren't clear, you can end up measuring the improper components in the wrong location at the inappropriate moment with poor exactitude or trustworthiness. Setting primary school objectives and purposes early in the workflow is important for keeping an eye on current and future difficulties in management. They can then choose those parameters that have the closest connections to those difficulties of leadership to measure framework qualities. But society has always set environmental specifications that are only about making money in the short term, like getting the most crops from plantations or the most timber from woods (Han et al., 2024). These goals could make healthy ecosystems less likely to endure in the long term. Organizations goals, and hence the choice of parameters, should be based on an understanding of how the administration of resources choices will affect both the near and long-term.

Lack of Scientific Rigor

It is hard to check the knowledge provided by these components because there aren't many good ways to find biological variables(Cao et al., 2023)V. We don't know how to interpret changes in factors is over time and space until standard methods for selecting and applying them are created. By creating and using conventional procedures for choosing biological variables, we can make sure that the elements we choose are relevant for managing challenges, that they can be repeated, and that there is no bias in the choice we make (Gao et al., 2023).

Technological Limitations

One of the biggest problems with climate surveillance will be getting passive information from sensors from mobile devices and gadgets that were wear as well as data from people's perceptions, into official information streams while completely safeguarding privacy and following the EU's GDPR. Donating data for science is growing more popular (Saheer et al., 2022). For example, Open Individuals gives academics, businesses, NGOs, and politicians access to Rods to new information sources and quantities that no one else possesses. The standard mobile devices and smart watches may now give you a lot of information, such as your location, motion individual carbon monoxide mobility the footprints, health data, and additional information (Internatio & D, n.d.). Technological innovation has definitely made it easier to keep an eye on things, yet some biological and environment interprets are still hard to see clearly. For example, monitoring methods give us useful information, but they might not show us how small-scale ecosystems work. Also, constraints on technology make it hard to get data that is current, which makes it hard for us to understand adjustments that happen quickly (Wellhausen et al., 2019). There is also a new environmentally focused technique for studying how people feel in cities that is connected to mobile devices (Chen et al., 2024). We might learn more about stressful cities by finding out how stress affects health. Wearable technology alongside additional low-cost sensors could help us learn more about things that can improve people's health, the availability of open spaces, city development, and steps to lower cities' greenhouse gases. Environmental study has mostly been done in the Global North, where people have time for unresolved pursuits and volunteer work. Still, because of technological advancements, the developing world's southern region has moved ahead of many important stages of economic growth. For example, in many places in which there are no had landlines, there currently are a lot of telephones (Gomiero, 2016). The information from a number of open ecological science devices, like this application, show that biological activities are far less prevalent in nations with poor infrastructure.

Interdisciplinary Collaboration

The researchers from different domains, such the biological sciences, atmospheric science, earth science, and social science, need to work together on environmental issues and climate change observing. It might be hard to communicate and work together well when people use different words, methods, and scientific agendas (Adewale, 2024). To get over these problems, it's important to find harmony and encourage collaborative work across disciplines. Another big challenge is the need to keep track of where data items and research results come from during the whole process, from collecting data to making sure it is of excellent craftsmanship analyzing it, simulations it, and eventually producing it (Kataria et al., 2023). the authenticity especially important for conclusions from science that are used to make administration and policy purchases since it can be hard to repeat field studies and processes because it's hard to recreate the conditions in which they were done. Researchers in the computer field are making a lot of progress when coming up with ways to collect verification knowledge. You can use programmed statistical applications like R and technological procedure structures like Keller University and Tavern to keep track of the data being analyzed and processed that ultimately produced a certain set of accomplishments.

FUTURE DIRECTIONS

Despite several years of adhere to lacked a solid theoretical base, climate change awareness has grown into a growing area that is important for individuals seeking to increase public participation and is quickly becoming its own distinct area of study. Pubic convictions, viewpoints, capacities for comprehension, and endorsements for policies across distinct nations have been assessed using a number of polls, both general and conducted for academic use. A few focus collaborations also exist. Furthermore, certain research have evaluated particular arguments, perspectives, and viewer reactions, and these investigations have produced valuable information; however, these research results are not publicly available (Guizilini et al., 2019). The media's depiction of warming temperatures has been the subject of multiple studies. The latest literature has expanded the scope of this type of study to include reports from outlets outside of Western European and American print media. The remaining knowledge or assumptions regarding information about climate change are largely derived from research in related disciplines, such as communicating risks, science storytelling, and communication through the press, campaigns, online marketing, rhetoric, among continuing. Because of this, environmentally friendly interactions are a thriving area that requires further study and validation of its practical applications (T.-D. Le, 2024). Here are some new things to think about while planning for the upcoming years of investigation and behavior

Key Elements of the Communication Process

The following areas need more studies audience-specific information and structuring; the affect of establishing for successful enthusiasm; the significance of multiple representatives for varying listeners; the potential, limitations, and most suitable applications of new media technologies for various forms of public enjoyment; the cause of environment on exchange of information; and the successful outcome of various marketing initiatives (Giri et al., 2020).

Communication Technologies and Modes

Concerning the efficient and moral dissemination of climate change data visualizations, there is a dearth of literature (Hou et al., 2023). Despite the growing importance of the performing arts in raising awareness about the impact of climate change and encouraging more people to become that are involved, their participation in this movement has yet to be well examined (Xie et al., 2024).

Communicating Mitigation and Adaptation

However, there is also a lack of initial studies on the public's understanding of adaptations, successful approaches for expressing the need for modification, and any noteworthy distinctions between relating minimizing them and adjusting (Moser, 2010). This is despite the fact that knowledge concerning efficient conversation may be applicable to numerous ecological strategies (Saeed et al., 2019).

Long-Term and Deeper Engagement

There are crucial queries to ask about how to communicate about a problem that "does not go away," demands long-term involvement, and may cause disappointment because mitigation efforts do not immediately lead to a "improving" setting, given that global warming is unchangeable on individuals time frames and cannot be "solved" simply or swiftly "Green fatigue" has already garnered a lot of coverage from the media, who are critical of the difficulties "newsworthiness" for being so ubiquitous(Yang et al., 2023). However, there is a lack of knowledge regarding the best ways to interact with a problem and maintain the engagement of society participants over a prolonged amount of time. Visionary and visioning play an important part in participation procedures that span many years, and this area of study may help us learn more about their function.

Mass Mobilization

Researchers have proposed that mass mobilization could be necessary due to the gravity and timeliness of worldwide warming (Shenoy et al., 2024). Nevertheless, the nature of such an Endeavour has not been investigated just yet. Relevant to such investigation would involve a thorough investigation of methods for conveying excitement without additionally frightening or paralyzing listeners.

Dialogic Forms of Communication

Potential, impact, usefulness, and limitation of a single direction exchange of data vs bilateral, interacting, dialogic interactions are completely distinct. While some see them as a threat to the idea of mass mobilization, another see them as a check on the growing technocracy in international decision-and policy-making on this problem (Zhang et al., 2023).To further understand how discussion might influence participation, taking decisions, freedom, and societal reactions to the effects of climate change, additional research findings are needed.

CONCLUSION

The changing climate is a problem that affects everyone and needs to be dealt with right away. There were a lot of articles written about how to deal with and adapt to changes in the climate. But we need new ways to look at the complexities of global warming to come up with better and more effective procedures for mitigation and mitigation of damage (Saeed et al., 2021). As technology has improved ML and other procedures have become increasingly common in many areas, including global warming. The goal of this research is to look into the most common ML methods that were previously used to deal with climate change. Unguided, semi-supervised, and self-supervised education is all great ways to get useful information from a lot of climate data that doesn't have labels. These methods are making climate modeling along with tracking more accessible by making it less reliant on annotations made by hands. This is leading to big improvements in weather prediction, tracking the environment, and worldwide warming evaluations. The future of climate computer science will depend on how well these methods keep getting better (X. Wang et al., 2023).

REFERENCES

Adewale, T. (2024). How AI Training Methods Are Evolving : From Supervised Learning to Reinforcement and Beyond. November.

Agastya, C., Ghebremusse, S., Anderson, I., Reed, C., Vahabi, H., & Todeschini, A. (2021). Self-supervised Contrastive Learning for Irrigation Detection in Satellite Imagery. 1. https://arxiv.org/abs/2108.05484

Akhare, R. (2023). Hybrid approach between Supervised and Unsupervised Learning: Self Supervised Learning. *Article in Journal of Xidian University*, *17*(8), 1–9. DOI: 10.37896/jxu17.8/026

Allmer, J. (2014). miRNomics. *MicroRNA Biology and Computational Analysis.*, *1107*, 333. DOI: 10.1007/978-1-62703-748-8

Alzubi, J., Nayyar, A., & Kumar, A. (2018). Machine Learning from Theory to Algorithms: An Overview. Journal of Physics: Conference Series, 1142(1), 0–15. https://doi.org/DOI: 10.1088/1742-6596/1142/1/012012

Ashok, A., Lambhate, D., & Singh, J. (2022). Self-Supervised Representations of Geolocated Weather Time Series-an Evaluation and Analysis. www.aaai.org

Astruc, G., Gonthier, N., Mallet, C., & Landrieu, L. (2024). OmniSat: Self-Supervised Modality Fusion for Earth Observation. https://doi.org/DOI: 10.1007/978-3-031-73390-1_24

Bahlali, A. R., & Bachir, A. (2025). Semi-Supervised Approach for Encrypted Traffic Anomaly Self-Supervised Learning Meets Custom Autoencoder : A Semi-Supervised Approach for Encrypted Traffic Anomaly Detection.

Cao, S., Joshi, D., Gui, L. Y., & Wang, Y. X. (2023). HASSOD: Hierarchical Adaptive Self-Supervised Object Detection. Advances in Neural Information Processing Systems, 36(NeurIPS).

ÇELİK, Ö. (2018). A Research on Machine Learning Methods and Its Applications. *Journal of Educational Technology and Online Learning*, *1*(3), 25–40. DOI: 10.31681/jetol.457046

Chen, H., Gouin-Vallerand, C., Bouchard, K., Gaboury, S., Ngankam, H. K., Lussier, M., Couture, M., Bier, N., & Giroux, S. (2024). Utilizing Self-Supervised Learning for Recognizing Human Activity in Older Adults through Labeling Applications in Real-World Smart Homes. September, 275–283. https://doi.org/DOI: 10.1145/3677525.3678671

Del Pup, F., & Atzori, M. (2023). Applications of Self-Supervised Learning to Biomedical Signals: A Survey. *IEEE Access : Practical Innovations, Open Solutions*, *11*(November), 144180–144203. DOI: 10.1109/ACCESS.2023.3344531

Gao, L., Wang, D., Zhuang, L., Sun, X., Huang, M., & Plaza, A. (2023). BS3LNet: A New Blind-Spot Self-Supervised Learning Network for Hyperspectral Anomaly Detection. *IEEE Transactions on Geoscience and Remote Sensing*, *61*. Advance online publication. DOI: 10.1109/TGRS.2023.3246565

Ghanbari, A., Shirdel, G. H., & Maleki, F. (2024). Semi-Self-Supervised Domain Adaptation: Developing Deep Learning Models with Limited Annotated Data for Wheat Head Segmentation. *Algorithms*, *17*(6). Advance online publication. DOI: 10.3390/a17060267

Giri, R., Tenneti, S. V, Cheng, F., Helwani, K., Isik, U., & Krishnaswamy, A. (2020). Detection and Classification of Acoustic Scenes and Events 2020 SELF-SUPERVISED CLASSIFICATION FOR DETECTING ANOMALOUS SOUNDS. November.

Gomiero, T. (2016). Soil degradation, land scarcity and food security: Reviewing a complex challenge. *Sustainability (Switzerland)*, *8*(3), 1–41. DOI: 10.3390/su8030281

Guizilini, V., Li, J., Rares, A., Pillai, S., & Gaidon, A. (2019). Robust Semi-Supervised Monocular Depth Estimation with Reprojected Distances. Proceedings of Machine Learning Research, 100(CoRL), 503–512.

Haasnoot, M., van 't Klooster, S., & van Alphen, J. (2018). Designing a monitoring system to detect signals to adapt to uncertain climate change. *Global Environmental Change*, *52*(July), 273–285. DOI: 10.1016/j.gloenvcha.2018.08.003

Han, H., Liu, Z., Li, J., & Zeng, Z. (2024). Challenges in remote sensing based climate and crop monitoring: Navigating the complexities using AI. *Journal of Cloud Computing (Heidelberg, Germany)*, *13*(1). Advance online publication. DOI: 10.1186/s13677-023-00583-8

Hashmi, E., Yayilgan, S. Y., Yamin, M. M., Abomhara, M., & Ullah, M. (2025). Self-supervised hate speech detection in Norwegian texts with lexical and semantic augmentations. Expert Systems with Applications, 264(April 2024), 125843. https://doi.org/DOI: 10.1016/j.eswa.2024.125843

Hou, Z., He, Y., Cen, Y., Liu, X., Dong, Y., Kharlamov, E., & Tang, J. (2023). GraphMAE2: A Decoding-Enhanced Masked Self-Supervised Graph Learner. ACM Web Conference 2023 - Proceedings of the World Wide Web Conference, WWW 2023, 737–746. https://doi.org/DOI: 10.1145/3543507.3583379

Hoyer, L., Dai, D., Wang, Q., Chen, Y., & Van Gool, L. (2023). Improving Semi-Supervised and Domain-Adaptive Semantic Segmentation with Self-Supervised Depth Estimation. *International Journal of Computer Vision, 131*(8), 2070–2096. DOI: 10.1007/s11263-023-01799-6

Hur, J., & Roth, S. (2020). Self-supervised monocular scene flow estimation. *Proceedings of the IEEE Computer Society Conference on Computer Vision and Pattern Recognition, i*, 7394–7403. https://doi.org/DOI: 10.1109/CVPR42600.2020.00742

Internatio, A., & D, L. V. D. Q. R. (n.d.). Review W of Eu an Stud Dies.

James, G., Witten, D., Hastie, T., & Tibshirani, R. (2000). An introduction to Statistical Learning. In Current medicinal chemistry (Vol. 7, Issue 10). https://doi.org/DOI: 10.1007/978-1-4614-7138-7

Jin, W., Wang, N., Tao, T., Shi, B., Bi, H., Zhao, B., Wu, H., Duan, H., & Yang, G. (2024). A veracity dissemination consistency-based few-shot fake news detection framework by synergizing adversarial and contrastive self-supervised learning. *Scientific Reports, 14*(1), 1–22. DOI: 10.1038/s41598-024-70039-9 PMID: 39174581

Kataria, S., Villalba, J., Moro-Velázquez, L., Thebaud, T., & Dehak, N. (2023). Self-FiLM: Conditioning GANs with self-supervised representations for bandwidth extension based speaker recognition. Proceedings of the Annual Conference of the International Speech Communication Association, INTERSPEECH, 2023-August, 4688–4692. https://doi.org/DOI: 10.21437/Interspeech.2023-2031

Khaertdinov, B., & Asteriadis, S. (2022). Temporal Feature Alignment in Contrastive Self-Supervised Learning for Human Activity Recognition. *2022 IEEE International Joint Conference on Biometrics, IJCB 2022*. https://doi.org/DOI: 10.1109/IJCB54206.2022.10007984

Le, T.-D. (2024). Boosting Transformer's Robustness and Efficacy in PPG Signal Artifact Detection with Self-Supervised Learning. Ml. https://arxiv.org/abs/2401.01013

Le, T. D., Macabiau, C., Albert, K., Jouvet, P., & Noumeir, R. (2024). A Novel Transformer-based Self-Supervised Learning Method to Enhance Photoplethysmogram Signal Artifact Detection. *IEEE Access : Practical Innovations, Open Solutions, 12*(November). Advance online publication. DOI: 10.1109/ACCESS.2024.3488595

LIU. D. (2022). Self-Supervised Learning Frameworks for Iot Applications. PhD Thesis.

Mahmoud, H., Wu, W., & Gaber, M. M. (2022). A Time-Series Self-Supervised Learning Approach to Detection of Cyber-physical Attacks in Water Distribution Systems. *Energies*, *15*(3), 1–18. DOI: 10.3390/en15030914

Moser, S. C. (2010). Communicating climate change: History, challenges, process and future directions. *Wiley Interdisciplinary Reviews: Climate Change*, *1*(1), 31–53. DOI: 10.1002/wcc.11

Myakala, P. K. (2025). Scaling AI with Limited Labeled Data : A Self-Supervised Learning Approach.

Nasteski, V. (2017). An overview of the supervised machine learning methods. *Horizons.B*, *4*(December), 51–62. DOI: 10.20544/horizons.b.04.1.17.p05

Qi, G. J., & Luo, J. (2022). Small Data Challenges in Big Data Era: A Survey of Recent Progress on Unsupervised and Semi-Supervised Methods. *IEEE Transactions on Pattern Analysis and Machine Intelligence*, *44*(4), 2168–2187. DOI: 10.1109/TPAMI.2020.3031898 PMID: 33074801

Saeed, A., Ozcelebi, T., & Lukkien, J. (2019). Multi-task Self-Supervised Learning for Human Activity Detection. *Proceedings of the ACM on Interactive, Mobile, Wearable and Ubiquitous Technologies*, *3*(2), 1–30. DOI: 10.1145/3328932

Saeed, A., Ungureanu, V., & Gfeller, B. (2021). Sense and Learn: Self-supervision for omnipresent sensors. *Machine Learning with Applications*, *6*(April), 100152. DOI: 10.1016/j.mlwa.2021.100152

Saheer, L. B., Shahawy, M., Approach, S., & Tree, U. (2022). Self-supervised Approach for Urban Tree Recognition on Lakshmi Babu Saheer, Mohamed Shahawy To cite this version : HAL Id : hal-03789042.

Seneviratne, S., Wijnands, J. S., Nice, K., Zhao, H., Godic, B., Mavoa, S., Vidanaarachchi, R., Stevenson, M., Garcia, L., Hunter, R. F., & Thompson, J. (2022). Urban feature analysis from aerial remote sensing imagery using self-supervised and semi-supervised computer vision. 1–35. https://arxiv.org/abs/2208.08047

Shenoy, J., Zhang, X. D., Tao, B., Mehrotra, S., Yang, R., Zhao, H., & Vasisht, D. (2024). Self-Supervised Learning across the Spectrum. *Remote Sensing*, *16*(18). Advance online publication. DOI: 10.3390/rs16183470

Shetty, S. H., Shetty, S., Singh, C., & Rao, A. (2022). Supervised machine learning: Algorithms and applications. *Fundamentals and Methods of Machine and Deep Learning: Algorithms, Tools, and Applications*, (January), 1–16. DOI: 10.1002/9781119821908.ch1

Shou, Y., Cao, X., & Meng, D. (2024). SpeGCL: Self-supervised Graph Spectrum Contrastive Learning without Positive Samples. 1–13. https://arxiv.org/abs/2410.10365

Souto, E. J. P., & Oliveira, H. A. B. F. (2022). Recognition Using Physiological Signals.

Souza, J. C. (2024). A Brief Overview of Deep Learning Approaches for Climate Data - Proceeding A Brief Overview of Deep Learning Approaches for Climate Data. April, 10–12. https://doi.org/DOI: 10.13140/RG.2.2.26670.52800

Tang, H., Bhatti, U. A., Li, J., Marjan, S., Baryalai, M., Assam, M., Ghadi, Y. Y., & Mohamed, H. G. (2023). A New Hybrid Forecasting Model Based on Dual Series Decomposition with Long-Term Short-Term Memory. *International Journal of Intelligent Systems, 2023*. Advance online publication. DOI: 10.1155/2023/9407104

Vera, P., Marie, M., Gaviria, J. A., Miroslav, T., Petr, H., Esther, B., Tudor, T., Nina, M., & Chawdhery, R. A. (2020). *Pr ep rin t n pe er re v Pr ep er* ed. 1–24.

Voigt, B., Fischer, O., Schilling, B., Krumnow, C., & Herta, C. (2023). Investigation of semi- and self-supervised learning methods in the histopathological domain. *Journal of Pathology Informatics*, *14*(January), 100305. DOI: 10.1016/j.jpi.2023.100305 PMID: 37025325

Wang, X., Yang, N., Liu, E., Gu, W., Zhang, J., Zhao, S., Sun, G., & Wang, J. (2023). Tree Species Classification Based on Self-Supervised Learning with Multisource Remote Sensing Images. *Applied Sciences (Switzerland)*, *13*(3). Advance online publication. DOI: 10.3390/app13031928

Wang, Y., Albrecht, C. M., Braham, N. A. A., Mou, L., & Zhu, X. X. (2022). Self-Supervised Learning in Remote Sensing: A review. *IEEE Geoscience and Remote Sensing Magazine*, *10*(4), 213–247. DOI: 10.1109/MGRS.2022.3198244

Wellhausen, L., Dosovitskiy, A., Ranftl, R., Walas, K., Cadena, C., & Hutter, M. (2019). Where should i walk(Predicting terrain properties from images via self-supervised learning. *IEEE Robotics and Automation Letters*, *4*(2), 1509–1516. DOI: 10.1109/LRA.2019.2895390

Xie, Y., Zhang, H., & Babar, M. A. (2024). LogSD: Detecting Anomalies from System Logs through Self-Supervised Learning and Frequency-Based Masking. Proceedings of the ACM on Software Engineering, 1(FSE), 2098–2120. https://doi.org/DOI: 10.1145/3660800

Yang, J., Chen, X., Zou, H., Wang, D., & Xie, L. (2023). AutoFi: Toward Automatic Wi-Fi Human Sensing via Geometric Self-Supervised Learning. *IEEE Internet of Things Journal, 10*(8), 7416–7425. DOI: 10.1109/JIOT.2022.3228820

Zhang, Y., Wang, J., Chen, Y., Yu, H., & Qin, T. (2023). Adaptive Memory Networks With Self-Supervised Learning for Unsupervised Anomaly Detection. *IEEE Transactions on Knowledge and Data Engineering*, *35*(12), 12068–12080. DOI: 10.1109/TKDE.2021.3139916

Chapter 9
Wood Wide Web:
Integrating Artificial Intelligence in Forest Communication

Arun Agrawal
https://orcid.org/0000-0001-7233-6660
Institute of Technology and Management, Gwalior, India

Nalini Santosh Jagtap
https://orcid.org/0000-0002-8477-9300
Pimpri Chinchwad College of Engineering and Research, India

Dipali Himmatrao Patil
JSPMs Rajarshi Shahu College of Engineering, India

Anju Lata Gajpal
Jansanchar Vishwavidhalaya, India

M. V. V. S. Prasad
GITAM Univesrity, India

Vinaydeep Brar
Dr. D.Y. Patil B-School, Pune, India

Deepak Gupta
https://orcid.org/0000-0003-3929-1362
Institute of Technology and Management, Gwalior, India

Gaurav Dubey
https://orcid.org/0009-0000-5973-3944
Institute of Technology and Management, Gwalior, India

ABSTRACT

This chapter delves into the emerging intersection of advanced computational techniques and the study of mycorrhizal networks in forest ecosystems, often referred to as the "World Wide Web of Wood." The chapter explores how tools such as machine learning, sensor networks, and predictive models can contribute to a deeper understanding of fungal-root communication, resource distribution, and forest resilience. By combining data-driven analyses with traditional ecological studies, the research reveals complex patterns of interactions that were previously imperceptible. The chapter proposes a new framework for mapping these underground networks using neural models, which has potential applications in conservation, climate adaptation,

DOI: 10.4018/979-8-3373-3501-8.ch009

and sustainable forestry. This interdisciplinary approach highlights how technological innovations can enrich ecological insights and support effective ecosystem management.

INTRODUCTION

There is a growing recognition that forests function as more than just a collection of individual trees, but rather as complex, interconnected communities facilitated by underground fungal networks (Simard et al., 1997). Mycorrhizal fungi form symbiotic relationships with tree roots, constituting an extensive underground network known as the "wood web" (Selosse et al., 2006). This biological infrastructure enables the transfer of carbon, nitrogen, water, defense signals, and other important resources between trees, fundamentally reshaping our understanding of forest ecosystems and plant-to-plant communication (Bever et al., 2010).

Currently, advances in data-driven technologies are revolutionizing ecological research, revealing complex patterns in complex data sets that were previously undetectable to humans (Christin et al., 2019). Combining these computational tools with insights into fungal-root networks opens new avenues for exploring forest dynamics and improving management practices.

This Chapter examines how advanced algorithms and predictive models can be utilized to:

1. Accurately map mycorrhizal network structures and behaviors
2. Model resource and signal flows between trees
3. Predict forest responses to environmental stresses
4. Guide conservation strategies to protect critical network hubs
5. Support sustainable forestry by preserving the integrity of these hidden systems

We emphasize both current applications and future conceptual models focused on practical solutions to enhance forest resilience amid climate change and human impacts.

BACKGROUND

Mycorrhizal Networks and Forest Communication

Mycorrhizal fungi exist in symbiotic relationships with nearly 90% of terrestrial plant species, primarily through ectomycorrhizal (EM) and arbuscular mycorrhizal

(AM) networks (Smith & Read, 2010). These mutually beneficial partnerships are fundamental to plant survival and ecosystem stability. Fungi obtain carbohydrates—products of photosynthesis—from their plant hosts, and in turn, they significantly enhance the plant's ability to absorb water and essential nutrients such as phosphorus and nitrogen from the soil.

In addition to nutrient exchange, these networks facilitate complex interplant communication. Pioneering research by Simard et al. (1997) showed that trees can transfer carbon to neighboring plants, even across species. This phenomenon suggests a form of cooperation within forest communities. Later studies expanded on this understanding, showing that mycorrhizal networks can also transmit biochemical signals, such as stress responses and allelochemicals, allowing plants to warn each other of pests, pathogens, or environmental stressors (Barto et al., 2011).

These large underground systems often revolve around "hub" trees (usually older, mature individuals) that act as central nodes in networks similar to human-designed communication systems. These scale-free networks are biologically resilient and efficient, supporting both ecological cooperation and competition (Beiler et al., 2010).

The spatial and temporal complexity of mycorrhizal interactions poses a major challenge to traditional ecological approaches. As a result, researchers are increasingly turning to advanced computational techniques, such as machine learning and network analysis, to reveal subtle patterns and dynamic relationships in these biological systems. These methods are enabling scientists to better understand forest communication, resilience, and the broader role of mycorrhizal networks in maintaining ecosystem health.

Artificial Intelligence in Ecological Research

In recent years, computational tools have become an integral part of ecological research, facilitating species identification, habitat mapping, and real-time ecosystem monitoring (Pimm et al., 2015). Among these tools, machine learning, especially deep learning, has shown extraordinary potential in extracting complex patterns from large, unstructured ecological datasets that are often difficult to process using traditional analytical methods (Christin et al., 2019). Convolutional neural networks (CNNs) have significantly advanced image-based species classification, enabling automated and accurate species identification from camera traps and satellite imagery (Norouzzadeh et al., 2018). Meanwhile, recurrent neural networks (RNNs) and their variants, such as long short-term memory (LSTM) networks, have demonstrated strong performance in modeling temporal patterns, making them ideal for predicting ecological trends and disturbances (Reichstein et al., 2019).

While these techniques have been primarily applied to aboveground ecological processes, such as wildlife monitoring and vegetation analysis, their applications in belowground ecosystems remain relatively unexplored. However, with the growing interest in understanding belowground networks, such as mycorrhizal communication systems, these advanced models are expected to become powerful tools for detecting hidden signals, simulating nonlinear interactions, and revealing the structure of complex belowground networks.

AI TECHNOLOGIES FOR MAPPING THE WOOD WIDE WEB

Sensing Technologies and Data Collection

Advanced sensing technologies form the backbone of computational approaches to studying mycorrhizal networks. These technologies can be broadly categorized into three key strategies:

1. In **situ sensor networks** involve the deployment of micro-devices directly into the soil to monitor key parameters such as moisture, temperature, chemical gradients, and electrical signals. These sensors form a dynamic network through wireless connections, collecting real-time, high-resolution spatiotemporal data (Cao et al., 2008). This continuous stream of data provides valuable insights into belowground ecological processes and is highly compatible with machine learning methods, which can detect hidden patterns and predict system behavior. The combination of sensor technology and computational models allows researchers to more effectively monitor the belowground environment, thereby gaining a deeper understanding of soil health, nutrient cycling, and mycorrhizal interactions.
2. **Molecular techniques** such as environmental DNA (eDNA) sampling and metabarcoding have become powerful tools for studying soil fungal communities. These methods allow researchers to collect and sequence genetic material directly from environmental samples, generating large datasets that capture biodiversity in great detail (Taberlet et al., 2012). When combined with computational analysis, especially deep learning models, these datasets can be mined to reveal species distributions, co-occurrence patterns, and potential network pathways that are often not discoverable through manual or traditional methods. Deep learning excels at processing complex, high-dimensional data, making it ideal for modeling belowground ecological networks and interactions. This combination opens new avenues for understanding the hidden dynamics of mycorrhizal systems and their role in forest ecosystems.
3. **Computer vision**-enhanced remote sensing techniques, such as hyperspectral imaging and lidar, provide non-invasive methods to detect subtle aboveground

indicators associated with belowground fungal activity (Asner & Martin, 2016). These indicators may include changes in canopy structure, leaf reflectance, or vegetation health that are associated with mycorrhizal presence and function. By capturing fine spectral and structural details over broad forest areas, these tools allow researchers to infer belowground processes without disturbing the ecosystem. When processed using intelligent algorithms such as machine learning and deep neural networks, these data can be transformed into detailed spatial maps of fungal networks. This approach can enable efficient and large-scale monitoring of mycorrhizal dynamics and their ecological impacts, thus bridging the gap between surface observations and belowground biological processes.

One notable example is the combination of soil microsensors and neural network analysis to detect carbon transfer between Douglas-fir seedlings, demonstrating the feasibility and effectiveness of real-time monitoring of resource exchanges in forest ecosystems.

Neural Networks for Network Mapping

The structural similarities between neural networks and mycorrhizal systems make computational models particularly useful for studying communication in underground forests. Several architectural strategies have emerged that are particularly promising:

1. **Graph neural networks (GNNs)** are powerful tools for analyzing complex interconnected systems such as mycorrhizal networks. They excel at identifying key structural features, such as central hub trees and frequently used fungal pathways, by modelling the relationships between nodes (representing trees, fungi, or soil patches) and their connections (Wu et al., 2022). By capturing both local and global patterns, GNNs are able to detect hidden interaction dynamics in forest ecosystems. Liu et al. (2021) applied GNNs to forest plot data and demonstrated that these models can accurately predict mycorrhizal connectivity based on factors such as spatial proximity. This approach provides a scalable and data-driven way to explore belowground symbiotic networks, allowing ecologists to map and monitor fungal interactions across diverse landscapes with unprecedented precision..
2. **Unsupervised learning techniques** including clustering algorithms and autoencoders, are valuable tools for revealing hidden patterns in ecological data without labeled inputs. In the study of mycorrhizal networks, these models have been used to analyze large, complex datasets and reveal naturally occurring structures in underground fungal systems (Xie et al., 2016). Clustering algorithms group similar network features based on common properties, while

autoencoders reveal key variations by reducing the dimensionality of the data. When applied to mycorrhizal systems, these techniques have successfully identified distinct structural patterns in different forest ecosystems, indicating that environmental factors and species composition shape fungal network structure in unique ways. This data-driven approach allows researchers to classify and compare mycorrhizal systems on a global scale, providing new insights into their ecological roles and adaptability.

3. **Bayesian networks** provide a powerful framework for modeling probabilistic relationships and assessing causal relationships in complex ecosystems. By representing variables as nodes and their dependencies as directed edges, these models can infer how changes in one factor affect another, even in the presence of uncertainty (Heckerman et al., 1995). In studies of intra-root communication in fungi, Bayesian networks are particularly useful for distinguishing true signaling (e.g., nutrient exchange or biochemical signaling) from coincidental correlations. This helps researchers identify the true pathways by which mycorrhizal fungi interact with their host plants, thereby elucidating the mechanisms of belowground communication. By integrating prior knowledge and observational data, Bayesian methods allow ecologists to move beyond descriptive analyses and build causal models that elucidate the functional dynamics of mycorrhizal systems.

Spatiotemporal Models of Network Dynamics.

Forest communication networks are dynamic systems that change and adapt over time rather than maintaining a fixed structure. A variety of computational methods have been developed to capture these changing interactions:

1. **Recurrent neural networks (RNNs)**, especially long short-term memory (LSTM) models, are well suited for analyzing time series data from soil sensors. These models can capture temporal dependencies and thus detect patterns of resource sharing and signaling events within mycorrhizal networks (Hochreiter & Schmidhuber, 1997). By processing continuous data streams (e.g., soil moisture, nutrient levels, and electrical signals), LSTMs can identify cycles, anomalies, and coordinated responses between plants and fungi. This capability allows for a deeper understanding of the timing and dynamics of belowground communication, providing valuable insights into how mycorrhizal systems respond to environmental changes.
2. **Agent-based models (ABMs)** are combined with reinforcement learning (RL) to simulate fungi and trees as autonomous decision-making agents that can adapt over time. Each agent interacts with its environment and learns the best strategy

through trial and error, guided by feedback signals (Railsback & Grimm, 2019). This hybrid approach allows researchers to explore how mycorrhizal networks evolve under different environmental conditions, such as nutrient availability or climate stress. By modeling individual behaviors and their collective outcomes, the ABM-RL framework can reveal emergent properties of fungus-plant interactions and predict how communication and resource-sharing strategies change in response to ecological stress.

3. **Digital twins** of forest environments integrate various data streams—such as sensor readings, remote sensing, and ecological models—to create a real-time virtual replica of a forest ecosystem (Jones et al., 2020). These dynamic models simulate the behavior of mycorrhizal networks and their interactions with trees and soils, allowing researchers to monitor the system's response to disturbances such as drought, disease, or logging. By continuously updating with real-time data, digital twins can accurately predict ecosystem resilience and recovery, making them a valuable tool for forest management and conservation planning. In one notable study, Zhang et al. (2021) applied a temporal convolutional neural network to soil condition data and achieved 83% accuracy in predicting fungal network growth, significantly outperforming traditional statistical methods.

APPLICATIONS IN FOREST ECOLOGY AND MANAGEMENT

Early Warning Systems for Forest Health

Using smart systems to monitor mycorrhizal networks is an emerging forest health assessment strategy that can provide critical early warnings before visible symptoms of stress emerge (Bueno et al., 2020). These systems rely on machine learning models (particularly neural networks) that are trained on large amounts of historical soil and sensor data. They enable researchers and forest managers to identify anomalies in patterns of belowground signals that often indicate:

1. **Pathogen spread** through fungal pathways before symptoms are evident on leaves or bark.
2. **Early drought responses**, detected via shifts in water and electrical signal transmission, long before wilting or discoloration appears.
3. **Nutrient imbalance signals** exchanged among interconnected tree systems that may reflect local soil deficiencies or competition.

To effectively implement this approach, the following guidelines are recommended:

- **Sensor deployment**: Install a dense in-situ network of soil sensors to measure moisture, temperature, conductivity, and nutrient levels at different depths.
- **Data integration**: Combine real-time data from sensors with historical baselines and remote sensing inputs to enhance model robustness.
- **Model training**: Use supervised and unsupervised neural networks (e.g., long short-term memory (LSTM), autoencoders) to identify temporal patterns and flag deviations.
- **Anomaly detection protocols**: Establish threshold criteria for pattern deviations to trigger alerts of potential forest stress or ecosystem disturbances.
- **Field validation**: Conduct controlled experiments to test model predictions, calibrate sensors, and confirm the biological relevance of detected anomalies.
- **Preventive action planning**: Develop rapid response strategies based on alerts, such as localized watering, disease mitigation, or nutrient adjustments.

Field studies in temperate forests have shown that such systems can detect disturbances up to three weeks before any visual signs appear (Reinsch et al., 2024). This early warning capability provides a powerful tool for proactive forest management, ecosystem resilience, and biodiversity conservation.

Climate Change Adaptation Strategies

Understanding the role of mycorrhizal networks in enhancing forest resilience is critical for developing effective climate adaptation strategies. These underground fungal root systems promote resource sharing and communication between trees, making forests more resilient to environmental stresses. Computational methods provide powerful tools to predict and optimize the performance of these networks under future climate scenarios..

1. **Predictive models**: Deep learning models trained on long-term climate, soil, and biological data can predict how rising temperatures and changing precipitation patterns will affect fungal connectivity, nutrient flows, and symbiotic efficiency (Gu et al., 2020). This enables proactive planning to mitigate disturbances.
2. **Optimization algorithms**: Tools such as genetic algorithms help identify the combinations of tree and fungal species that are best suited for reforestation or assisted migration projects. These strategies aim to maintain or enhance network stability in a changing environment (Johnson et al., 2020).
3. **Scenario simulations**: Agent-based models (ABMs) simulate the behavior of individual trees and fungi, allowing researchers to explore the outcomes of different forest management interventions, such as selective planting, thinning, or species mixing, and assess their long-term effects on the health of mycorrhizal

networks (Lehmann et al., 2021).In a practical example, Martinez-Garcia et al. (2022) applied reinforcement learning to optimize tree planting layouts. Their model promoted robust mycorrhizal interactions and led to a 27% improvement in seedling survival during drought simulations, demonstrating the real-world potential of AI-guided forest design.

Conservation of Keystone Network Components

Not all Wood Wide Web nodes contribute equally to the overall health and resilience of forest ecosystems. Using computational network analysis, researchers can identify key components, such as hub trees (which maintain connections with numerous other trees) and key fungal species that play a central role in facilitating nutrient exchange and signaling. These network elements function like major connection points in human-designed systems, meaning their disruption could lead to the fragmentation of the entire network.

Identifying and protecting these key nodes has strategic advantages, especially when conservation resources are limited. By focusing on key bottlenecks where fungal and tree connectivity is most vulnerable, forest managers can prioritize interventions that will deliver the greatest ecological benefits.

In an important study, Chen et al. (2021) demonstrated that protecting just 15% of trees identified as hubs through network centrality metrics was sufficient to preserve 78% of the overall communication capacity within forest mycorrhizal networks. This targeted conservation approach highlights the value of combining ecological knowledge with computational tools to enhance forest resilience, optimize resource allocation, and guide reforestation strategies.

Sustainable Forestry Practices

Commercial forestry has much to gain from integrating computational approaches that both maintain the integrity of mycorrhizal networks and enable sustainable timber harvests. These networks are essential to forest health, supporting nutrient cycling, water distribution, and tree regeneration. Disruption of these networks during harvesting can harm the long-term productivity of forests. However, emerging technologies now offer data-driven solutions to mitigate these risks.

A key application is the use of selective harvesting algorithms that can identify individual trees that can be removed with minimal impact on the overall mycorrhizal network structure (Wang et al., 2022). In addition, machine learning models can help optimize the spatial distribution of retained trees to ensure that core connections between fungi and tree species remain intact (Thompson et al., 2022). In addition, predictive systems can generate planting recommendations to accelerate the reestablishment of fungal connections after harvesting operations (Nguyen et al., 2021).

These computational strategies not only protect ecological functions, but also provide economic benefits. Recent economic modeling studies have shown that incorporating these tools into forest management can increase the long-term value of timber production by 18-23%, primarily due to increased regeneration rates and greater resilience of forests to changing environmental conditions (Harrison & López, 2022).

By aligning commercial objectives with ecological sustainability, these approaches signal a shift toward smarter forestry practices that balance profit with conservation.

NOVEL FRAMEWORK: THE FOREST NETWORK INTELLIGENCE SYSTEM (FNIS)

Based on existing technologies and research needs, we propose a comprehensive framework for applying artificial intelligence to forest communication networks. The Forest Network Intelligence System (FNIS) integrates multiple artificial intelligence approaches into a unified architecture, aiming to monitor, analyze, and support the management of the Wood World Wide Web.

System Architecture

The Forest Network Intelligence System (FNIS) is a novel multi-layered architecture designed to monitor, analyze, and manage the dynamics of mycorrhizal communication networks in forest ecosystems. It integrates real-time environmental perception, advanced artificial intelligence (AI), and decision support capabilities, and is divided into four functional layers:

1. **Sensing layer:** This foundational layer consists of a distributed network of soil sensors, regular molecular sampling units, and remote sensing platforms such as drones and satellites. These tools collect multimodal data such as temperature, soil moisture, chemical composition, and biomarkers to capture spatial and temporal changes in the fungus-root interface. The combination of local and large-scale sensing ensures a holistic understanding of the ecological conditions below and above ground.
2. **Processing layer:** To reduce latency and data overload, edge computing devices are deployed near the data collection point. These units perform initial filtering, noise reduction, and feature extraction to prepare data for efficient transmission. This distributed processing model enables fast response and saves bandwidth.

3. **Analysis layer:** This layer is the core of FNIS and hosts various AI algorithms, including graph neural networks (GNNs) for network topology analysis, convolutional neural networks (CNNs) for image-based fungal detection, and recurrent neural networks (RNNs) for interpreting time series sensor data. Together, these algorithms reveal network patterns, anomalies, and key attributes, providing a deep understanding of the ecological environment.
4. **4. Interface layer:** Ultimately, the processed insights will be transformed into a user-friendly format through visual dashboards, decision support tools, and automated alerts. These outputs will help forest managers and researchers make proactive, data-driven decisions to maintain forest health and resilience..

The FNIS framework exemplifies how intelligent systems can bridge ecological complexity and practical forest management in the age of AI.

Implementation Challenges and Solutions

Several challenges must be addressed for successful implementation of the FNIS framework:

Implementing intelligent monitoring systems in forest ecosystems presents several challenges, but recent innovations offer promising solutions:

1. **Power constraints**: Forest sensor networks often operate in remote areas where energy is scarce. Low-power machine learning algorithms combined with energy harvesting technologies such as solar and thermoelectric generators can significantly reduce energy consumption while maintaining performance, enabling long-term autonomous deployment.
2. **Data integration**: Given the diversity of forest types and environmental conditions, integrating data from multiple locations is critical to gain generalizable insights. Federated learning allows decentralized model training across different forest locations, minimizing the need for large-scale data transfer and enhancing privacy and efficiency.
3. **Model interpretability**: While neural networks can detect complex patterns, their decisions can be opaque. Combining attention mechanisms with explainable artificial intelligence (XAI) techniques makes it possible to trace how specific inputs affect outputs, making predictions more transparent and actionable for forest managers.
4. **Ecosystem impact**: The introduction of technology into natural ecosystems must minimize disturbance. The use of biodegradable sensors and low-impact deployment methods (e.g., drone placement or shallow soil burial) can help

reduce environmental disturbance and align research activities with conservation ethics..

These advancements support scalable, responsible AI applications for forest resilience and conservation.

Validation and Testing

We propose the design and deployment of a prototype Forest Network Intelligence System (FNIS) at a temperate forest research site, encompassing a 20-hectare mixed-age forest dominated by Douglas fir (*Pseudotsuga menziesii*) and western hemlock (*Tsuga heterophylla*). The FNIS architecture is intended to integrate in situ sensors, remote sensing platforms, molecular sampling techniques, and AI-driven analytics to monitor and interpret belowground communication between fungal networks and tree roots.

The proposed system aims to achieve the following:

1. **Mycorrhizal Prediction Using AI:** Leverage machine learning models to predict mycorrhizal connections with an expected accuracy of approximately 87%, based on non-invasive, above-ground features such as canopy structure, spatial distribution, and tree health indicators.
2. **Validation Through Isotope Labeling:** Incorporate experimental isotope tracing to confirm carbon transfer events. The system is designed to use real-time data from soil sensors and molecular sampling to track physiological exchanges within fungal-root networks.
3. **Temporal Pattern Analysis:** Utilize time-series analysis to detect and characterize daily rhythms in belowground communication, with particular attention to variations in signaling and nutrient flow linked to environmental variables like light cycles and temperature fluctuations.

FNIS is expected to lead to novel ecological insights and practical forestry applications. By revealing hidden dynamics in forest ecosystems, the system provides actionable intelligence for precision forestry, enabling early detection of tree stress, mapping of key fungal pathways, and predicting resource allocation patterns. This represents a transformative shift toward forest management as a smart, connected system rather than a collection of individual trees. The continued development and refinement of the FNIS framework is expected to advance ecological science and sustainable forest management practices.

ETHICAL CONSIDERATIONS AND FUTURE DIRECTIONS

Ethical Dimensions

The integration of artificial intelligence (AI) into natural systems, particularly forest ecosystems, presents several **ethical considerations** that must guide responsible research and application.

1. **Non-disruption Principle**: Technologies used to monitor mycorrhizal networks must prioritize minimal ecological interference. This includes using biodegradable materials, non-invasive sensors, and low-impact deployment methods to ensure that the act of observation does not alter the very processes being studied.
2. **Data Openness vs. Forest Protection**: While open data sharing enhances scientific collaboration, revealing the exact locations of critical network hubs or keystone species may expose these components to threats such as poaching, illegal logging, or commercial exploitation. Ethical frameworks must weigh the benefits of transparency against the need to safeguard vulnerable ecosystems.
3. **Indigenous Knowledge Integration**: Forests have long been understood through the lens of traditional ecological knowledge (TEK) held by Indigenous communities. AI systems should respectfully integrate TEK, recognizing its value in understanding ecological relationships, and ensure proper attribution and informed consent. Collaborations must be co-designed with Indigenous stakeholders, ensuring that benefits are shared and cultural perspectives are honored.

Addressing these ethical issues is essential to ensure that AI supports sustainable forest stewardship while respecting ecological integrity and cultural knowledge systems.

Technological Horizons

Future Developments in AI Applications for Forest Network Analysis

Advancing the application of AI in studying mycorrhizal and forest communication networks will rely on several **emerging technologies** that promise to overcome current limitations and open new frontiers:

1. **Quantum Computing Approaches**: Traditional computing struggles with the massive complexity of ecological networks, especially those with dynamic, non-linear interactions. Quantum computing offers the potential to simulate

entangled relationships and multidimensional variables simultaneously, enabling researchers to model mycorrhizal interactions at ecosystem scales with vastly improved speed and precision.

2. **Biological-Digital Interfaces**: Future systems may incorporate bio-hybrid sensors capable of directly interfacing with chemical signaling pathways in fungal networks. These interfaces could detect real-time fluctuations in allelochemicals, defense signals, or hormonal exchanges, offering a more granular understanding of plant-fungal communication without relying solely on proxies like carbon flux or electrical signal.
3. **Autonomous Ecological Monitoring Systems**: By integrating machine learning with robotics, future platforms may adapt their sampling strategies in real-time based on ongoing network activity. For instance, mobile sensor units could reposition themselves to areas showing unusual signal patterns or anticipated stress responses, improving data relevance and efficiency.

Together, these innovations could significantly enhance our ability to **predict, protect, and manage** forest ecosystems in the face of accelerating environmental change.

Research Priorities

To advance this field, we identify five priority research directions:

1. Standardizing data collection protocols to enable cross-site machine learning applications
2. Developing transfer learning approaches that can apply insights across different forest types
3. Creating open-source software platforms specifically designed for mycorrhizal network analysis
4. Establishing long-term monitoring sites with integrated AI systems
5. Investigating potential applications for urban forest management and restoration ecology

CONCLUSION

The integration of computational intelligence with the study of forest communication networks marks a transformative step toward understanding ecosystem dynamics. Using neural models to analyze the Wood Wide Web provides researchers with unprecedented insights into the intricate relationships that sustain forest communities. These insights support the development of more effective conservation methods, climate adaptation strategies, and sustainable forest management.

The technologies and frameworks discussed here reveal a powerful synergy between artificial neural networks and their biological counterparts in mycorrhizal systems. As advances continue in both areas, we expect increasingly sophisticated applications that will enhance our understanding of forests as interconnected systems rather than mere collections of individual trees.

Future work should prioritize standardizing approaches, broadening the geographic and ecological reach of computational applications, and creating accessible tools for forest managers. By combining computational and natural intelligence, we can achieve a fuller understanding of forest communication and develop more effective ways to protect these critical ecosystems amid global change.

REFERENCES

Asner, G. P., & Martin, R. E. (2016). Spectranomics: Emerging science and conservation opportunities at the interface of biodiversity and remote sensing. *Global Ecology and Conservation*, *8*, 212–219.

Barto, E. K., Hilker, M., Müller, F., Mohney, B. K., Weidenhamer, J. D., & Rillig, M. C. (2011). The fungal fast lane: Common mycorrhizal networks extend bioactive zones of allelochemicals in soils. *PLoS One*, *6*(11), e27195. PMID: 22110615

Bever, J. D., Dickie, I. A., Facelli, E., Facelli, J. M., Klironomos, J., Moora, M., & Zobel, M. (2010). Rooting theories of plant community ecology in microbial interactions. *Trends in Ecology & Evolution*, *25*(8), 468–478. PMID: 20557974

Bueno, C. G., Pickles, B. J., Truong, C., & Watts-Williams, S. J. (2020). Mycorrhizas for a sustainable world. *The New Phytologist*, *225*(3), 1065–1069. PMID: 31894588

Cao, H., Gerhold, K., Mayers, J. R., Wiest, M. M., Watkins, S. M., & Hotamisligil, G. S. (2008). Identification of a lipokine, a lipid hormone linking adipose tissue to systemic metabolism. *Cell*, *134*(6), 933–944. PMID: 18805087

Chen, T., Pan, H., Lu, M., Hang, J., Lam, C. K. C., Yuan, C., & Pearlmutter, D. (2021). Effects of tree plantings and aspect ratios on pedestrian visual and thermal comfort using scaled outdoor experiments. *The Science of the Total Environment*, *801*, 149527. PMID: 34416606

Christin, S., Hervet, É., & Lecomte, N. (2019). Applications for deep learning in ecology. *Methods in Ecology and Evolution*, *10*(10), 1632–1644.

Gu, S., Kelly, B., & Xiu, D. (2020). Empirical asset pricing via machine learning. *Review of Financial Studies*, *33*(5), 2223–2273.

Heckerman, D., Geiger, D., & Chickering, D. M. (1995). Learning Bayesian networks: The combination of knowledge and statistical data. *Machine Learning*, *20*, 197–243.

Hochreiter, S., & Schmidhuber, J. (1997). Long short-term memory. *Neural Computation*, *9*(8), 1735–1780. PMID: 9377276

Johnson, F., Valderrama, A., Valle, C., Crawford, B., Soto, R., & Ñanculef, R. (2020). Automating configuration of convolutional neural network hyperparameters using genetic algorithm. *IEEE Access : Practical Innovations, Open Solutions*, *8*, 156139–156152.

Jones, J. A., Wei, X., Archer, E., Bishop, K., Blanco, J. A., Ellison, D., ... & Creed, I. F. (2020). Forest-water interactions under global change. Forest-water interactions, 589-624.

Lehmann, J., Cowie, A., Masiello, C. A., Kammann, C., Woolf, D., Amonette, J. E., & Whitman, T. (2021). Biochar in climate change mitigation. *Nature Geoscience*, *14*(12), 883–892.

Liu, X., Jin, W., Ma, Y., Li, Y., Liu, H., Wang, Y., & Tang, J. (2021, July). Elastic graph neural networks. In *International Conference on Machine Learning* (pp. 6837-6849). PMLR.

Nguyen, T., Novak, R., Xiao, L., & Lee, J. (2021). Dataset distillation with infinitely wide convolutional networks. *Advances in Neural Information Processing Systems*, *34*, 5186–5198.

Norouzzadeh, M. S., Nguyen, A., Kosmala, M., Swanson, A., Palmer, M. S., Packer, C., & Clune, J. (2018). Automatically identifying, counting, and describing wild animals in camera-trap images with deep learning. *Proceedings of the National Academy of Sciences of the United States of America*, *115*(25), E5716–E5725. PMID: 29871948

Pimm, S. L., Alibhai, S., Bergl, R., Dehgan, A., Giri, C., Jewell, Z., & Loarie, S. (2015). Emerging technologies to conserve biodiversity. *Trends in Ecology & Evolution*, *30*(11), 685–696. PMID: 26437636

Railsback, S. F., & Grimm, V. (2019). *Agent-based and individual-based modeling: a practical introduction*. Princeton university press.

Reichstein, M., Camps-Valls, G., Stevens, B., Jung, M., Denzler, J., Carvalhais, N., & Prabhat, F. (2019). Deep learning and process understanding for data-driven Earth system science. *Nature*, *566*(7743), 195–204. PMID: 30760912

Reinsch, S., Robinson, D. A., van Soest, M. A., Keith, A. M., Parry, S., & Tye, A. M. (2024). Temperate soils exposed to drought—Key processes, impacts, indicators, and unknowns. *Land (Basel)*, *13*(11), 1759.

Selosse, M. A., Richard, F., He, X., & Simard, S. W. (2006). Mycorrhizal networks: Des liaisons dangereuses? *Trends in Ecology & Evolution*, *21*(11), 621–628. PMID: 16843567

Simard, S. W., Perry, D. A., Jones, M. D., Myrold, D. D., Durall, D. M., & Molina, R. (1997). Net transfer of carbon between ectomycorrhizal tree species in the field. *Nature*, *388*(6642), 579–582.

Smith, S. E., & Read, D. J. (2010). *Mycorrhizal symbiosis*. Academic press.

Wang, Y., He, X., & Yu, F. (2022). Non-host plants: Are they mycorrhizal networks players? *Plant Diversity*, *44*(2), 127–134. PMID: 35505991

Wu, S., Sun, F., Zhang, W., Xie, X., & Cui, B. (2022). Graph neural networks in recommender systems: A survey. *ACM Computing Surveys*, *55*(5), 1–37.

Xie, J., Girshick, R., & Farhadi, A. (2016, June). Unsupervised deep embedding for clustering analysis. In *International conference on machine learning* (pp. 478-487). PMLR.

Zhang, Z., Yu, X., Rong, X., & Iwata, M. (2021). Spatial-temporal neural network for P300 detection. *IEEE Access : Practical Innovations, Open Solutions*, *9*, 163441–163455.

Compilation of References

Aamir, M., Bazai, S., Bhatti, U. A., Dayo, Z. A., Liu, J., & Zhang, K. (2023). Applications of machine learning in medicine: Current trends and prospects. In *2023 Global Conference on Wireless and Optical Technologies (GCWOT)* (pp. 1-4). IEEE.

Aamir, M., & Bhatti, U. A. (2025). Remote sensing for climate monitoring and agriculture. *Computers, Materials & Continua*, *75*(3).

Aamir, M., Gao, J., Zhang, Y., Tang, H., & Bhatti, U. A. (2024). Overview of Underwater Object Detection Based on Image Processing Techniques. In *7th International Conference on Pattern Recognition and Artificial Intelligence*.

Aamir, M., Li, Z., Bazai, S., Wagan, R. A., Bhatti, U. A., Nizamani, M. M., & Akram, S. (2021). Spatiotemporal change of air-quality patterns in Hubei province—A pre-to post-COVID-19 analysis using path analysis and regression. *Atmosphere*, *12*(10), 1338.

Aamir, M., Rahman, Z., Abro, W. A., Bhatti, U. A., Dayo, Z. A., & Ishfaq, M. (2023). A progressive approach to generic object detection: A two-stage framework for image recognition. *Computers, Materials & Continua*, *75*(3).

Aamir, M., Rahman, Z., Abro, W. A., Bhatti, U. A., Dayo, Z. A., & Ishfaq, M. (2023). Brain tumor classification utilizing deep features derived from high-quality regions in MRI images. *Biomedical Signal Processing and Control*, *85*, 104988.

Aarthi, R., & Vinayagasundaram, B. (2022). Effective management of class imbalance problem in climate data analysis using a hybrid of deep learning and data level sampling. *Journal of Intelligent & Fuzzy Systems*, *43*(4), 4187–4199. DOI: 10.3233/jifs-210666

Abdul, S., Adeghe, E. P., Adegoke, B. O., Adegoke, A. A., & Udedeh, E. H. (2024). AI-enhanced healthcare management during natural disasters: Conceptual insights. *Engineering Science & Technology Journal*, *5*(5), 1794–1816.

Abouhalima, M., das Neves, L., Taveira-Pinto, F., & Rosa-Santos, P. (2024). Machine learning in Coastal Engineering: Applications, challenges, and perspectives. *Journal of Marine Science and Engineering*, *12*(4), 638.

Adewale, T. (2024). How AI Training Methods Are Evolving : From Supervised Learning to Reinforcement and Beyond. November.

Adriano, B., Yokoya, N., Xia, J., Miura, H., Liu, W., Matsuoka, M., & Koshimura, S. (2020). Learning from Multimodal and Multitemporal Earth Observation Data for Building Damage Mapping (No. arXiv:2009.06200). arXiv. https://doi.org// arXiv.2009.06200DOI: 10.48550

Agastya, C., Ghebremusse, S., Anderson, I., Reed, C., Vahabi, H., & Todeschini, A. (2021). Self-supervised Contrastive Learning for Irrigation Detection in Satellite Imagery. 1. https://arxiv.org/abs/2108.05484

Agbehadji, I., Mabhaudhi, T., Botai, J., & Masinde, M. (2023). A systematic review of existing early warning systems' challenges and opportunities in cloud computing early warning systems. *Climate (Basel)*, *11*(9), 188. DOI: 10.3390/cli11090188

Aggarwal, L., & Papay, F. (2022). Applications of multispectral and hyperspectral imaging in dermatology. *Experimental Dermatology*. Advance online publication. DOI: 10.1111/exd.14624 PMID: 35671110

Ahasnain, A., Hashmi, M. Z., Khan, S., Bhatti, U. A., Min, X., Yue, Y., & He, Y. (2024). Predicting ambient PM2.5 concentrations via time series models in Anhui Province, China. *Environmental Monitoring and Assessment*, *196*(5), 487. PMID: 38687422

Akhare, R. (2023). Hybrid approach between Supervised and Unsupervised Learning: Self Supervised Learning. *Article in Journal of Xidian University*, *17*(8), 1–9. DOI: 10.37896/jxu17.8/026

Akindote, O. J., Adegbite, A. O., Dawodu, S. O., Omotosho, A., Anyanwu, A., & Maduka, C. P. (2023). Comparative review of big data analytics and GIS in healthcare decision-making. *World Journal of Advanced Research and Reviews*, *20*(3), 1293–1302.

Al Shafian, S., & Hu, D. (2024). Integrating Machine Learning and Remote Sensing in Disaster Management: A Decadal Review of Post-Disaster Building Damage Assessment. *Buildings (Basel, Switzerland)*, *14*(8), 8. Advance online publication. DOI: 10.3390/buildings14082344

Alam, F., Ofli, F., & Imran, M. (2018). Crisismmd: Multimodal twitter datasets from natural disasters. Proceedings of the International AAAI Conference on Web and Social Media, 12(1). https://ojs.aaai.org/index.php/ICWSM/article/view/14983

Alam, F., Ofli, F., Imran, M., Alam, T., & Qazi, U. (2020). Deep Learning Benchmarks and Datasets for Social Media Image Classification for Disaster Response. *2020 IEEE/ACM International Conference on Advances in Social Networks Analysis and Mining (ASONAM)*, 151–158. https://doi.org/DOI: 10.1109/ASONAM49781.2020.9381294

Alam, F., Qazi, U., Imran, M., & Ofli, F. (2021). HumAID: Human-Annotated Disaster Incidents Data from Twitter with Deep Learning Benchmarks. *Proceedings of the International AAAI Conference on Web and Social Media*, 15, 933–942. https://doi.org/DOI: 10.1609/icwsm.v15i1.18116

Alam, M. A., Sohel, A., Uddin, M. M., & Siddiki, A. (2024). Big Data and Chronic Disease Management Through Patient Monitoring And Treatment With Data Analytics. Academic Journal on Artificial Intelligence, Machine Learning. *Data Science and Management Information Systems*, *1*(01), 77–94.

Albahri, A. S., Khaleel, Y. L., Habeeb, M. A., Ismael, R. D., Hameed, Q. A., Deveci, M., Homod, R. Z., Albahri, O. S., Alamoodi, A. H., & Alzubaidi, L. (2024). A systematic review of trustworthy artificial intelligence applications in natural disasters. *Computers & Electrical Engineering*, *118*, 109409.

Alemayehu, M., Workie, S., Belew, M., Agimas, M., Derseh, N., Abuhay, H., & Bazezew, A. (2024). Knowledge towards the health impacts of climate change and associated factors among undergraduate health sciences students in amhara region: A multi-centered study. *Frontiers in Environmental Health*, *3*. Advance online publication. DOI: 10.3389/fenvh.2024.1363395

Algiriyage, N., Prasanna, R., Stock, K., Doyle, E. E. H., & Johnston, D. (2021). Multi-source Multimodal Data and Deep Learning for Disaster Response: A Systematic Review. *SN Computer Science*, *3*(1), 92. DOI: 10.1007/s42979-021-00971-4 PMID: 34870241

Alhatami, E., Bhatti, U. A., Huang, M. X., & Feng, S. L. (2023). Review and enhancement of discrete cosine transform (DCT) for medical image fusion. *Proceedings of International Conference on Information Technology and ….*

Alhatami, E., Bhatti, U. A., Mengxing, H., & Feng, S. (2024). Advanced fuzzy denoising technique for agricultural remote sensing: Modified partition filter for suppressing impulsive noise. *IEEE Access : Practical Innovations, Open Solutions*, *12*, 6721–6735.

Ali, A. H., & Thakkar, R. (2023). Climate changes through data science: Understanding and mitigating environmental crisis. *Mesopotamian Journal of Big Data, 2023*, 125–137.

Allmer, J. (2014). miRNomics. *MicroRNA Biology and Computational Analysis., 1107*, 333. DOI: 10.1007/978-1-62703-748-8

Al-Mhiqani, M. N., Ahmad, R., Zainal Abidin, Z., Yassin, W., Hassan, A., Abdulkareem, K. H., & Yunos, Z. (2020). A review of insider threat detection: Classification, machine learning techniques, datasets, open challenges, and recommendations. *Applied Sciences (Basel, Switzerland), 10*(15), 5208.

Alzubi, J., Nayyar, A., & Kumar, A. (2018). Machine Learning from Theory to Algorithms: An Overview. Journal of Physics: Conference Series, 1142(1), 0–15. https://doi.org/DOI: 10.1088/1742-6596/1142/1/012012

Anand, A. (2025). Machine learning and artificial intelligence in environmental change prediction. *Interantional Journal of Scientific Research in Engineering and Management, 09*(01), 1–9. DOI: 10.55041/ijsrem40801

Anderegg, W., Prall, J., Harold, J., & Schneider, S. (2010). Expert credibility in climate change. *Proceedings of the National Academy of Sciences of the United States of America, 107*(27), 12107–12109. DOI: 10.1073/pnas.1003187107 PMID: 20566872

An, S., Kim, H., & Kim, S. (2021). Rate-dependent hysteresis of the atlantic meridional overturning circulation system and its asymmetric loop. *Geophysical Research Letters, 48*(1). Advance online publication. DOI: 10.1029/2020gl090132

Ashok, A., Lambhate, D., & Singh, J. (2022). Self-Supervised Representations of Geolocated Weather Time Series-an Evaluation and Analysis. www.aaai.org

Asif, M., & Bhatti, U. A. (2025). Evaluating the effectiveness of green finance mechanisms, digitalization, globalization, and renewable energy consumption in advancing environmental sustainability in G20 countries. *Energy & Environment.*

Asif, M., Song, H., Zia, M. A., Ali, S., Awwad, E. M., Ghadi, Y. Y., Al-Razgan, M., Sadiq, T., & Bhatti, U. A. (2024). Shedding light on diagnostic precision: GANs for low light endoscopy image enhancements. Research. https://doi.org/DOI: 10.21203/rs.3.rs-421332s1/v1

Asner, G. P., & Martin, R. E. (2016). Spectranomics: Emerging science and conservation opportunities at the interface of biodiversity and remote sensing. *Global Ecology and Conservation, 8*, 212–219.

Astruc, G., Gonthier, N., Mallet, C., & Landrieu, L. (2024). OmniSat: Self-Supervised Modality Fusion for Earth Observation. https://doi.org/DOI: 10.1007/978-3-031-73390-1_24

Avalon-Cullen, C., Caudill, C., Newlands, N. K., & Enenkel, M. (2023). Big data, small island: Earth observations for improving flood and landslide risk assessment in Jamaica. *Geosciences*, *13*(3), 64.

Bahlali, A. R., & Bachir, A. (2025). Semi-Supervised Approach for Encrypted Traffic Anomaly Self-Supervised Learning Meets Custom Autoencoder : A Semi-Supervised Approach for Encrypted Traffic Anomaly Detection.

Bai, Z., Zhang, Y., Bhatti, U. A., & Liu, J. (2024). Feature Enhanced fusion of U-Net and VIT Transformer for Brain Tumor MRI Image Segmentation. *IEEE Journal of Biomedical and Health Informatics*.

Baloch, L., Bazai, S. U., Marjan, S., Aftab, F., & Aslam, S. (2023). A Review of Big Data Trends and Challenges in Healthcare. 14(May), 1320–1333. https://doi .org/DOI: 10.14716/ijtech.v14i6.6643

Barry, J. P., & Dayton, P. K. (1991). Physical heterogeneity and the organization of marine communities. In *Ecological heterogeneity* (pp. 270–320). Springer New York.

Barto, E. K., Hilker, M., Müller, F., Mohney, B. K., Weidenhamer, J. D., & Rillig, M. C. (2011). The fungal fast lane: Common mycorrhizal networks extend bioactive zones of allelochemicals in soils. *PLoS One*, *6*(11), e27195. PMID: 22110615

Basit, A., Siddique, M. A., Bashir, S., Naseer, E., & Sarfraz, M. S. (2024). Deep Learning-Based Detection of Oil Spills in Pakistan's Exclusive Economic Zone from January 2017 to December 2023. *Remote Sensing*, *16*(13), 13. Advance online publication. DOI: 10.3390/rs16132432

Bazai, S. U., & Jang-Jaccard, J. (2019). SparkDA: RDD-based high-performance data anonymization technique for Spark platform. In International conference on network and system security (pp. 646-662). Cham: Springer International Publishing.

Bazai, S. U., & Jang-Jaccard, J. (2020). In-memory data anonymization using scalable and high performance rdd design. *Electronics (Basel)*, *9*(10), 1732.

Bazai, S. U., Jang-Jaccard, J., & Alavizadeh, H. (2021). A novel hybrid approach for multi-dimensional data anonymization for apache spark. *ACM Transactions on Privacy and Security*, *25*(1), 1–25.

Bazai, S. U., Jang-Jaccard, J., & Wang, R. (2017). Anonymizing k-NN classification on MapReduce. *International Conference on Mobile Networks and Management*, 364–377.

Beechie, T., Pess, G., Roni, P., & Giannico, G. (2008). Setting river restoration priorities: A review of approaches and a general protocol for identifying and prioritizing actions. *North American Journal of Fisheries Management, 28*(3), 891–905.

Belcore, E., Latella, M., Piras, M., & Camporeale, C. (2024). Enhancing precision in coastal dunes vegetation mapping: Ultra-high resolution hierarchical classification at the individual plant level. *International Journal of Remote Sensing, 45*(13), 4527–4552.

Benedetti, P., Ienco, D., Gaetano, R., Ose, K., Pensa, R. G., & Dupuy, S. (2018). M^3\ textFusion: A Deep Learning Architecture for Multiscale Multimodal Multitemporal Satellite Data Fusion. *IEEE Journal of Selected Topics in Applied Earth Observations and Remote Sensing, 11*(12), 4939–4949. DOI: 10.1109/JSTARS.2018.2876357

Bengio, Y. (2012, June). Deep learning of representations for unsupervised and transfer learning. In Proceedings of ICML workshop on unsupervised and transfer learning (pp. 17-36). JMLR Workshop and Conference Proceedings.

Bever, J. D., Dickie, I. A., Facelli, E., Facelli, J. M., Klironomos, J., Moora, M., & Zobel, M. (2010). Rooting theories of plant community ecology in microbial interactions. *Trends in Ecology & Evolution, 25*(8), 468–478. PMID: 20557974

Bhatti, M. A., Zeeshan, Z., Syam, M. S., Bhatti, U. A., Khan, A., Ghadi, Y. Y., & Alsenan, S. (2024a). Advanced plant disease segmentation in precision agriculture using optimal dimensionality reduction with fuzzy C-means clustering and deep learning. IEEE Journal of Selected Topics in Applied Earth Observations and Remote Sensing.

Bhatti, M. A., Zeeshan, Z., Syam, M. S., Bhatti, U. A., Khan, A., Ghadi, Y. Y., & Alsenan, S. (2024d). Advanced plant disease segmentation in precision agriculture using optimal dimensionality reduction with fuzzy C-means clustering and deep learning. IEEE Journal of Selected Topics in Applied Earth Observations and Remote Sensing.

Bhatti, U. A., Ali, S., & Aslam, U. (2018). Data-driven switched-gain observation and control. *Nonlinear Dynamics*, •••, 1–19.

Bhatti, U. A., Bhatti, M. A., Tang, H., Syam, M. S., Awwad, E. M., Sharaf, M., & Ghadi, Y. Y. (2024c). Global production patterns: Understanding the relationship between greenhouse gas emissions, agriculture greening and climate variability. *Environmental Research*. PMID: 38169167

Bhatti, U. A., Hao, T., Khan, A., Ghadi, Y. Y., Bhatti, M. A., & Khan, K. A. (2024b). Investigating the nexus between energy, socio-economic factors and environmental pollution: A geo-spatial multi-regression approach. *Gondwana Research*.

Bhatti, U. A., Hashmi, M. Z., Sun, Y., Masud, M., & Nizamani, M. M. (2023). Artificial intelligence applications in reduction of carbon emissions: Step towards sustainable environment. [). Frontiers Media SA.]. *Frontiers in Environmental Science*, *11*, 1183620.

Bhatti, U. A., & Hasnain, A. (2025). Modeling and assessing environmental change using remote sensing and machine learning. *Computers, Materials & Continua*.

Bhatti, U. A., Huang, M., Wang, H., Zhang, Y., Mehmood, A., & Di, W. (2018). Recommendation system for immunization coverage and monitoring. *Human Vaccines & Immunotherapeutics*, *14*(1), 165–171. PMID: 29068748

Bhatti, U. A., Huang, M., Wu, D., Zhang, Y., Mehmood, A., & Han, H. (2019). Recommendation system using feature extraction and pattern recognition in clinical care systems. *Enterprise Information Systems*, *13*(3), 329–351.

Bhatti, U. A., Li, D., & Tang, H. (2025). Oceanic biometric recognition algorithm based on generalized zero-shot learning. In *Deep Learning for Earth Observation and Climate Monitoring* (pp. 249–262). CRC Press.

Bhatti, U. A., Li, J., Huang, M., & Ahmad, R. M. (2024). Remote-sensing image classification: A comprehensive review and applications. In *Deep Learning for Multimedia Processing Applications* (pp. 18–47). CRC Press.

Bhatti, U. A., Li, J., & Nawaz, S. A. (2023). *Deep Learning for Multimedia Processing Applications: Volume One: Image Security and Intelligent Systems for Multimedia Processing*. CRC Press.

Bhatti, U. A., Li, J., & Zhang, Y. (2024). Enhancing lung cancer diagnosis with data fusion and mobile edge computing using DenseNet and CNN. *IEEE Transactions on Biomedical Engineering*, •••, 1–4.

Bhatti, U. A., & Liu, J. (2019). *Medical image watermarking algorithm based on DCT-DFRFT*. Innovation in Medicine and Healthcare Systems, and Multimedia.

Bhatti, U. A., Marjan, S., Wahid, A., Syam, M. S., Huang, M., Tang, H., & Hasnain, A. (2023). The effects of socioeconomic factors on particulate matter concentration in China's: New evidence from spatial econometric model. *Journal of Cleaner Production*, *417*, 137969.

Bhatti, U. A., Masud, M., Bazai, S. U., & Tang, H. (2023). Investigating AI-based smart precision agriculture techniques. [). Frontiers Media SA.]. *Frontiers in Plant Science*, *14*, 1237783. PMID: 37496855

Bhatti, U. A., Mengxing, H., Li, J., Bazai, S. U., & Aamir, M. (2024). *Deep learning for multimedia processing applications: Volume two: Signal processing and pattern recognition*. CRC Press.

Bhatti, U. A., Mengxing, H., Neira-Molin, H., Marjan, S., Baryalai, M., Hao, T., & Wu, G. (2023). MFFCG–Multi feature fusion for hyperspectral image classification using graph attention network. *Expert Systems with Applications*, •••, 120496.

Bhatti, U. A., Ming-Quan, Z., Qing-Song, H., Ali, S., Hussain, A., Yuhuan, Y., & Yu, Z. (2021). Advanced color edge detection using Clifford algebra in satellite images. *IEEE Photonics Journal*, *13*(2), 1–20.

Bhatti, U. A., Song, Z., & Hasnain, M. A. (2023). Predicting the impact of change in air quality patterns due to COVID-19 lockdown policies in multiple urban cities of Henan: A deep learning approach. *Atmosphere*, *14*(5), 902.

Bhatti, U. A., Tang, H., Wu, G., Marjan, S., & Hussain, A. (2023). Deep learning with graph convolutional networks: An overview and latest applications in computational intelligence. *International Journal of Intelligent Systems*, *2023*, 278.

Bhatti, U. A., Yan, Y., Zhou, M., Ali, S., Hussain, A., Qingsong, H., & Yu, Z. (2021). Time series analysis and forecasting of air pollution particulate matter (PM2.5): An SARIMA and factor analysis approach. *IEEE Access : Practical Innovations, Open Solutions*, *9*, 41019–41031.

Bhatti, U. A., Yu, K., Mamyrbayev, O. Z., Aitkazina, A., Hao, T., & Zhumazhan, N. O. (2024). Recommendations for Healthcare: An Interpretable Approach Using Deep Learning. *7th International Conference on Pattern Recognition and Artificial Intelligence*.

Bhatti, U. A., Yu, Z., Chanussot, J., Zeeshan, Z., Yuan, L., Luo, W., & Nawaz, S. A.. (2021). Local similarity-based spatial–spectral fusion hyperspectral image classification with deep CNN and Gabor filtering. *IEEE Transactions on Geoscience and Remote Sensing*, *60*, 1–15.

Bhatti, U. A., Yu, Z., Yuan, L., Zeeshan, Z., Nawaz, S. A., Bhatti, M., & Mehmood, A. (2020). Geometric algebra applications in geospatial artificial intelligence and remote sensing image processing. *IEEE Access : Practical Innovations, Open Solutions*, *8*, 155783–155796.

Bhatti, U. A., Zeeshan, Z., Nizamani, M. M., Bazai, S., Yu, Z., & Yuan, L. (2022). Assessing the change of ambient air quality patterns in Jiangsu Province of China pre-to post-COVID-19. *Chemosphere*, *288*, 132569. PMID: 34655644

Bhatti, U. A., Zhang, Y., Feng, W., & Huang, M. (2016). Research on the smartphone based eHealth systems for strengthening healthcare organization. *International Conference on Smart Health*, 91-101.

Bhatti, U., Bazai, S., Hussain, S., Fakhar, S., Ku, C., Marjan, S., Por, Y., & Jing, L. (2023). Deep learning-based trees disease recognition and classification using hyperspectral data. *Computers, Materials & Continua*, *77*, 681–697.

Brown, K. M., Hambidge, C. H., & Brownett, J. M. (2016). Progress in operational flood mapping using satellite synthetic aperture radar (SAR) and airborne light detection and ranging (LiDAR) data. *Progress in Physical Geography*, *40*(2), 196–214.

Bueno, C. G., Pickles, B. J., Truong, C., & Watts-Williams, S. J. (2020). Mycorrhizas for a sustainable world. *The New Phytologist*, *225*(3), 1065–1069. PMID: 31894588

Cai, J., Chatoux, H., Boust, C., & Mansouri, A. (2021). Extending the unmixing methods to multispectral images. *Color and Imaging Conference*, 29(1), 311-316. DOI: 10.2352/issn.2169-2629.2021.29.311

Camilleri, S., De Giglio, M., Stecchi, F., & Pérez-Hurtado, A. (2017). Land use and land cover change analysis in predominantly man-made coastal wetlands: Towards a methodological framework. *Wetlands Ecology and Management*, *25*, 23–43.

Cao, S., Joshi, D., Gui, L. Y., & Wang, Y. X. (2023). HASSOD: Hierarchical Adaptive Self-Supervised Object Detection. Advances in Neural Information Processing Systems, 36(NeurIPS).

Cao, H., Gerhold, K., Mayers, J. R., Wiest, M. M., Watkins, S. M., & Hotamisligil, G. S. (2008). Identification of a lipokine, a lipid hormone linking adipose tissue to systemic metabolism. *Cell*, *134*(6), 933–944. PMID: 18805087

Cao, P., Li, J., Liu, J., Li, Y., & Bhatti, U. A. (2023). Robust Zero Watermarking Algorithm for Medical Images Based on STAR-TEBLID-DCT. *International KES Conference on Innovation in Medicine and Healthcare*, 231-245.

Carlton, J., Perry-Hill, R., Huber, M., & Prokopy, L. (2015). The climate change consensus extends beyond climate scientists. *Environmental Research Letters, 10*(9), 094025. DOI: 10.1088/1748-9326/10/9/094025

Catthoor, F., & Danckaert, K. (2002). *Data access and storage management for embedded programmable processors*. Springer Science & Business Media.

ÇELİK, Ö. (2018). A Research on Machine Learning Methods and Its Applications. *Journal of Educational Technology and Online Learning, 1*(3), 25–40. DOI: 10.31681/jetol.457046

Chammas, S., Wang, M., Schneider, T., Ihme, M., Chen, Y., & Anderson, J. (2023). Accelerating large-eddy simulations of clouds with tensor processing units. *Journal of Advances in Modeling Earth Systems, 15*(10). Advance online publication. DOI: 10.1029/2023ms003619

Chander, B., & Gopalakrishnan, K. (2023). Data clustering using unsupervised machine learning. In *Statistical Modeling in Machine Learning* (pp. 179–204). Academic Press.

Chen, F., Zhang, W., Fawzi, M., Mfarrej, B., & Hamzah, M. (2024). Ecotoxicology and Environmental Safety Breathing in danger : Understanding the multifaceted impact of air pollution on health impacts. 280(December 2023). https://doi.org/ DOI: 10.1016/j.ecoenv.2024.116532

Chen, H., Gouin-Vallerand, C., Bouchard, K., Gaboury, S., Ngankam, H. K., Lussier, M., Couture, M., Bier, N., & Giroux, S. (2024). Utilizing Self-Supervised Learning for Recognizing Human Activity in Older Adults through Labeling Applications in Real-World Smart Homes. September, 275–283. https://doi.org/DOI: 10.1145/3677525.3678671

Chen, L., Zhang, Y., Xu, B., Shao, K., Yan, J., & Bhatti, U. A. (2024). A lot-based VMD-CNN-BiGRU indoor mariculture water quality prediction method including attention mechanism. *International Journal of High Speed Electronics and Systems*. Advance online publication. DOI: 10.1142/S0129156425400105

Chen, S., & Lin, Z. (2025). Research on ecological environment monitoring and protection applications based on remote sensing technology. *Academic Journal of Science and Technology, 14*(1), 55–57. DOI: 10.54097/k03eem97

Chen, T., Pan, H., Lu, M., Hang, J., Lam, C. K. C., Yuan, C., & Pearlmutter, D. (2021). Effects of tree plantings and aspect ratios on pedestrian visual and thermal comfort using scaled outdoor experiments. *The Science of the Total Environment, 801*, 149527. PMID: 34416606

Christie, G., Fendley, N., Wilson, J., & Mukherjee, R. (2018). Functional map of the world. Proceedings of the IEEE Conference on Computer Vision and Pattern Recognition, 6172–6180. https://openaccess.thecvf.com/content_cvpr_2018/html/Christie_Functional_Map_of_CVPR_2018_paper.html

Christin, S., Hervet, É., & Lecomte, N. (2019). Applications for deep learning in ecology. *Methods in Ecology and Evolution*, *10*(10), 1632–1644.

Civco, D. L., Hurd, J. D., Wilson, E. H., Song, M., & Zhang, Z. (2002, April). A comparison of land use and land cover change detection methods. In *ASPRS-ACSM Annual Conference* (Vol. 21, pp. 18-33).

Cobb, C. (2024). The politics of climate denialism and the secondary denialism of economics. *American Journal of Economics and Sociology*, *84*(1), 69–88. DOI: 10.1111/ajes.12606

Cracknell, M. J., & Reading, A. M. (2014). Geological mapping using remote sensing data: A comparison of five machine learning algorithms, their response to variations in the spatial distribution of training data and the use of explicit spatial information. *Computers & Geosciences*, *63*, 22–33.

CRED. (n.d.). EM-DAT - The international disaster database. Retrieved May 31, 2025, from https://www.emdat.be/

Cui, W., Li, J., Liu, J., Fang, Y., Xiao, X., Bhatti, U. A., & Nawaz, S. A. (2021). A zero-watermarking algorithm for medical images based on Gabor-DCT. Cyberspace Safety and Security: 12th International Symposium, CSS 2020.

Dale, V. H. (1997). The relationship between land-use change and climate change. *Ecological Applications*, *7*(3), 753–769.

Datta, D., & Deb, S. (2012). Analysis of coastal land use/land cover changes in the Indian Sunderbans using remotely sensed data. *Geo-Spatial Information Science*, *15*(4), 241–250.

Daudt, R. C., Le Saux, B., Boulch, A., & Gousseau, Y. (2019). Multitask learning for large-scale semantic change detection. *Computer Vision and Image Understanding*, *187*, 102783.

Dee, D., Uppala, S., Simmons, A., Berrisford, P., Poli, P., Kobayashi, S., & Vitart, F. (2011). The era-interim reanalysis: Configuration and performance of the data assimilation system. *Quarterly Journal of the Royal Meteorological Society*, *137*(656), 553–597. DOI: 10.1002/qj.828

Del Pup, F., & Atzori, M. (2023). Applications of Self-Supervised Learning to Biomedical Signals: A Survey. *IEEE Access : Practical Innovations, Open Solutions, 11*(November), 144180–144203. DOI: 10.1109/ACCESS.2023.3344531

Denker, J., Schwartz, D., Wittner, B., Solla, S., Howard, R., Jackel, L., & Hopfield, J. (1987). Large automatic learning, rule extraction, and generalization. *Complex Systems, 1*(5), 877–922.

Dewitte, S. (2020). Editorial for special issue "earth radiation budget". *Remote Sensing, 12*(20), 3379. DOI: 10.3390/rs12203379

Doan, T.-N., & Le-Thi, D.-N. (2025). A Novel Deep Learning Model for Flood Detection from Synthetic Aperture Radar Images. *Journal of Advances in Information Technology, 16*(1), 57–70. DOI: 10.12720/jait.16.1.57-70

Dong, F., Li, J., Bhatti, U. A., & Li, Y. (2023). Robust Zero-Watermarking Algorithm for Medical Images Based on Direction Gradient Histogram and DCT. Artificial Intelligence and Security: 6th International Conference, ICAIS.

Dubey, S. R., Singh, S. K., & Chaudhuri, B. B. (2022). Activation Functions in Deep Learning: A Comprehensive Survey and Benchmark.

Ebel, P., Saha, S., & Zhu, X. X. (2021). FUSING MULTI-MODAL DATA FOR SUPERVISED CHANGE DETECTION. The International Archives of the Photogrammetry, Remote Sensing and Spatial Information Sciences, XLIII-B3-2021, 243–249. XXIV ISPRS Congress <q>Imaging today, foreseeing tomorrow</q>, Commission III - 2021 edition, 5–9 July 2021. https://doi.org/DOI: 10.5194/isprs-archives-XLIII-B3-2021-243-2021

Emami, P., & Marzban, A. (2023). The synergy of artificial intelligence (ai) and geographic information systems (gis) for enhanced disaster management: Opportunities and challenges. *Disaster Medicine and Public Health Preparedness, 17.* Advance online publication. DOI: 10.1017/dmp.2023.174 PMID: 37694385

Eshbayev, O., Mirzaliev, S., Rozikov, R., Kuzikulova, D., & Шакирова, Г. (2022). Nlp and ml based approach of increasing the efficiency of environmental management operations and engineering practices. *IOP Conference Series. Earth and Environmental Science, 1045*(1), 012058. DOI: 10.1088/1755-1315/1045/1/012058

Essamlali, I., Nhaila, H., & El Khaili, M. (2024). Supervised Machine Learning Approaches for Predicting Key Pollutants and for the Sustainable Enhancement of Urban Air Quality. *Systematic Reviews.*

Estes, J. E., & Loveland, T. R. (1999). Toward the use of remote sensing and other data to delineate functional types in terrestrial and aquatic systems. In *Developments in Atmospheric Science* (Vol. 24, pp. 125–150). Elsevier.

Fanelli, S., Pratici, L., Salvatore, F. P., Donelli, C. C., & Zangrandi, A. (2023). Big data analysis for decision-making processes: Challenges and opportunities for the management of health-care organizations. *Management Research Review*, *46*(3), 369–389.

Fang, Y., Liu, J., Li, J., Cheng, J., Hu, J., Yi, D., Xiao, X., & Bhatti, U. A. (2023). Robust zero-watermarking algorithm for medical images based on SIFT and Bandelet-DCT. *Multimedia Tools and Applications*, *81*(12), 16863–16879.

Fan, Y., Li, J., Bhatti, U. A., Shao, C., Gong, C., Cheng, J., & Chen, Y. (2021). A Multi-Watermarking Algorithm for Medical Images Using Inception V3 and DCT. *Computers, Materials & Continua*, *74*(1), 1–13.

Fan, Y., Zeng, C., & Bhatti, U. A. (2021). Robust zero-watermarking algorithm for medical images based on dual-tree complex wavelet transform and discrete cosine transform. *Journal of Medical Imaging and Health Informatics*, *9*(1), 188–194.

Feng, S., Liu, Q., Patel, A., Bazai, S. U., Jin, C., Kim, J. S., Sarrafzadeh, M., Azzollini, D., Yeoh, J., & Kim, E. (2022). Automated pneumothorax triaging in chest X-rays in the New Zealand population using deep-learning algorithms. *Journal of Medical Imaging and Radiation Oncology*, *66*(8), 1035–1043. PMID: 35224858

Fer, I., Gardella, A., Shiklomanov, A., Campbell, E., Cowdery, E., Kauwe, M., & Dietze, M. (2020). Beyond ecosystem modeling: A roadmap to community cyberinfrastructure for ecological data-model integration. *Global Change Biology*, *27*(1), 13–26. DOI: 10.1111/gcb.15409 PMID: 33075199

Fernández-Montes De Oca, A. I., Ghilardi, A., Kauffer, E., Mas, J. F., Sánchez-Cordero, V., & Gallardo-Cruz, J. A. (2025). Building a Methodological Reference Framework for Quantifying Tropical Deforestation with Remote Sensing. *Sustainability*, *17*(4), 1394. DOI: 10.3390/su17041394

Fomalont, E. B. (1999). Image analysis. In Synthesis Imaging in Radio Astronomy II (Vol. 180, p. 301).

Galvan, L. P. C., Bhatti, U. A., Campo, C. C., & Stanojevic, S. (2022). Toward sustainable development: The nexus between CO2 emission, economic growth, trade openness, and the middle-income trap in Latin American countries. *Frontiers in Environmental Science*, •••, 1003.

Gambín, Á. F., Angelats, E., González, J. S., Miozzo, M., & Dini, P. (2021). Sustainable marine ecosystems: Deep learning for water quality assessment and forecasting. *IEEE Access : Practical Innovations, Open Solutions*, *9*, 121344–121365.

Ganesan, D., Ratnasamy, S., Wang, H., & Estrin, D. (2004). Coping with irregular spatio-temporal sampling in sensor networks. *Computer Communication Review*, *34*(1), 125–130.

Gao, J., Li, D., Zhang, Y., & Bhatti, U. A. (2024). EF-VPT-Net: Enhanced Feature-Based Vision Patch Transformer Network for Accurate Brain Tumor Segmentation in Magnetic Resonance Imaging. *IEEE Journal of Biomedical and Health Informatics*.

Gao, J., Zhang, Y., Geng, X., Tang, H., & Bhatti, U. A. (2022). PE-Transformer: Path enhanced transformer for improving underwater object detection. *Expert Systems with Applications*, •••, 123253.

Gao, L., Wang, D., Zhuang, L., Sun, X., Huang, M., & Plaza, A. (2023). BS3LNet: A New Blind-Spot Self-Supervised Learning Network for Hyperspectral Anomaly Detection. *IEEE Transactions on Geoscience and Remote Sensing*, *61*. Advance online publication. DOI: 10.1109/TGRS.2023.3246565

Gao, M., & Bhatti, U. A. (2024). *Deep Learning Based Applications for Multimedia Processing Applications* (Vol. 1 and 2). Taylor & Francis.

Garson, G. D. (2021). *Data analytics for the social sciences: Applications in R*. Routledge.

Gerland, P. (1996, June). Socio-economic data and GIS: datasets, databases, indicators, and data integration issues. In UNEP/CGIAR (Consultative Group on International Agricultural Research), Arendal III Workshop on Use of GIS in Agricultural Research Management. Norway.

Ghanbari, A., Shirdel, G. H., & Maleki, F. (2024). Semi-Self-Supervised Domain Adaptation: Developing Deep Learning Models with Limited Annotated Data for Wheat Head Segmentation. *Algorithms*, *17*(6). Advance online publication. DOI: 10.3390/a17060267

Ghazanchaei, E., Allahbakhshi, K., Khorasani-Zavareh, D., Aghazadeh-Attari, J., & Mohebbi, I. (2023). Challenges in providing care for patients with chronic diseases during disasters: A qualitative study with focus on diabetes and chronic respiratory diseases in Iran. *Tanaffos*, *22*(1), 83. PMID: 37920319

Gholami, S., Robinson, C., Ortiz, A., Yang, S., Margutti, J., Birge, C., Dodhia, R., & Ferres, J. L. (2022). On the Deployment of Post-Disaster Building Damage Assessment Tools using Satellite Imagery: A Deep Learning Approach. *2022 IEEE International Conference on Data Mining Workshops (ICDMW)*, 1029–1036. https://doi.org/DOI: 10.1109/ICDMW58026.2022.00134

Giri, R., Tenneti, S. V, Cheng, F., Helwani, K., Isik, U., & Krishnaswamy, A. (2020). Detection and Classification of Acoustic Scenes and Events 2020 SELF-SUPERVISED CLASSIFICATION FOR DETECTING ANOMALOUS SOUNDS. November.

Glorot, X., & Bordes, A. (2011). Deep Sparse Rectifier. *Neural Networks*, *15*, 315–323.

Gómez, N. I. M., Arévalo, J. L. D., & Jiménez, P. A. L. (2021). Air quality and urban sustainable development : The application of machine learning tools. *International Journal of Environmental Science and Technology*, *18*(4), 1029–1046. DOI: 10.1007/s13762-020-02896-6

Gomiero, T. (2016). Soil degradation, land scarcity and food security: Reviewing a complex challenge. *Sustainability (Switzerland)*, *8*(3), 1–41. DOI: 10.3390/su8030281

Goodfellow, I., Pouget-Abadie, J., Mirza, M., Xu, B., Warde-Farley, D., Ozair, S., & Bengio, Y. (2020). Generative adversarial networks. *Communications of the ACM*, *63*(11), 139–144. DOI: 10.1145/3422622

Goundar, S., & Rayani, P. K. (2021). *Applications of Big Data in Large-and Small-Scale Systems. IGI Global.Kuh, D., & Shlomo, Y. Ben. (2004). A life course approach to chronic disease epidemiology (Issue 2)*. Oxford university press.

Green, E. P., Mumby, P. J., Edwards, A. J., & Clark, C. D. (1996). A review of remote sensing for the assessment and management of tropical coastal resources. *Coastal Management*, *24*(1), 1–40.

Guild, L. S., Cohen, W. B., & Kauffman, J. B. (2004). Detection of deforestation and land conversion in Rondonia, Brazil using change detection techniques. *International Journal of Remote Sensing*, *25*(4), 731–750.

Guizilini, V., Li, J., Rares, A., Pillai, S., & Gaidon, A. (2019). Robust Semi-Supervised Monocular Depth Estimation with Reprojected Distances. Proceedings of Machine Learning Research, 100(CoRL), 503–512.

Guo, Q., Ren, M., Wu, S., & Sun, Y. (2022). Applications of artificial intelligence in the field of air pollution : A bibliometric analysis.

Gupta, A. K., & Nair, S. S. (2011, May). Environmental knowledge for disaster risk management. In *Abstract Volume of International Conference* (pp. 10-11).

Gu, S., Kelly, B., & Xiu, D. (2020). Empirical asset pricing via machine learning. *Review of Financial Studies, 33*(5), 2223–2273.

Haasnoot, M., van 't Klooster, S., & van Alphen, J. (2018). Designing a monitoring system to detect signals to adapt to uncertain climate change. *Global Environmental Change, 52*(July), 273–285. DOI: 10.1016/j.gloenvcha.2018.08.003

Haider, A., Hussain, L., Tareen, A. W., Bazai, S. U., Aslam, S., Neo, M., & Amphawan, A. (2024). A comparative study of machine learning techniques for accurate disease prediction using symptom-based diagnosis. In AIP Conference Proceedings (Vol. 3153, No. 1). AIP Publishing.

Hamza, M., Bazai, S. U., Ghafoor, M. I., Ullah, S., Akram, S., & Khan, M. S. (2023). Generative adversarial networks (gans) video framework: A systematic literature review. In *2023 International Conference on Energy, Power, Environment, Control, and Computing (ICEPECC)* (pp. 1-5). IEEE.

Hamza, M., Bazai, S. U., Ghafoor, M. I., Ullah, S., Akram, S., & Khan, M. S. (2023). Generative adversarial networks (GANs) video framework: A systematic literature review. In *2023 International Conference on Energy, Power, Environment, Control, and Computing (ICEPECC)* (pp. 1-5). IEEE.

Hamza, M., Bazai, S. U., Ghafoor, M. I., Ullah, S., Akram, S., & Khan, M. S. (2023). Generative adversarial networks (GANS) video framework: A systematic literature review. In *2023 International Conference on Energy, Power, Environment, Control, and Computing (ICEPECC)* (pp. 1-5). IEEE.

Han, H., Huang, M., Zhang, Y., & Bhatti, U. A. (2018). An architecture of secure health information storage system based on blockchain technology. Cloud Computing and Security: 4th International Conference, ICCCS 2018, Haikou, China, June 8-10, 2018, Revised Selected Papers, Part II 4, 578–588.

Hanekamp, J., & Bergkamp, L. (2016). The 'best available science' and the paris agreement on climate change. *European Journal of Risk Regulation, 7*(1), 42–48. DOI: 10.1017/s1867299x00005365

Han, H., Bazai, S. U., Bhatti, M. A., Basit, A., Wahid, A., Bhatti, U. A., Ghadi, Y. Y., & Algarni, A. (2024). Hybrid Climate Forecasting: Variational Mode Decomposition and Convolutional Neural Network with Long-Term Short Memory. *Polish Journal of Environmental Studies, 33*(2).

Han, H., Hasnain, A., Bhatti, U. A., Yue, Y., He, Y., & Wei, G., & ur Rahman, W. (2024a). Assessing the spatio-temporal impact of the COVID-19 pandemic lockdown on air quality in Jiangsu Province, China. *Environment, Development and Sustainability*, •••, 1–21.

Han, H., Liu, Z., Li, J., & Zeng, Z. (2024). Challenges in remote sensing based climate and crop monitoring: Navigating the complexities using AI. *Journal of Cloud Computing (Heidelberg, Germany)*, *13*(1). Advance online publication. DOI: 10.1186/s13677-023-00583-8

Han, H., Zeeshan, Z., Assam, M., Khan, F. U., Khan, W., Asif, M., Bhatti, U. A., Hasnain, A., Awwad, E. M., & Sarhan, N. (2024c). *Deep learning techniques for enhanced mangrove land use and land change from remote sensing imagery: A blue carbon perspective*. Big Data Research., DOI: 10.1016/j.bdr.2024.100478

Han, H., Zeeshan, Z., Talpur, B. A., Sadiq, T., Bhatti, U. A., & Awwad, E. M. (2024b). Studying long-term relationship between carbon emissions, soil, and climate change: Insights from a global Earth modeling framework. *International Journal of Applied Earth Observation and Geoinformation*, *130*, 118049.

Hansen, M. C., DeFries, R. S., Townshend, J. R., & Sohlberg, R. (2000). Global land cover classification at 1 km spatial resolution using a classification tree approach. *International Journal of Remote Sensing*, *21*(6-7), 1331–1364.

Haq, S. U., Bazai, S. U., Fatima, A., Marjan, S., Yang, J., Yee, L., & Ku, C. S. (2023). Reseek-arrhythmia: Empirical evaluation of ResNet architecture for detection of arrhythmia. *Diagnostics (Basel)*, *13*(18), 2867. PMID: 37761234

Harker-Schuch, I., Mills, F., Lade, S., & Colvin, R. (2018). Opinions of 12 to 13-year-olds in austria and australia on the worry, cause and imminence of climate change. DOI: 10.1101/333237

Hashmi, E., Yayilgan, S. Y., Yamin, M. M., Abomhara, M., & Ullah, M. (2025). Self-supervised hate speech detection in Norwegian texts with lexical and semantic augmentations. Expert Systems with Applications, 264(April 2024), 125843. https://doi.org/DOI: 10.1016/j.eswa.2024.125843

Hasnain, A., Hashmi, M. Z., & Bhatti, U. A. (2024). Prediction of particulate matter pollution using a long short-term memory model in Zhejiang Province, China. *International Journal of Wireless and Mobile Computing*, *23*(2), 163–172.

Hasnain, A., Sheng, Y., Hashmi, M. Z., Bhatti, U. A., Hussain, A., Hameed, M., & Yu, Z. (2022). Time series analysis and forecasting of air pollutants based on Prophet forecasting model in Jiangsu province, China. *Frontiers in Environmental Science*, *10*, 945628.

Hasnain, M. Z., & Bhatti, U. A. (2022). Improved multi-step dissolved oxygen prediction in aquaculture using adaptive temporal convolution. *Computers and Electronics in Agriculture*.

Hasnaoui, Y., Tachi, S. E., Bouguerra, H., & Yaseen, Z. M. (2025). Transfer learning-based deep learning models for flood and erosion detection in coastal area of Algeria. *Earth Science Informatics*, *18*(2), 380. DOI: 10.1007/s12145-025-01866-1

He, B., & Bhatti, U. A. (2024). Smart cities and smart networks: AI applications in urban geography and industrial communication. *International Journal of High Speed Electronics and Systems*. Advance online publication. DOI: 10.1142/S0129156424401220

Heckerman, D., Geiger, D., & Chickering, D. M. (1995). Learning Bayesian networks: The combination of knowledge and statistical data. *Machine Learning*, *20*, 197–243.

Hino, M., Benami, E., & Brooks, N. (2018). Machine learning for environmental monitoring. *Nature Sustainability*, *1*(10), 583–588. DOI: 10.1038/s41893-018-0142-9

Hirschmugl, M., Gallaun, H., Dees, M., Datta, P., Deutscher, J., Koutsias, N., & Schardt, M. (2017). Methods for mapping forest disturbance and degradation from optical earth observation data: A review. *Current Forestry Reports*, *3*, 32–45.

Hmielowski, J., Feldman, L., Myers, T., Leiserowitz, A., & Maibach, E. (2013). An attack on science? media use, trust in scientists, and perceptions of global warming. *Public Understanding of Science (Bristol, England)*, *23*(7), 866–883. DOI: 10.1177/0963662513480091 PMID: 23825287

Hochreiter, S., & Schmidhuber, J. (1997). Long short-term memory. *Neural Computation*, *9*(8), 1735–1780. PMID: 9377276

Hong, D., Gao, L., Yokoya, N., Yao, J., Chanussot, J., Du, Q., & Zhang, B. (2021). More Diverse Means Better: Multimodal Deep Learning Meets Remote-Sensing Imagery Classification. *IEEE Transactions on Geoscience and Remote Sensing*, *59*(5), 4340–4354. DOI: 10.1109/TGRS.2020.3016820

Hou, Z., He, Y., Cen, Y., Liu, X., Dong, Y., Kharlamov, E., & Tang, J. (2023). GraphMAE2: A Decoding-Enhanced Masked Self-Supervised Graph Learner. ACM Web Conference 2023 - Proceedings of the World Wide Web Conference, WWW 2023, 737–746. https://doi.org/DOI: 10.1145/3543507.3583379

Hoyer, L., Dai, D., Wang, Q., Chen, Y., & Van Gool, L. (2023). Improving Semi-Supervised and Domain-Adaptive Semantic Segmentation with Self-Supervised Depth Estimation. *International Journal of Computer Vision*, *131*(8), 2070–2096. DOI: 10.1007/s11263-023-01799-6

Huang, M., Han, H., Wang, H., Li, L., Zhang, Y., & Bhatti, U. A. (2018). A clinical decision support framework for heterogeneous data sources. *IEEE Journal of Biomedical and Health Informatics*, *22*(6), 1824–1833. PMID: 29994279

Huang, M., Zhang, X. S., Bhatti, U. A., Wu, Y. Y., Zhang, Y., & Ghadi, Y. Y. (2024). An interpretable approach using hybrid graph networks and explainable AI for intelligent diagnosis recommendations in chronic disease care. *Biomedical Signal Processing and Control*, *91*, 105913.

Hur, J., & Roth, S. (2020). Self-supervised monocular scene flow estimation. *Proceedings of the IEEE Computer Society Conference on Computer Vision and Pattern Recognition, i*, 7394–7403. https://doi.org/DOI: 10.1109/CVPR42600.2020.00742

Hurt, J. A., Scott, G. J., Anderson, D. T., & Davis, C. H. (2018). Benchmark Meta-Dataset of High-Resolution Remote Sensing Imagery for Training Robust Deep Learning Models in Machine-Assisted Visual Analytics. *2018 IEEE Applied Imagery Pattern Recognition Workshop (AIPR)*, 1–9. https://doi.org/DOI: 10.1109/AIPR.2018.8707433

Hussain, S. M., Braydich-Stolle, L. K., Schrand, A. M., Murdock, R. C., Yu, K. O., Mattie, D. M., & Terrones, M. (2009). Toxicity evaluation for safe use of nanomaterials: Recent achievements and technical challenges. *Advanced Materials*, *21*(16), 1549–1559

Iftikhar, A., Purvis, L., Giannoccaro, I., & Wang, Y. (2023). The impact of supply chain complexities on supply chain resilience: The mediating effect of big data analytics. *Production Planning and Control*, *34*(16), 1562–1582.

Igwama, G. T., Olaboye, J. A., Maha, C. C., Ajegbile, M. D., & Abdul, S. (2024). Big data analytics for epidemic forecasting: Policy Frameworks and technical approaches. *International Journal of Applied Research in Social Sciences*, *6*(7), 1449–1460.

Ijaz, H., Ahmad, R., Ahmed, R., Ahmed, W., Kai, Y., & Jun, W. (2023). A UAV-Assisted Edge Framework for Real-Time Disaster Management. *IEEE Transactions on Geoscience and Remote Sensing*, *61*, 1–13. DOI: 10.1109/TGRS.2023.3306151

Inman, D. L., & Brush, B. M. (1973). The Coastal Challenge: Fragile ribbons which border our land require more understanding, new technology, and resolute planning. *Science*, *181*(4094), 20–32. PMID: 17769816

Internatio, A., & D, L. V. D. Q. R. (n.d.). Review W of Eu an Stud Dies.

James, G., Witten, D., Hastie, T., & Tibshirani, R. (2000). An introduction to Statistical Learning. In Current medicinal chemistry (Vol. 7, Issue 10). https://doi.org/ DOI: 10.1007/978-1-4614-7138-7

Jaspal, R., Nerlich, B., & Koteyko, N. (2012). Contesting science by appealing to its norms. *Science Communication*, *35*(3), 383–410. DOI: 10.1177/1075547012459274

Jenifer, A. E., Aparna, A., Sudha, N., & Kumar, A. (2022). AgriFloodNet: A dual patch CNN architecture for mapping flooded agricultural lands via bi-temporal multi-sensor images. *Geocarto International*, *37*(26), 13618–13637. DOI: 10.1080/10106049.2022.2082549

Jiao, Z. (2024). The application of remote sensing techniques in ecological environment monitoring. *Highlights in Science Engineering and Technology*, *81*, 449–455. DOI: 10.54097/7dqegz64

Jin, W., Wang, N., Tao, T., Shi, B., Bi, H., Zhao, B., Wu, H., Duan, H., & Yang, G. (2024). A veracity dissemination consistency-based few-shot fake news detection framework by synergizing adversarial and contrastive self-supervised learning. *Scientific Reports*, *14*(1), 1–22. DOI: 10.1038/s41598-024-70039-9 PMID: 39174581

Johnson, F., Valderrama, A., Valle, C., Crawford, B., Soto, R., & Ñanculef, R. (2020). Automating configuration of convolutional neural network hyperparameters using genetic algorithm. *IEEE Access : Practical Innovations, Open Solutions*, *8*, 156139–156152.

Jones, J. A., Wei, X., Archer, E., Bishop, K., Blanco, J. A., Ellison, D., ... & Creed, I. F. (2020). Forest-water interactions under global change. Forest-water interactions, 589-624.

Joyce, K. E., Samsonov, S. V., Levick, S. R., Engelbrecht, J., & Belliss, S. (2014). Mapping and monitoring geological hazards using optical, LiDAR, and synthetic aperture RADAR image data. *Natural Hazards*, *73*, 137–163.

Joysoyal, R., Uddin, S. S., Islam, T., Sarker, S. K., Li, L., Ahsan, F., & Bhatti, U. A. (2024). Blockchain for sustainable city transformation: A review on Bangladesh. *Engineering Reports*, •••, 12948.

Kang, G. K., Gao, J. Z., Chiao, S., Lu, S., & Xie, G. (2018). Air Quality Prediction. *Big Data and Machine Learning Approaches.*, *9*(1). Advance online publication. DOI: 10.18178/ijesd.2018.9.1.1066

Kataria, S., Villalba, J., Moro-Velázquez, L., Thebaud, T., & Dehak, N. (2023). Self-FiLM: Conditioning GANs with self-supervised representations for bandwidth extension based speaker recognition. Proceedings of the Annual Conference of the International Speech Communication Association, INTERSPEECH, 2023-August, 4688–4692. https://doi.org/DOI: 10.21437/Interspeech.2023-2031

Khaertdinov, B., & Asteriadis, S. (2022). Temporal Feature Alignment in Contrastive Self-Supervised Learning for Human Activity Recognition. *2022 IEEE International Joint Conference on Biometrics, IJCB 2022*. https://doi.org/DOI: 10.1109/IJCB54206.2022.10007984

Khan, S. M., Shafi, I., Butt, W. H., Diez, I. de la T., Flores, M. A. L., Galán, J. C., & Ashraf, I. (2023). A systematic review of disaster management systems: Approaches, challenges, and future directions. *Land (Basel)*, *12*(8), 1514.

Khatami, R., Mountrakis, G., & Stehman, S. V. (2016). A meta-analysis of remote sensing research on supervised pixel-based land-cover image classification processes: General guidelines for practitioners and future research. *Remote Sensing of Environment*, *177*, 89–100.

Kılınç, M. (2024). *Information Systems. December. Kılınç, M. (2024).* Data-Driven Analytical Techniques in Geographic Information Systems.

Kior, A., Sukhov, V., & Sukhova, E. (2021). Application of reflectance indices for remote sensing of plants and revealing actions of stressors. *Photonics*, *8*(12), 582. DOI: 10.3390/photonics8120582

Kior, A., Yudina, L., Zolin, Y., Sukhov, V., & Sukhova, E. (2024). Rgb imaging as a tool for remote sensing of characteristics of terrestrial plants: A review. *Plants*, *13*(9), 1262. DOI: 10.3390/plants13091262 PMID: 38732477

Kizilay, F., Narman, M. R., Song, H., Narman, H. S., Cosgun, C., & Alzarrad, A. (2024). Evaluating fine tuned deep learning models for real-time earthquake damage assessment with drone-based images. *AI in Civil Engineering*, *3*(1), 15. DOI: 10.1007/s43503-024-00034-6

Klemas, V. V. (2001). Remote sensing of landscape-level coastal environmental indicators. *Environmental Management*, *27*, 47–57. PMID: 11083908

Kommineni, H. P., & International, M. (2024). Data-Driven Business Intelligence in Energy Distribution : Analytics and Environment- Focused Approaches. December. https://doi.org/DOI: 10.18034/gdeb.v13i1.779

Krämer, M., & Senner, J. (2015). A modular software architecture for processing of big geospatial data in the cloud. *Computers & Graphics*, *49*, 69–81. DOI: 10.1016/j.cag.2015.02.005

Krause, J. R., Watson, E. B., Wigand, C., & Maher, N. (2020). Are tidal salt marshes exposed to nutrient pollution more vulnerable to sea level rise? *Wetlands*, *40*(5), 1539–1548. PMID: 35068652

Kumar, A., Jain, R., Gupta, M., & Islam, S. M. N. (2023). *6G-enabled IoT and AI for smart healthcare: challenges, impact, and analysis*. CRC Press.

Kurniawan, M. H., Handiyani, H., Nuraini, T., Hariyati, R. T. S., & Sutrisno, S. (2024). A systematic review of artificial intelligence-powered (AI-powered) chatbot intervention for managing chronic illness. *Annals of Medicine*, *56*(1), 2302980. PMID: 38466897

Kyrkou, C., & Theocharides, T. (2020). EmergencyNet: Efficient Aerial Image Classification for Drone-Based Emergency Monitoring Using Atrous Convolutional Feature Fusion. *IEEE Journal of Selected Topics in Applied Earth Observations and Remote Sensing*, *13*, 1687–1699. DOI: 10.1109/JSTARS.2020.2969809

Lapetina, A., & Sheng, Y. P. (2015). Simulating complex storm surge dynamics: Three-dimensionality, vegetation effect, and onshore sediment transport. *Journal of Geophysical Research. Oceans*, *120*(11), 7363–7380.

Lary, D. J., Alavi, A. H., Gandomi, A. H., & Walker, A. L. (2016). Machine learning in geosciences and remote sensing. *Geoscience Frontiers*, *7*(1), 3–10.

Latella, M., Luijendijk, A., Moreno-Rodenas, A. M., & Camporeale, C. (2021). Satellite image processing for the coarse-scale investigation of sandy coastal areas. *Remote Sensing*, *13*(22), 4613.

Lee, S., Hyun, C., & Lee, M. (2023). Machine Learning Big Data Analysis of the Impact of Air Pollutants on Rhinitis-Related Hospital Visits. 1–17.

Lee, S. Y., Dunn, R. J. K., Young, R. A., Connolly, R. M., Dale, P. E. R., Dehayr, R., & Welsh, D. T. (2006). Impact of urbanization on coastal wetland structure and function. *Austral Ecology*, *31*(2), 149–163.

Lehmann, J., Cowie, A., Masiello, C. A., Kammann, C., Woolf, D., Amonette, J. E., & Whitman, T. (2021). Biochar in climate change mitigation. *Nature Geoscience*, *14*(12), 883–892.

Lemos, M., & Rood, R. (2010). Climate projections and their impact on policy and practice. *Wiley Interdisciplinary Reviews: Climate Change*, *1*(5), 670–682. DOI: 10.1002/wcc.71

Le, T. D., Macabiau, C., Albert, K., Jouvet, P., & Noumeir, R. (2024). A Novel Transformer-based Self-Supervised Learning Method to Enhance Photoplethysmogram Signal Artifact Detection. *IEEE Access : Practical Innovations, Open Solutions*, *12*(November). Advance online publication. DOI: 10.1109/ACCESS.2024.3488595

Le, T.-D. (2024). Boosting Transformer's Robustness and Efficacy in PPG Signal Artifact Detection with Self-Supervised Learning. Ml. https://arxiv.org/abs/2401.01013

Li, J., & Bhatti, U. A. (2021). Robust zero watermarking algorithm for medical images based on Tetrolet-DCT. Cyberspace Safety and Security: 12th International Symposium, CSS 2020.

Li, D., & Bhatti, U. A. (2024). A Sustainable Computing-Based Multi-Objective Feeding Scheduling Model for Low-Carbon Factory Aquaculture Using Enhanced NSGA-II in Agricultural Consumer Electronics. *IEEE Transactions on Consumer Electronics*.

Li, D., Bhatti, U. A., & Zhang, Y. (2024). MSTFNet: A Mamba and Dual Swin-Transformer Fusion Network for Remote Sensing Image Classification for Precision Agriculture Land Processing. *Environmental Earth Sciences*, *83*(23), 1–16.

Li, D., Bhatti, U. A., & Zhang, Y. (2025). *Robust multi-watermarking algorithm for medical images based on SqueezeNet transfer learning*. Deep Learning for Multimedia Processing Applications.

Li, D., Yu, M., & Bhatti, U. A. (2024). Enhanced feature-based network for accurate segmentation of MRI images. *IEEE Journal of Biomedical and Health Informatics*.

Li, D., Zhang, H., & Bhatti, U. A. (2022). A robust zero-watermarking algorithm based on PHTs-DCT for encrypted medical images. *Innovation in Medicine and Healthcare*, *5*(2), 102–113.

Li, D., Zhang, Y., & Bhatti, U. A. (2024). Color correction methods for underwater image enhancement: A systematic literature review. *PLoS One*, *20*(3), e0317306. PMID: 40063649

Li, D., Zhang, Y., & Bhatti, U. A. (2025). *A robust color image zero-watermarking based on stationary wavelet transform and Daisy descriptor*. Deep Learning for Multimedia Processing Applications.

Li, D., Zhang, Y., & Bhatti, U. A. (2025). CSTFNet: A CNN and Dual Swin-Transformer Fusion Network for Remote Sensing Hyperspectral Data Fusion and Classification of Coastal Areas. *IEEE Journal of Selected Topics in Applied Earth Observations and Remote Sensing*.

Li, H., Zeng, Y., & Bhatti, U. A. (2024). Bacterial Inhibition Zone Recognition Method Based on YOLO. In *7th International Conference on Pattern Recognition and Artificial Intelligence*.

Li, J., Bhatti, U. A., Shoukat, M. U., & Nawaz, S. A. (2020). Image fusion based on discrete cosine transform with high compression. *International Conference on Signal and Image Processing*.

Li, J., Bhatti, U. A., Shoukat, M. U., & Nawaz, S. A. (2020). Medical image hybrid watermark algorithm based on frequency domain processing and inception v3. *Advanced Intelligent Systems*, •••, 2400654.

Li, J., Hong, D., Gao, L., Yao, J., Zheng, K., Zhang, B., & Chanussot, J. (2022). Deep learning in multimodal remote sensing data fusion: A comprehensive review. *International Journal of Applied Earth Observation and Geoinformation*, *112*, 102926. DOI: 10.1016/j.jag.2022.102926

Li, J., Nawaz, S. A., Liu, J., & Bhatti, U. A. (2021). A robust digital image watermarking algorithm based on IWT-SVD. *Indian Journal of Engineering*, *17*(47), 110–116.

Li, J., Nawaz, S. A., Shoukat, M. U., & Bhatti, U. A. (2020). Improved watermarking algorithm based on SURF and SVD with wavelet transformation against geometric attacks. *Proceedings of the 2020 3rd International Conference on Image and Graphics*.

Li, J., Yu, Z., Yuan, L., Nawaz, S. A., & Bhatti, U. A. (2022). SURF based security of remote sensing images by encrypted watermark. *Proceedings of the 7th International Conference on Signal and Image Processing (ICSIP)*.

Li, K., & Lam, N. S. (2018). A spatial dynamic model of population changes in a vulnerable coastal environment. *International Journal of Geographical Information Science*, *32*(4), 685–710.

Li, N., Liu, C., & Pfeifer, N. (2019). Improving LiDAR classification accuracy by contextual label smoothing in post-processing. *ISPRS Journal of Photogrammetry and Remote Sensing*, *148*, 13–31.

Li, T., Li, J., Liu, J., Huang, M., Chen, Y. W., & Bhatti, U. A. (2022). Robust watermarking algorithm for medical images based on log-polar transform. *EURASIP Journal on Wireless Communications and Networking*, *2022*(1), 24.

LIU. D. (2022). Self-Supervised Learning Frameworks for Iot Applications. PhD Thesis.

Liu, J., Bhatti, U. A., & Zhang, Y. (2024). EF-UV: Feature Enhanced fusion of U-Net and VIT Transformer for Brain Tumor MRI Image Segmentation. *IEEE Journal of Biomedical and Health Informatics.*

Liu, J., Li, J., & Bhatti, U. A. (2024). Medical Image Hybrid Watermark Algorithm Based on Frequency Domain Processing and Inception v3. *Advanced Intelligent Systems*, •••, 2400654.

Liu, J., Li, J., Ma, J., Sadiq, N., Bhatti, U. A., & Ai, Y. (2019). A robust multi-watermarking algorithm for medical images based on DTCWT-DCT and Henon map. *Applied Sciences (Basel, Switzerland)*, *9*(4), 700.

Liu, J., Li, J., Zhang, K., Bhatti, U. A., & Ai, Y. (2019). Zero-watermarking algorithm for medical images based on dual-tree complex wavelet transform and discrete cosine transform. *Journal of Medical Imaging and Health Informatics*, *9*(1), 188–194.

Liu, X., Jin, W., Ma, Y., Li, Y., Liu, H., Wang, Y., & Tang, J. (2021, July). Elastic graph neural networks. In *International Conference on Machine Learning* (pp. 6837-6849). PMLR.

Li, W., & Hsu, C. Y. (2020). Automated terrain feature identification from remote sensing imagery: A deep learning approach. *International Journal of Geographical Information Science*, *34*(4), 637–660.

Li, W., Sun, K., Zhao, H., Li, W., Wei, J., & Gao, S. (2022). Extracting buildings from high-resolution remote sensing images by deep ConvNets equipped with structural-cue-guided feature alignment. *International Journal of Applied Earth Observation and Geoinformation*, *113*, 102970. DOI: 10.1016/j.jag.2022.102970

Li, X., Hao, X., Yin, L., Liu, L., Ma, Y., Yang, R., & Song, Q. (2021). Validation of gps-based monitoring and remote sensing of ice-shelf and ice-sheet movement changes. *Sensors (Basel)*, *21*(23), 7822. DOI: 10.3390/s21237822 PMID: 34883824

Li, X., Hou, B., Tang, H., Talpur, B. A., Zeeshan, Z., Bhatti, U. A., Liao, J., Liu, J., & Zhang, Y. (2024). Abnormal crops image data acquisition strategy by exploiting edge intelligence and dynamic-static synergy in smart agriculture. *IEEE Journal of Selected Topics in Applied Earth Observations and Remote Sensing.*

Li, X., Wang, S., & Bhatti, U. A. (2024). State-of-the-Art Approach: Boosting IoT Sensor Network Security with Visual Secret Sharing. *IEEE Smart World Congress (SWC).*

Li, Y., & Bhatti, U. A. (2024). A robust watermarking algorithm for medical images based on K-means and DCT. *International Journal of Wireless and Mobile Computing, 23*(2), 163–172.

Li, Y., Zhang, M., & Bhatti, U. A. (2025). *Robust zero watermarking algorithm for encrypted medical volume data based on PJFM and 3D-DCT*. Deep Learning for Multimedia Processing Applications.

Li, Z., Shen, H., Weng, Q., Zhang, Y., Dou, P., & Zhang, L. (2022). Cloud and cloud shadow detection for optical satellite imagery: Features, algorithms, validation, and prospects. *ISPRS Journal of Photogrammetry and Remote Sensing, 188*, 89–108.

Loganathan, A., Sumithra, P., & Deneshkumar, V. (2022)... *Estimation of Air Quality Index Using Multiple Linear Regression.*, (December). Advance online publication. DOI: 10.12691/aees-10-12-3

Lokmic-Tomkins, Z., Bhandari, D., Bain, C., Borda, A., Kariotis, T. C., & Reser, D. (2023). Lessons learned from natural disasters around digital health technologies and delivering quality healthcare. *International Journal of Environmental Research and Public Health, 20*(5), 4542. PMID: 36901559

Long, K., Shen, C., Tian, C., Zhang, K., Bhatti, U. A., Kong, D. F., Feng, S., & Cheng, H. (2023). Single UWB anchor aided PDR heading and step length correcting indoor localization system. *IEEE Access : Practical Innovations, Open Solutions, 9*, 11511–11522.

Lorenz, S., Dessai, S., Forster, P., & Paavola, J. (2016). Adaptation planning and the use of climate change projections in local government in england and germany. *Regional Environmental Change, 17*(2), 425–435. DOI: 10.1007/s10113-016-1030-3 PMID: 32269501

Mahmoud, A. S., Mohamed, S. A., El-Khoriby, R. A., AbdelSalam, H. M., & El-Khodary, I. A. (2023). Oil Spill Identification based on Dual Attention UNet Model Using Synthetic Aperture Radar Images. *Photonirvachak (Dehra Dun), 51*(1), 121–133. DOI: 10.1007/s12524-022-01624-6

Mahmoud, H., Wu, W., & Gaber, M. M. (2022). A Time-Series Self-Supervised Learning Approach to Detection of Cyber-physical Attacks in Water Distribution Systems. *Energies, 15*(3), 1–18. DOI: 10.3390/en15030914

Mayer-Schönberger, V., & Cukier, K. (2013). *Big data: A revolution that will transform how we live, work, and think*. Houghton Mifflin Harcourt.

McCright, A., & Dunlap, R. (2011). Cool dudes: The denial of climate change among conservative white males in the united states. *Global Environmental Change, 21*(4), 1163–1172. DOI: 10.1016/j.gloenvcha.2011.06.003

Meliadou, A., Santoro, F., Nader, M. R., Abou Dagher, M., Al Indary, S., & Abi Salloum, B. (2012). Prioritising coastal zone management issues through fuzzy cognitive mapping approach. *Journal of Environmental Management, 97*, 56–68. PMID: 22325583

Meng, X., Shen, H., Li, H., Zhang, L., & Fu, R. (2019). Review of the pansharpening methods for remote sensing images based on the idea of meta-analysis: Practical discussion and challenges. *Information Fusion, 46*, 102–113.

Merkle, N., Bahmanyar, R., Henry, C., Azimi, S. M., Yuan, X., Schopferer, S., Gstaiger, V., Auer, S., Schneibel, A., Wieland, M., & Kraft, T. (2023). Drones4Good: Supporting Disaster Relief Through Remote Sensing and AI. *2023 IEEE/CVF International Conference on Computer Vision Workshops (ICCVW)*, 3772–3776. https://doi.org/DOI: 10.1109/ICCVW60793.2023.00407

Minelli, M., Chambers, M., & Dhiraj, A. (2012). *Big data, big analytics: emerging business intelligence and analytic trends for today's businesses*. John Wiley & Sons.

Mintz, D. (2012). *Technical Assistance Document for the Reporting of Daily Air Quality – the Air Quality Index*. AQI.

Mitsuda, Y., & Ito, S. (2011). A review of spatial-explicit factors determining spatial distribution of land use/land-use change. *Landscape and Ecological Engineering, 7*, 117–125.

Monem, R., & Bastawissy, A. (2016). Dira: A framework of data integration using data quality. *International Journal of Data Mining & Knowledge Management Process, 6*(2), 37–58. DOI: 10.5121/ijdkp.2016.6204

Monteleoni, C., Schmidt, G., & McQuade, S. (2013). Climate informatics: Accelerating discovering in climate science with machine learning. *Computing in Science & Engineering, 15*(5), 32–40. DOI: 10.1109/mcse.2013.50

Mookambiga, A., & Gomathi, V. (2016). Comprehensive review on fusion techniques for spatial information enhancement in hyperspectral imagery. *Multidimensional Systems and Signal Processing, 27*(4), 863–889. DOI: 10.1007/s11045-016-0415-2

Moser, S. C. (2010). Communicating climate change: History, challenges, process and future directions. *Wiley Interdisciplinary Reviews: Climate Change, 1*(1), 31–53. DOI: 10.1002/wcc.11

Murala, D. K., Panda, S. K., & Dash, S. P. (2023). MedMetaverse: Medical care of chronic disease patients and managing data using artificial intelligence, blockchain, and wearable devices state-of-the-art methodology. *IEEE Access : Practical Innovations, Open Solutions*, *11*, 138954–138985.

Mutmainah, O., Hakim, A., & Syam, M. (2022). Validity of physics teaching materials based on stem to improve climate literacy of high school students. *Jurnal Pendidikan Fisika Dan Teknologi*, *8*(2), 208–216. DOI: 10.29303/jpft.v8i2.4053

Myakala, P. K. (2025). Scaling AI with Limited Labeled Data : A Self-Supervised Learning Approach.

Nabeel, S. M., Bazai, S. U., Alasbali, N., Liu, Y., Ghafoor, M. I., Khan, R., & Por, L. Y. (2024). Optimizing lung cancer classification through hyperparameter tuning. *Digital Health*, *10*, 20552076241249661. PMID: 38698834

Najafabadi, M. M., Villanustre, F., Khoshgoftaar, T. M., Seliya, N., Wald, R., & Muharemagic, E. (2015). Deep learning applications and challenges in big data analytics. *Journal of Big Data*, *2*, 1–21.

Nasteski, V. (2017). An overview of the supervised machine learning methods. *Horizons.B*, *4*(December), 51–62. DOI: 10.20544/horizons.b.04.1.17.p05

Nawaz, S. A., Li, J., & Bhatti, U. A., SU, B., & A Mehmood. (2021). Hybrid medical image zero watermarking via discrete wavelet transform-ResNet101 and discrete cosine transform. *Computers & Electrical Engineering*, *112*, 108985.

Nawaz, S. A., Li, J., Bhatti, U. A., Bazai, S. U., Zafar, A., Bhatti, M. A., Mehmood, A., & Ain, Q. ul, & Shoukat, M. U. (2021). A hybrid approach to forecast the COVID-19 epidemic trend. *PLoS One*, *16*(10), e0256971. PMID: 34606503

Nawaz, S. A., Li, J., Bhatti, U. A., Shoukat, M. U., & Ahmad, R. M. (2022). AI-based object detection latest trends in remote sensing, multimedia and agriculture applications. *Frontiers in Plant Science*, *13*, 1041514. PMID: 37082514

Nawaz, S. A., Li, J., Bhatti, U. A., Shoukat, M. U., Ahmad, R. M., & Hussain, S. (2023). AI-based object detection latest trends in remote sensing, multimedia, and agriculture applications. *Frontiers in Plant Science*, •••, 4273. PMID: 37082514

Neal, S. (2013). Transition culture: Politics, localities and ruralities. *Journal of Rural Studies*, *32*, 60–69.

Nguyen, T., Novak, R., Xiao, L., & Lee, J. (2021). Dataset distillation with infinitely wide convolutional networks. *Advances in Neural Information Processing Systems*, *34*, 5186–5198.

Nicholls, R. J. (2011). Planning for the impacts of sea level rise. *Oceanography (Washington, D.C.)*, *24*(2), 144–157.

Ni, M., Wu, Q., Li, G., & Li, D. (2025). Remote sensing technology for observing tree mortality and its influences on carbon–water dynamics. *Forests*, *16*(2), 194. DOI: 10.3390/f16020194

Noor, S., Bazai, S. U., Ghafoor, M. I., Marjan, S., Akram, S., & Ali, F. (2023). Generative adversarial networks for anomaly detection: a systematic literature review. In 2023 4th International Conference on Computing, Mathematics and Engineering Technologies (iCoMET) (pp. 1-6). IEEE.

Noor, R., Wahid, A., Bazai, S. U., Khan, A., Fang, M., Bhatti, U. A., & Ghadi, Y. Y. (2024). DLGAN: Undersampled MRI reconstruction using deep learning based generative adversarial network. *Biomedical Signal Processing and Control*, *93*, 106218.

Noor, R., Wahid, A., Bazai, S. U., Khan, A., Fang, M., Syam, M. S., & Bhatti, U. A. (2024). DLGAN: Undersampled MRI reconstruction using deep learning-based generative adversarial network. *Biomedical Signal Processing and Control*, *93*, 106218.

Norouzzadeh, M. S., Nguyen, A., Kosmala, M., Swanson, A., Palmer, M. S., Packer, C., & Clune, J. (2018). Automatically identifying, counting, and describing wild animals in camera-trap images with deep learning. *Proceedings of the National Academy of Sciences of the United States of America*, *115*(25), E5716–E5725. PMID: 29871948

Novikarany, R. (2025). Benefits of remote sensing for environmental monitoring. *International Journal of Integrative Sciences*, *4*(2), 445–454. DOI: 10.55927/ijis.v4i2.19

Ogugua, J. O., Onwumere, C., Arowoogun, J. O., Anyanwu, E. C., Odilibe, I. P., & Akomolafe, O. (2024). Data science in public health: A review of predictive analytics for disease control in the USA and Africa. *World Journal of Advanced Research and Reviews*, *21*(1), 2753–2769.

Otukei, J. R., & Blaschke, T. (2010). Land cover change assessment using decision trees, support vector machines and maximum likelihood classification algorithms. *International Journal of Applied Earth Observation and Geoinformation*, *12*, S27–S31.

Ouchra, H., Belangour, A., & Erraissi, A. (2022). A comparative study on pixel-based classification and object-oriented classification of satellite image. *International Journal of Engineering Trends and Technology*, *70*(8), 206–215.

Pal, M., & Mather, P. M. (2003). An assessment of the effectiveness of decision tree methods for land cover classification. *Remote Sensing of Environment, 86*(4), 554–565.

Park, S., Jeong, S., Park, Y., Kim, S., Lee, D., Mo, Y., & Park, K. (2021). Phenological analysis of sub-alpine forest on jeju island, south korea, using data fusion of landsat and modis products. *Forests, 12*(3), 286. DOI: 10.3390/f12030286

Peng, M., Liu, Y., & Bhatti, U. A. (2023). Crop monitoring using remote sensing land use and land change data: Comparative analysis of deep learning methods using pre-trained CNN models. *Big Data Research, 36*, 100448.

Peng, M., Liu, Y., & Bhatti, U. A. (2024). Deep learning-based tree disease recognition and classification using hyperspectral data. *Computers, Materials & Continua, 77*(1), 112–121.

Peng, M., Liu, Y., Qadri, I. A., Bhatti, U. A., Ahmed, B., Sarhan, N. M., & Awwad, E. M. (2024a). Advanced image segmentation for precision agriculture using CNN-GAT fusion and fuzzy C-means clustering. *Computers and Electronics in Agriculture, 226*, 109431.

Pimm, S. L., Alibhai, S., Bergl, R., Dehgan, A., Giri, C., Jewell, Z., & Loarie, S. (2015). Emerging technologies to conserve biodiversity. *Trends in Ecology & Evolution, 30*(11), 685–696. PMID: 26437636

Qi, G. J., & Luo, J. (2022). Small Data Challenges in Big Data Era: A Survey of Recent Progress on Unsupervised and Semi-Supervised Methods. *IEEE Transactions on Pattern Analysis and Machine Intelligence, 44*(4), 2168–2187. DOI: 10.1109/TPAMI.2020.3031898 PMID: 33074801

Quinn, J. A., Nyhan, M. M., Navarro, C., Coluccia, D., Bromley, L., & Luengo-Oroz, M. (2018). Humanitarian applications of machine learning with remote-sensing data: Review and case study in refugee settlement mapping. *Philosophical Transactions. Series A, Mathematical, Physical, and Engineering Sciences, 376*(2128), 20170363. DOI: 10.1098/rsta.2017.0363 PMID: 30082308

Rahman, Z., Abro, W. A., Aamir, M., Bhatti, U. A., Dayo, Z. A., & Ishfaq, M. (2023). Object recognition using machine learning techniques. *Computers, Materials & Continua.*

Railsback, S. F., & Grimm, V. (2019). *Agent-based and individual-based modeling: a practical introduction.* Princeton university press.

Rane, N. (2023). Integrating leading-edge artificial intelligence (AI), internet of things (IOT), and big data technologies for smart and sustainable architecture, engineering and construction (AEC) industry: Challenges and future directions. Engineering and Construction (AEC) Industry: Challenges and Future Directions (September 24, 2023).

Rao, A., Jung, J., Silva, V., Molinario, G., & Yun, S.-H. (2023). Earthquake building damage detection based on synthetic-aperture-radar imagery and machine learning. *Natural Hazards and Earth System Sciences*, *23*(2), 789–807. DOI: 10.5194/nhess-23-789-2023

Rautela, K. S., & Goyal, M. K. (2024). Transforming air pollution management in India with AI and machine learning technologies. *Scientific Reports*, •••, 1–19. DOI: 10.1038/s41598-024-71269-7 PMID: 39223178

Reichstein, M., Camps-Valls, G., Stevens, B., Jung, M., Denzler, J., Carvalhais, N., & Prabhat, F. (2019). Deep learning and process understanding for data-driven Earth system science. *Nature*, *566*(7743), 195–204. PMID: 30760912

Reinsch, S., Robinson, D. A., van Soest, M. A., Keith, A. M., Parry, S., & Tye, A. M. (2024). Temperate soils exposed to drought—Key processes, impacts, indicators, and unknowns. *Land (Basel)*, *13*(11), 1759.

Rhind, D. (1992). Data access, charging and copyright and their implications for geographical information systems. *International Journal of Geographical Information Systems*, *6*(1), 13–30.

Rittle-Johnson, B., Star, J. R., & Durkin, K. (2009). The importance of prior knowledge when comparing examples: Influences on conceptual and procedural knowledge of equation solving. *Journal of Educational Psychology*, *101*(4), 836.

Rodriguez-Galiano, V., Sanchez-Castillo, M., Chica-Olmo, M., & Chica-Rivas, M. J. O. G. R. (2015). Machine learning predictive models for mineral prospectivity: An evaluation of neural networks, random forest, regression trees and support vector machines. *Ore Geology Reviews*, *71*, 804–818.

Rudner, T. G. J., Rußwurm, M., Fil, J., Pelich, R., & Bischke, B. (n.d.). *Rapid Computer Vision-Aided Disaster Response via Fusion of Multiresolution*. Multisensor, and Multitemporal Satellite Imagery.

Saeed, A., Ozcelebi, T., & Lukkien, J. (2019). Multi-task Self-Supervised Learning for Human Activity Detection. *Proceedings of the ACM on Interactive, Mobile, Wearable and Ubiquitous Technologies*, *3*(2), 1–30. DOI: 10.1145/3328932

Saeed, A., Ungureanu, V., & Gfeller, B. (2021). Sense and Learn: Self-supervision for omnipresent sensors. *Machine Learning with Applications*, *6*(April), 100152. DOI: 10.1016/j.mlwa.2021.100152

Saheer, L. B., Shahawy, M., Approach, S., & Tree, U. (2022). Self-supervised Approach for Urban Tree Recognition on Lakshmi Babu Saheer, Mohamed Shahawy To cite this version : HAL Id : hal-03789042.

Sain, S. (2022). Data science and climate risk analytics. *Environmetrics*, *34*(2). Advance online publication. DOI: 10.1002/env.2749

Sánchez-Arcilla, A., García-León, M., Gracia, V., Devoy, R., Stanica, A., & Gault, J. (2016). Managing coastal environments under climate change: Pathways to adaptation. *The Science of the Total Environment*, *572*, 1336–1352. PMID: 26875603

Sarala, R. (2025). Machine learning and spatio-temporal patterns in climate change. *Journal of Information Systems Engineering & Management*, *10*(15s), 77–85. DOI: 10.52783/jisem.v10i15s.2431

Sarathchandra, D., & Haltinner, K. (2019). Trust/distrust judgments and perceptions of climate science: A research note on skeptics' rationalizations. *Public Understanding of Science (Bristol, England)*, *29*(1), 53–60. DOI: 10.1177/0963662519886089 PMID: 31691642

Sarathchandra, D., & Haltinner, K. (2021). A survey instrument to measure skeptics' (dis)trust in climate science. *Climate (Basel)*, *9*(2), 18. DOI: 10.3390/cli9020018

Sattar, A., Assam, M., Alahmadi, T. J., Bhatti, U. A., Tang, H., & Aamir, M. (2024). Remote sensing based advance image captioning improved feature attention. In *7th International Conference on Pattern Recognition and Artificial Intelligence*.

Schindler, K. (2012). An overview and comparison of smooth labeling methods for land-cover classification. *IEEE Transactions on Geoscience and Remote Sensing*, *50*(11), 4534–4545.

Sebestyén, V., Czvetkó, T., & Abonyi, J. (2021). The Applicability of Big Data in Climate Change Research. *The Importance of System of Systems Thinking.*, *9*(March), 1–26. DOI: 10.3389/fenvs.2021.619092

Selosse, M. A., Richard, F., He, X., & Simard, S. W. (2006). Mycorrhizal networks: Des liaisons dangereuses? *Trends in Ecology & Evolution*, *21*(11), 621–628. PMID: 16843567

Seneviratne, S., Wijnands, J. S., Nice, K., Zhao, H., Godic, B., Mavoa, S., Vidanaarachchi, R., Stevenson, M., Garcia, L., Hunter, R. F., & Thompson, J. (2022). Urban feature analysis from aerial remote sensing imagery using self-supervised and semi-supervised computer vision. 1–35. https://arxiv.org/abs/2208.08047

Sesnie, S. E., Finegan, B., Gessler, P. E., Thessler, S., Ramos Bendana, Z., & Smith, A. M. (2010). The multispectral separability of Costa Rican rainforest types with support vector machines and Random Forest decision trees. *International Journal of Remote Sensing*, *31*(11), 2885–2909.

Shao, K., Li, D., Tang, H., Zhang, Y., Xu, B., & Bhatti, U. A. (2024). Adaptive Temporal Convolutional Network (Atcn) and Internally-Optimized Transformer for Dissolved Oxygen Prediction in Aquaculture. Available at *SSRN* 5039613.

Shen, L., Lu, Y., Chen, H., Wei, H., Xie, D., Yue, J., Chen, R., Lv, S., & Jiang, B. (2021). S2Looking: A satellite side-looking dataset for building change detection. *Remote Sensing*, *13*(24), 5094.

Shenoy, J., Zhang, X. D., Tao, B., Mehrotra, S., Yang, R., Zhao, H., & Vasisht, D. (2024). Self-Supervised Learning across the Spectrum. *Remote Sensing*, *16*(18). Advance online publication. DOI: 10.3390/rs16183470

Shetty, S. H., Shetty, S., Singh, C., & Rao, A. (2022). Supervised machine learning: Algorithms and applications. *Fundamentals and Methods of Machine and Deep Learning: Algorithms, Tools, and Applications*, (January), 1–16. DOI: 10.1002/9781119821908.ch1

Shoukat, M. U., Nawaz, S. A., Raza, M., Bhatti, U. A., & Mehmood, A. (2021). A robust watermarking algorithm for encrypted face recognition based on DWT-DCT and tent map. Cloud Computing and Security: 4th International Conference, ICCCS 2018.

Shou, Y., Cao, X., & Meng, D. (2024). SpeGCL: Self-supervised Graph Spectrum Contrastive Learning without Positive Samples. 1–13. https://arxiv.org/abs/2410.10365

Shu, C., & Sun, L. (2020). Automatic target recognition method for multitemporal remote sensing image. *Open Physics*, *18*(1), 170–181. DOI: 10.1515/phys-2020-0015

Simard, S. W., Perry, D. A., Jones, M. D., Myrold, D. D., Durall, D. M., & Molina, R. (1997). Net transfer of carbon between ectomycorrhizal tree species in the field. *Nature*, *388*(6642), 579–582.

Simboura, N., Pavlidou, A., Bald, J., Tsapakis, M., Pagou, K., Zeri, C., & Panayotidis, P. (2016). Response of ecological indices to nutrient and chemical contaminant stress factors in Eastern Mediterranean coastal waters. *Ecological Indicators*, *70*, 89–105.

Simon, P. (2013). *Too big to ignore: the business case for big data* (Vol. 72). John Wiley & Sons.

Singh, G. (2023). Environmental monitoring with machine learning. *Epra International Journal of Multidisciplinary Research (Ijmr), 208-212*. Advance online publication. DOI: 10.36713/epra13330

Singh, S., Sharma, P., Pal, N., Sarma, D. K., Tiwari, R., & Kumar, M. (2024). Holistic one health surveillance framework: Synergizing environmental, animal, and human determinants for enhanced infectious disease management. *ACS Infectious Diseases*, *10*(3), 808–826. PMID: 38415654

Smith, S. E., & Read, D. J. (2010). *Mycorrhizal symbiosis*. Academic press.

Snezhana, D. (2023). Applying artificial intelligence (AI) for mitigation climate change consequences of the natural disasters. Dineva, S.(2023). Applying Artificial Intelligence (AI) for Mitigation Climate Change Consequences of the Natural Disasters. *Research Journal of Ecology and Environmental Sciences*, *3*(1), 1–8.

Souto, E. J. P., & Oliveira, H. A. B. F. (2022). Recognition Using Physiological Signals.

Souza, J. C. (2024). A Brief Overview of Deep Learning Approaches for Climate Data - Proceeding A Brief Overview of Deep Learning Approaches for Climate Data. April, 10–12. https://doi.org/DOI: 10.13140/RG.2.2.26670.52800

Sreekesh, S., Kaur, N., & Sreerama Naik, S. R. (2020). An Obia and rule algorithm for coastline extraction from high-and medium-resolution multispectral remote sensing images. *Remote Sensing in Earth Systems Sciences*, *3*, 24–34.

Srivastava, P. K., Han, D., Rico-Ramirez, M. A., Bray, M., & Islam, T. (2012). Selection of classification techniques for land use/land cover change investigation. *Advances in Space Research*, *50*(9), 1250–1265.

Stawicki, S. P., Firstenberg, M. S., Galwankar, S. C., Izurieta, R., & Papadimos, T. (2021). *Contemporary Developments and Perspectives in International Health Security* (Vol. 1). BoD–Books on Demand.

Storch, H., & Bray, D. (2017). Models, manifestation and attribution of climate change. *Meteorology Hydrology and Water Management*, *5*(1), 47–52. DOI: 10.26491/mhwm/67388

Subramaniam, S., Raju, N., Ganesan, A., Rajavel, N., & Dixit, S. (2022). Artificial Intelligence Technologies for Forecasting Air Pollution and Human Health : A Narrative Review. 1–36.

Sun, S., Xue, Q., Xing, X., Zhao, H., & Zhang, F. (2024). Remote Sensing Image Interpretation for Coastal Zones: A Review. *Remote Sensing*, *16*(24), 4701.

Sun, Y., Khan, M. A., & Bhatti, U. A. (2024). Advanced plant disease segmentation in precision agriculture using optimal dimensionality reduction with fuzzy C-means clustering and deep learning. *IEEE Journal of Selected Topics in Applied Earth Observations and Remote Sensing*, *17*, 121–131.

Taj, G., & Navi, M. (2020). Air quality monitoring and assessment using. *IoT*. Advance online publication. DOI: 10.1088/1757-899X/955/1/012006

Tang, H., Li, D., & Bhatti, U. A. (2025). Robust Zero-Watermarking Algorithm for Medical Images Based on AGAST-LATCH and DCT. In Deep Learning for Multimedia Processing Applications.

Tang, H., Bhatti, U. A., Li, J., Marjan, S., Baryalai, M., Assam, M., Ghadi, Y. Y., & Mohamed, H. G. (2023). A New Hybrid Forecasting Model Based on Dual Series Decomposition with Long-Term Short-Term Memory. *International Journal of Intelligent Systems*, *2023*. Advance online publication. DOI: 10.1155/2023/9407104

Tang, H., Li, J., & Bhatti, U. A. (2025). *Robust zero watermarking algorithm for medical images based on BRISK and DCT*. Deep Learning for Multimedia Processing Applications.

Tan, L., Guo, J., Mohanarajah, S., & Zhou, K. (2020). Can we detect trends in natural disaster management with artificial intelligence? a review of modeling practices. *Natural Hazards*, *107*(3), 2389–2417. DOI: 10.1007/s11069-020-04429-3

Teh, D., & Rana, T. (2023). The use of Internet of Things, Big Data analytics and artificial intelligence for attaining UN's SDGs. In *Handbook of big data and analytics in accounting and auditing* (pp. 235–253). Springer.

Thota, S., Chitta, S., Vangoor, V. K. R., Ravi, C. S., & Bonam, V. S. M. (2023). Few-ShotLearning in Computer Vision: Practical Applications and Techniques. *Human-Computer Interaction*, *3*(1).

Toimil, A., Losada, I. J., Nicholls, R. J., Dalrymple, R. A., & Stive, M. J. (2020). Addressing the challenges of climate change risks and adaptation in coastal areas: A review. *Coastal Engineering*, *156*, 103611.

Toker, A., Kondmann, L., Weber, M., Eisenberger, M., Camero, A., Hu, J., Hoderlein, A. P., Şenaras, Ç., Davis, T., & Cremers, D. (2022). Dynamicearthnet: Daily multi-spectral satellite dataset for semantic change segmentation. Proceedings of the IEEE/CVF Conference on Computer Vision and Pattern Recognition, 21158–21167. https://openaccess.thecvf.com/content/CVPR2022/html/Toker_DynamicEarthNet_Daily_Multi-Spectral_Satellite_Dataset_for_Semantic_Change_Segmentation_CVPR_2022_paper.html

Topbaş, A., Jamil, A., Hameed, A. A., Ali, S. M., Bazai, S., & Shah, S. A. (2021). Sentiment analysis for covid-19 tweets using recurrent neural network (rnn) and bidirectional encoder representations (bert) models. 2021 International Conference on Computing, Electronic and Electrical Engineering (ICE Cube), 1–6.

Topbaş, A., Jamil, A., Hameed, A. A., Ali, S. M., Bazai, S., & Shah, S. A. (2021). Sentiment analysis for covid-19 tweets using recurrent neural network (RNN) and bidirectional encoder representations (BERT) models. 2021 International Conference on Computing, Electronic and Electrical Engineering (ICE Cube), 1–6.

Usama, M., Qadir, J., Raza, A., Arif, H., Yau, K. L. A., Elkhatib, Y., & Al-Fuqaha, A. (2019). Unsupervised machine learning for networking: Techniques, applications and research challenges. *IEEE Access : Practical Innovations, Open Solutions*, *7*, 65579–65615.

Uyar, F., & Karamustafaoğlu, O. (2024). Polar science awareness of science teachers. *International E-Journal of Educational Studies*, *8*(16), 32–44. DOI: 10.31458/iejes.1386430

Vera, P., Marie, M., Gaviria, J. A., Miroslav, T., Petr, H., Esther, B., Tudor, T., Nina, M., & Chawdhery, R. A. (2020). *Pr ep rin t n pe er re v Pr ep er* ed. 1–24.

Vögt, V., Harrs, J., Reinhart, V., Hollenbach, P., Bühler, M., & Tewes, T. (2023). Implementing agile data workflows to unlock climate-resilient urban planning. *Climate (Basel)*, *11*(9), 174. DOI: 10.3390/cli11090174

Voigt, B., Fischer, O., Schilling, B., Krumnow, C., & Herta, C. (2023). Investigation of semi- and self-supervised learning methods in the histopathological domain. *Journal of Pathology Informatics*, *14*(January), 100305. DOI: 10.1016/j.jpi.2023.100305 PMID: 37025325

Wallace, J. F., Caccetta, P. A., & Kiiveri, H. T. (2004). Recent developments in analysis of spatial and temporal data for landscape qualities and monitoring. *Austral Ecology*, *29*(1), 100–107.

Wang, B., Li, J., & Bhatti, U. A. (2024). Robust Zero-Watermarking Algorithm for Medical Images Based on Direction Gradient Histogram and DCT. *Computers, Materials & Continua.*

Wang, X., Li, J., & Bhatti, U. A. (2024). A robust watermarking algorithm for medical images based on DWT-Daisy and Discrete Cosine Transform. In *International KES Conference on Innovation in Medicine and Healthcare.*

Wang, X., Yang, N., Liu, E., Gu, W., Zhang, J., Zhao, S., Sun, G., & Wang, J. (2023). Tree Species Classification Based on Self-Supervised Learning with Multisource Remote Sensing Images. *Applied Sciences (Switzerland), 13*(3). Advance online publication. DOI: 10.3390/app13031928

Wang, Y., Albrecht, C. M., Braham, N. A. A., Mou, L., & Zhu, X. X. (2022). Self-Supervised Learning in Remote Sensing: A review. *IEEE Geoscience and Remote Sensing Magazine, 10*(4), 213–247. DOI: 10.1109/MGRS.2022.3198244

Wang, Y., He, X., & Yu, F. (2022). Non-host plants: Are they mycorrhizal networks players? *Plant Diversity, 44*(2), 127–134. PMID: 35505991

Wang, Y., & Yésou, H. (2018). Remote sensing of floodpath lakes and wetlands: A challenging frontier in the monitoring of changing environments. *Remote Sensing, 10*(12), 1955. DOI: 10.3390/rs10121955

Wellhausen, L., Dosovitskiy, A., Ranftl, R., Walas, K., Cadena, C., & Hutter, M. (2019). Where should i walk(Predicting terrain properties from images via self-supervised learning. *IEEE Robotics and Automation Letters, 4*(2), 1509–1516. DOI: 10.1109/LRA.2019.2895390

Whitcraft, A. K., Vermote, E. F., Becker-Reshef, I., & Justice, C. O. (2015). Cloud cover throughout the agricultural growing season: Impacts on passive optical earth observations. *Remote Sensing of Environment, 156*, 438–447.

Wiguna, S., Adriano, B., Mas, E., & Koshimura, S. (2024). Assessment of Deep Learning Models Trained Using Global Remote Sensing Imagery in Real-Context Emergency Response. IGARSS 2024 - 2024 IEEE International Geoscience and Remote Sensing Symposium, 1736–1740. https://doi.org/DOI: 10.1109/IGARSS53475.2024.10641821

Wu, D. D., Lambert, J., Olson, D. L., Ding, S., Wu, D., Zhao, L., & Li, X. (2022). Smart Healthcare Engineering Management and Risk Analytics. https://doi.org/DOI: 10.1007/978-981-19-2560-3

Wu, G., Bhatti, U. A., Nawaz, S. A., Hasnain, A., Bhatti, M. A., Fahim, A., & Kaleri, A. A. (2022). COVID-19 and Air Pollution : Air Quality Impact in 13 Cities of the Jiangsu Province of China. 31(5), 4907–4916. https://doi.org/DOI: 10.15244/pjoes/149714

Wu, H., Zhang, Y., & Bhatti, U. A. (2025). Climate change air quality monitoring using sentimental 2 dataset. In Deep Learning for Earth Observation and Climate Monitoring.

Wu, D., Huang, M., Zhang, Y., Bhatti, U. A., & Chen, Q. (2018). Strategy for assessment of disaster risk using typhoon hazards modeling based on chlorophyll-a content of seawater. *EURASIP Journal on Wireless Communications and Networking, 2018*, 1–12.

Wu, G., Tang, H., & Bhatti, U. A. (2024). Remote sensing and GIS-based applications in coastal zone management and environmental monitoring. *Environmental Monitoring and Assessment, 207*, 142.

Wu, H., Gao, J., & Bhatti, U. A. (2024). An advanced system approach for disaster risk assessment. *EURASIP Journal on Wireless Communications and Networking*.

Wu, S., Sun, F., Zhang, W., Xie, X., & Cui, B. (2022). Graph neural networks in recommender systems: A survey. *ACM Computing Surveys, 55*(5), 1–37.

Wu, X., Li, J., Tu, R., Cheng, J., Bhatti, U. A., & Ma, J. (2022). Contourlet-DCT based multiple robust watermarkings for medical images. *Multimedia Tools and Applications, 78*, 8463–8480.

Wu, X., Zhang, Z., Xiong, S., Zhang, W., Tang, J., Li, Z., An, B., & Li, R. (2023). A Near-Real-Time Flood Detection Method Based on Deep Learning and SAR Images. *Remote Sensing, 15*(8), 2046. DOI: 10.3390/rs15082046

Wu, Y., & Bhatti, U. A. (2024). Applications of machine learning in medicine: Current trends and prospects. *Global Conference on Wireless and Optical Technologies (GCWOT).*

Wu, Y., Zhang, L., Bhatti, U. A., & Huang, M. (2024). Interpretable machine learning for personalized medical recommendations: A LIME-based approach. *Diagnostics (Basel), 13*(16), 2681. PMID: 37627940

Xia, G.-S., Hu, J., Hu, F., Shi, B., Bai, X., Zhong, Y., Zhang, L., & Lu, X. (2017a). AID: A Benchmark Data Set for Performance Evaluation of Aerial Scene Classification. *IEEE Transactions on Geoscience and Remote Sensing, 55*(7), 3965–3981. DOI: 10.1109/TGRS.2017.2685945

Xie, Y., Zhang, H., & Babar, M. A. (2024). LogSD: Detecting Anomalies from System Logs through Self-Supervised Learning and Frequency-Based Masking. Proceedings of the ACM on Software Engineering, 1(FSE), 2098–2120. https://doi.org/DOI: 10.1145/3660800

Xie, J., Girshick, R., & Farhadi, A. (2016, June). Unsupervised deep embedding for clustering analysis. In *International conference on machine learning* (pp. 478-487). PMLR.

Xiong, Z., Wang, Y., Zhang, F., & Zhu, X. X. (2024). One for all: Toward unified foundation models for earth vision. IGARSS 2024-2024 IEEE International Geoscience and Remote Sensing Symposium, 2734–2738. https://ieeexplore.ieee.org/abstract/document/10641637/?casa_token=PKATjEH_jZQAAAAA:UdZ0Jci0u8hIJuB5PXAGg_9XTbDtpfWsUS8jAyQwTgXAnHRyYuhucbJE_ZeNwHUqW2chDw

Xiong, J., Li, J., Bhatti, U. A., & Shoukat, M. U. (2020). Medical Image Watermarking Algorithm based on DCT. *Journal of Medical Imaging and Health Informatics.*

Xu, B., He, B., & Bhatti, U. A. (2023). AI-based smart agriculture systems for remote sensing applications in coastal regions. *Frontiers in Plant Science*, *14*, 124455.

Xu, B., Liu, Y., & Bhatti, U. A. (2023). Robust watermarking method in wavelet domain based on sift features. *Proceedings of the 2020 3rd International Conference on Artificial Intelligence.*

Xu, S., He, X., Cao, X., & Hu, J. (2022). Damaged Building Detection with Improved Swin-Unet Model. *Wireless Communications and Mobile Computing*, *2022*(1), 2124949. DOI: 10.1155/2022/2124949

Yang, G., Li, J., & Bhatti, U. A. (2024). A zero-watermarking algorithm for medical images based on Gabor-DCT. Cyberspace Safety and Security: 12th International Symposium.

Yang, J., Chen, X., Zou, H., Wang, D., & Xie, L. (2023). AutoFi: Toward Automatic Wi-Fi Human Sensing via Geometric Self-Supervised Learning. *IEEE Internet of Things Journal*, *10*(8), 7416–7425. DOI: 10.1109/JIOT.2022.3228820

Yang, M., Li, J., & Bhatti, U. A. (2025). A novel convolution kernel-based robust watermarking scheme applied in medical image. *International Journal of Wireless and Mobile Computing*, *22*(3-4), 290–299.

Yang, M., Li, J., Bhatti, U. A., Shao, C., & Chen, Y. W. (2024). Robust Zero-Watermarking Algorithm for Medical Images Based on Non-Subsampled Shearlet Transform and Schur Decomposition. *Computers, Materials & Continua*, *75*(3).

Yang, M., Li, J., Bhatti, U. A., & Yuan, Y. (2024). Robust Zero Watermarking Algorithm for Medical Images Based on Non-Subsampled Shearlet Transform and Schur Decomposition. *Computers, Materials & Continua, 75*(3).

Yang, M., Li, J., Bhatti, U. A., Yuan, Y., & Zhang, Q. (2024). A Novel Robust Watermarking Algorithm for Encrypted Medical Images Based on Non-Subsampled Shearlet Transform and Schur Decomposition. Deep Learning for Multimedia Processing Applications: Volume One. *Image*.

Yuan, Y., Li, J., & Bhatti, U. A. (2024). *Robust Zero Watermarking Algorithm for Medical Images Based on STAR-TEBLID-DCT*. Innovation in Medicine and Healthcare.

Yu, C., Tan, J., Cheng, Y., & Mi, X. (2024). Data analysis and preprocessing techniques for air quality prediction : A survey. *Stochastic Environmental Research and Risk Assessment, 38*(6), 2095–2117. DOI: 10.1007/s00477-024-02693-4

Yu, M., Zeng, Z., Tang, H., Bhatti, U. A., & Fahim, A. (2024). Research on smartphone-based eHealth systems for strengthening healthcare organization. *International Conference on Smart Health*, 91-101.

Yu, M., Zhang, Y., & Bhatti, U. A. (2024). Research on Cross-Platform Motion Control System Based on gSOAP. In *7th International Conference on Pattern Recognition and Artificial Intelligence.*

Yum, S.-G., & Das Adhikari, M. (2023). Suitable site selection for the development of solar based smart hydrogen energy plant in the Gangwon-do region, South Korea using big data: A geospatial approach. *International Journal of Hydrogen Energy, 48*(93), 36295–36313.

Zaland, Z., Bazai, S. U., Marjan, S., & Ashraf, M. (2021). Three-tier password security algorithm for online databases. In 2021 2nd International Informatics and Software Engineering Conference (IISEC) (pp. 1-6). IEEE.

Zayed, R. (2024). Big Data AI System for Air Quality Prediction. January 2022.

Zeeshan, Z., Li, J., & Bhatti, U. A. (2024). Robust multi-watermarking algorithm for medical images based on GoogLeNet and Henon map. *Computers, Materials & Continua, 75*(1), 565–586.

Zeng, Y., Tang, H., Gao, J., & Bhatti, U. A. (2024). A Domain Knowledge Graph Construction Method Based on Joint Extraction of GlobalPointer. *7th International Conference on Pattern Recognition and Artificial Intelligence.*

Zhang, M., Li, J., & Bhatti, U. A. (2024). Secure data storage and processing architectures for climate Internet of Things systems. In Deep Learning for Earth Observation and Climate Monitoring.

Zhang, Q., & Bhatti, U. A. (2025). Forecasting the carbon emissions in African countries using EMD-GCN. IEEE 6th International Conference on Pattern Recognition and Artificial Intelligence.

Zhang, Q., Li, J., Liu, J., Bhatti, U. A., & Chen, Y. W. (2023). Robust Zero-Watermarking Algorithm for Medical Images Based on SUSAN-DCT. 2023 26th ACIS International Winter Conference on Software Engineering.

Zhang, X., Ma, Z., Liu, Y., & Bhatti, U. A. (2024). Robust Zero-Watermarking Algorithm for Medical Images Based on Maximally Stable Extremal Regions and DCT. IEEE 6th International Conference on Pattern Recognition and Artificial Intelligence.

Zhang, B., Wu, Y., Zhao, B., Chanussot, J., Hong, D., Yao, J., & Gao, L. (2022). Progress and challenges in intelligent remote sensing satellite systems. *IEEE Journal of Selected Topics in Applied Earth Observations and Remote Sensing*, *15*, 1814–1822.

Zhang, C., Filella, I., Garbulsky, M., & Peñuelas, J. (2016). Affecting factors and recent improvements of the photochemical reflectance index (pri) for remotely sensing foliar, canopy and ecosystemic radiation-use efficiencies. *Remote Sensing*, *8*(9), 677. DOI: 10.3390/rs8090677

Zhang, C., Iqbal, I., Bhatti, U. A., Liu, J., Awwad, E. M., & Sarhan, N. (2024a). ResNet50 in remote sensing and agriculture: Evaluating image captioning performance for high spectral data. *Environmental Earth Sciences*, *83*(23), 1–16.

Zhang, H., & Bhatti, U. A. (2024). *Deep learning for remote sensing and climate monitoring*. Deep Learning for Earth Observation and Climate Monitoring.

Zhang, J., Li, Y., & Bhatti, U. A. (2024). Dynamic Segmentation Enhanced Asynchronous Non-linear Federated Learning for Fault Diagnosis Using Incremental Aggregation Mechanism. *IEEE Smart World Congress (SWC)*.

Zhang, K., Zhang, Y., & Bhatti, U. A. (2024). A domain knowledge graph construction method based on joint extraction of GlobalPointer. In *7th International Conference on Pattern Recognition and Artificial Intelligence*.

Zhang, L., Wang, H., & Zhang, B., Xu, Li, J., & Bhatti, U. A. (2024). A New ANN Technique for Short-Term Wind Speed Prediction Based on SCADA System Data in Turkey. *Atmosphere*, *14*(10).

Zhang, Q., Li, J., Liu, J., Bhatti, U. A., & Chen, Y. W. (2023). Robust Zero-Watermarking Algorithm for Medical Images Based on GFTT-KAZE and DCT. 2023 26th ACIS International Winter Conferenceon. *Software Engineering*.

Zhang, S., Xu, H., & Bhatti, U. A. (2023). Hybrid climate forecasting: Variational mode decomposition and convolutional neural network with long-term short memory. *Polish Journal of Environmental Studies*, *32*(1), 15–25.

Zhang, W., Tang, P., & Zhao, L. (2021). Fast and accurate land-cover classification on medium-resolution remote-sensing images using segmentation models. *International Journal of Remote Sensing*, *42*(9), 3277–3301.

Zhang, X., & Bhatti, U. A. (2023). Image fusion based on discrete cosine transform for medical data. *Journal of Medical Imaging and Health Informatics*.

Zhang, Y., Chen, J., Ma, X., Wang, G., Bhatti, U. A., & Huang, M. (2024b). Interactive medical image annotation using improved Attention U-net with compound geodesic distance. *Expert Systems with Applications*, *237*, 121282. DOI: 10.1016/j.eswa.2023.121282

Zhang, Y., Li, D., & Bhatti, U. A. (2025). Practical PID Protocol of T–S Fuzzy Positive Multi-Agent Systems Under Attacks. *International Journal of Fuzzy Systems*.

Zhang, Y., Wang, J., Chen, Y., Yu, H., & Qin, T. (2023). Adaptive Memory Networks With Self-Supervised Learning for Unsupervised Anomaly Detection. *IEEE Transactions on Knowledge and Data Engineering*, *35*(12), 12068–12080. DOI: 10.1109/TKDE.2021.3139916

Zhang, Y., Xu, B., & Bhatti, U. A. (2025). Optimizing co-simulation with the age of information in intelligent cyber physical systems. *Internet of Things : Engineering Cyber Physical Human Systems*, *31*, 101550.

Zhang, Y., Xu, S., Chen, H., Bhatti, U. A., & Huang, M. (2024c). Context-aware environment online monitoring for safety autonomous vehicle systems: An automata-theoretic approach. *Journal of Cloud Computing (Heidelberg, Germany)*, *13*(6). Advance online publication. DOI: 10.1186/s13677-023-00567-8

Zhang, Z., Yu, X., Rong, X., & Iwata, M. (2021). Spatial-temporal neural network for P300 detection. *IEEE Access : Practical Innovations, Open Solutions*, *9*, 163441–163455.

Zhang, Z., Zeng, H., & Bhatti, U. A. (2024). A novel myocarditis detection combining deep reinforcement learning and an improved differential evolution algorithm. *CAAI Transactions on Intelligence Technology*.

Zha, Y., Yue, Y., Zhang, Y., & Bhatti, U. A. (2023). Simulation optimization of AOA estimation algorithm based on MIMO UWB communication system. *Proceedings of the 2nd International Conference on Information Technologies.*

Zhu, X. X., Tuia, D., Mou, L., Xia, G. S., Zhang, L., Xu, F., & Fraundorfer, F. (2017). Deep learning in remote sensing: A comprehensive review and list of resources. *IEEE Geoscience and Remote Sensing Magazine*, *5*(4), 8–36.

Zielinski, O., Busch, J. A., Cembella, A. D., Daly, K. L., Engelbrektsson, J., Hannides, A. K., & Schmidt, H. (2009). Detecting marine hazardous substances and organisms: Sensors for pollutants, toxins, and pathogens. *Ocean Science*, *5*(3), 329–349.

About the Contributors

Li Dekai is a researcher and scholar specializing in hyperspectral image processing and remote sensing, with a Ph.D. in Information and Communication Engineering from Hainan University. His academic pursuits are focused on innovative techniques in image processing, including spectral-spatial feature fusion for ecological applications and advanced medical image watermarking algorithms. Dr. Li has co-authored multiple high-impact publications in journals such as Electronics and IET Biometrics, showcasing robust methodologies in image encryption, watermarking, and metamaterials. His research contributions also include patented innovations in medical image processing, employing advanced deep learning models like DarkNet53 and Vision Transformers. Throughout his career, Dr. Li has actively contributed to several projects, such as developing hyperspectral image classification methods for mangrove species and pioneering robust watermarking techniques for medical images. His expertise bridges the gap between theory and practical applications, advancing the fields of remote sensing and secure image processing.

Uzair Aslam Bhatti is a prominent researcher and academic specializing in remote sensing and hyperspectral image processing. He currently serves as an Associate Researcher and Professor at Hainan University, where his work focuses on image security and intelligent systems for medical image processing. Dr. Bhatti has a Ph.D. in Information and Communication Engineering from Hainan University, with a specialization in artificial intelligence, machine learning, and data mining. His notable research includes environmental change analysis using high-resolution remote sensing data and the development of graph-based classification techniques for hyperspectral images. Dr. Bhatti has contributed extensively to scientific literature, authoring several high-impact publications on topics such as environmental monitoring, carbon emissions reduction, and advanced deep learning applications in remote sensing. In addition to his research, Dr. Bhatti has played key roles in various large-scale projects, such as integrating space-ground-device networks for ecological applications and exploring sustainable healthcare systems. His expertise bridges the fields of remote sensing, artificial intelligence, and data security, making significant contributions to advancing environmental and medical applications.

Sibghat Ullah Bazai is a renowned academic and researcher specializing in cybersecurity, remote sensing, and related technologies. He has made significant contributions to image classification, GIS applications, and environmental monitoring. As Chairperson of Computer Engineering at BUITEMS, Quetta, Pakistan, he has advanced academic excellence and fostered research initiatives. With a Ph.D. in Information Technology (Cybersecurity) from Massey University, New Zealand, Dr. Bazai has extensive experience in applying machine learning and deep learning to geospatial datasets. He has co-authored high-impact papers in top-tier journals and conferences, covering topics such as medical image anomaly detection, GIS-based urban property tax systems, and high-performance computing for environmental applications. His expertise in remote sensing data has established him as a leader in sustainable development research.

Mir Muhammad Nizamani is a dedicated researcher specializing in hyperspectral remote sensing and ecological studies. With a Ph.D. from Hainan University, Dr. Nizamani's work focuses on the spatial and ecological patterns of urban plant diversity in cities like Karachi, Haikou, and Sanya. His expertise includes applying advanced statistical and machine learning techniques to analyze and model ecological data. Dr. Nizamani has published extensively in high-impact journals, with over 45 scientific papers exploring topics like urban forestry, plant diversity, and the impacts of urbanization on ecological systems. His innovative approaches to remote sensing and ecological modeling contribute to understanding sustainable urban development and biodiversity conservation. Through a strong foundation in plant science and genetics, combined with advanced data analysis skills, Dr. Nizamani continues to make impactful contributions to remote sensing and environmental research.

***

A. Saran Kumar currently serves as an Assistant Professor at Coimbatore Institute of Technology in Coimbatore, India. He earned his Ph.D. in Information and Communication Engineering from PSG College of Technology, Coimbatore. With 8 years of experience, he has an extensive academic record, with over 10 papers indexed in Scopus, 4 papers in SCI-indexed journals, and 5 papers in various international journals. Additionally, he has presented more than 25 research papers at various national and international conferences. His research contributions also include 6 filed patents, with 3 patents already published. He has completed 15 courses on NPTEL and 5 courses on Coursera. He is a life member of IAENG, AMIEE and ISRD.

Muhammad Aamir is an Associate Professor at College of Computer Science and Artificial Intelligence, Huanggang Normal University in China, where he has been teaching and conducting research for the past five years. His work centers on applying artificial intelligence and computer vision to solve practical problems, especially in the area of healthcare. Whether it's building models to help interpret medical images or exploring ways to make AI more efficient and accessible, his goal is always to create technology that has a real impact. Throughout his career, Dr. Aamir has contributed to several research projects and published in respected international journals. He's especially interested in developing intelligent systems that are not only accurate, but also lightweight and adaptable to real-world needs, qualities that are critical in fields like medicine and public health. Beyond his research, Dr. Aamir is actively involved in the academic community. He regularly reviews articles for international journals, has served as a guest editor for special issues, and participates in conferences as a speaker and session chair. He also enjoys mentoring students and collaborating with researchers from different disciplines and backgrounds. Dr. Aamir brings a thoughtful, hands-on approach to his work and remains committed to advancing AI in ways that are ethical, inclusive, and deeply connected to solving real-world challenges.

Aruna R. Shet is an Assistant Professor in the Department of Commerce at New Horizon College, Kasturinagar, Bengaluru. With 15 years of teaching experience, she specializes in Financial Management, Advanced Accounting, Cost Accounting and more. She holds a M.Com degree. She has presented and published several papers in reputed journals. She is committed to imparting quality education to students. Her expertise and passion for teaching have earned her a respected position in her field.

Sandeep C. S is an accomplished academician and researcher with over 16 years of experience in teaching, research, and administration in Electronics and Communication Engineering. He serves as Associate Professor and Head of the ECE Department at Jawaharlal College of Engineering and Technology, Kerala. He is also engaged in postdoctoral research at the University of Louisiana, USA, focusing on AI, machine learning, and biomedical signal processing. He holds a Ph.D. from the University of Kerala, with research on advanced biomedical imaging. He earned his M.E. and B.E. from Anna University, Chennai, and an MBA in HRM from IGNOU. A Senior Member of IEEE and a recognized research supervisor under

APJ Abdul Kalam Technological University, he has authored over 60 research papers and contributed to books, chapters, and patents. His interdisciplinary projects span healthcare, computing, and intelligent systems. He continues to contribute actively to academic innovation and global research collaborations.

Gaurav Dubey is an accomplished academician with over 17 years of experience in computer science education. He serves as an Assistant Professor at ITM, Gwalior, and is currently pursuing a Ph.D. His expertise spans artificial intelligence, machine learning, and wireless sensor networks, with numerous publications in reputed Scopus-indexed journals and IEEE/Springer conferences. Dubey is also a published author with multiple books and book chapters, and holds patents and copyrights in AI and IoT innovations. He has actively contributed to academic administration, accreditation processes, and faculty development programs. Recognized with the Best Faculty Award, he continues to mentor students and contribute to research, bridging the gap between academia and applied technology.

Deepak Gupta is a distinguished academician with over 20 years of teaching experience and expertise in Computer Science and Engineering. He holds a Ph.D. in Computer Engineering and has an extensive research portfolio, including many international journal papers, multiple conference proceedings, and several books. His research interests span machine learning, artificial intelligence, and data mining. Currently, Dr. Gupta serves as an Associate Professor and Head of Department of CSE (Emerging Branches) at ITM Gwalior, MP, India. He has also contributed significantly to the field through patents and copyrights and actively engages as a reviewer, conference chair and editor.

Qazi Mudassar Ilyas received his Ph.D. in Information and Communication Engineering from Huazhong University of Science and Technology, China, in 2005, following his BS and MS degrees from the University of Agriculture, Faisalabad, Pakistan. With over two decades of experience in teaching, research, and academic leadership across Pakistan, Saudi Arabia, and Brunei Darussalam, he has made substantial contributions to higher education and applied research. Currently an Associate Professor at Sultan Sharif Ali Islamic University, Brunei Darussalam, Dr. Ilyas's work focuses on the emerging applications of AI and machine learning. His professional portfolio encompasses curriculum design, outcome-based education, academic accreditation, and quality assurance. He has supervised numerous student research projects, published extensively in peer-reviewed journals, and actively contributed to initiatives advancing digital transformation, innovation, and community engagement.

Nalini Santosh Jagtap working as Associate Professor in Department of Information Technology at Pimpri Chinchwad College of Engineering and Research, SPPU, Pune, India. Her current research interests include image classification, CBIR, image enhancements and Natural Language Processing. She has more than 20 research papers to her credit.

Urooj Khalid holds a BS degree in Computer Science from Balochistan University of Information Technology, Engineering, and Management Sciences (BUITEMS), Quetta, where she was awarded a merit scholarship. She is also an alumnus of the Global UGRAD Semester Exchange Program, USA (2019). She is pursuing her MS degree at BUITEMS, Quetta, with research interests in process mining, cloud computing, deep learning, and machine learning.

Gaurav Kumar is working as Associate Professor in the department of School of Computer Applications, Lovely Professional University, Phagwara, Punjab, India. His educational qualification is PhD(Computer Applications), UGC-NET, M.Tech, M.Phil(Computer Science), MCA with distinction. His research areas include Machine Learning, Learning Analytics, Artificial Intelligence, Human-Computer Interaction, Computer vision etc.. He is also a member of different reputed organizations like IEEE, ISCA, CSTA. He has published more than 11 peer-reviewed research articles(Scopus/WoS/UGC-Care) including 05 Scopus indexed and 03 WoS Core Collection indexed articles along with 04 UGC-Care journal articles and 1 Copyright.

Fateh Bahadur Kunwar is a distinguished academician and researcher with over 16 years of experience in higher education, currently serving as Assistant Professor (Senior Scale) at Geetanjali Institute of Technical Studies, Udaipur. He holds a Ph.D. in Computer Science & Engineering, an M.Tech. from KNIT Sultanpur (autonomous government institute), and an MCA from Dr. R. M. L. Avadh University, Ayodhya. Kunwar has published extensively in reputed journals and conferences including IEEE and Springer, authored several books and test series, and holds multiple national patents. He has guided numerous B.Tech. and M.Tech. students in research and development projects. His research interests span health informatics, machine learning, deep learning, explainable AI, biometrics, image processing, and soft computing. Actively engaged in the academic community, he frequently delivers expert talks, contributes to conferences, and serves as a reviewer and editorial board member, reflecting his commitment to advancing knowledge, innovation, and interdisciplinary collaboration in computer science

Zokir Mamadiyarov is working in the Department of Finance and Tourism at Termez University of Economics and Service, Termez, Uzbekistan.

Abid Mehmood received the M.Sc. degree in computer science from Quaid-i-Azam University, Pakistan, in 2001, and the Ph.D. degree in computer science from Universiti Teknologi Malaysia, in 2014. With a strong background in software development, he has led numerous high-impact enterprise projects, demonstrating expertise in both technical execution and strategic project management. He is currently an Associate Professor of artificial intelligence with Dakota State University, USA. His research interests include data science and analytics, with a focus on machine learning and deep learning, particularly in areas, such as image processing, pattern recognition, e-commerce, and software engineering.

M. Sankar holds a Ph.D. in Management and currently serves as a Professor in the Department of Management at Acharya Institute of Graduate Studies, Bangalore. With over 13 years of experience in management education and academic administration, he specializes in human resources, psychology, industrial relations, and counselling. He has published more than 22 research articles in reputed journals (Scopus, UGC-CARE, ABDC, Web of Science), authored two books, and holds three patents along with one registered copyright.

Alamgir Naushad is an Assistant Professor in the School of Avionics and Electrical Engineering (SAEE), College of Aeronautical Engineering (CAE) with expertise in wireless communications, image processing, deep learning, and AI applications. He completed his PhD from GIK Institute, Pakistan (2019), and previously served as Department Head (2020-2023), overseeing faculty, research, and policy development. Dr. Naushad teaches Information Security, Computational Logic, and Parallel and Distributed Computing, and advises students on careers in communications, networking, and AI-driven applications, including remote sensing and climate change monitoring. His research interests intersect with deep learning applications in remote sensing, making him a relevant contributor to the field.

Dipali Himmatrao Patil is an Assistant Professor, at the Department of Information Technology, JSPM's Rajarshi Shahu College of Engineering, Savitribai Phule Pune University, Maharashtra, India. She is having teaching experience of more than 24 years at Engineering College. Her research interests include Natural Language Processing, Machine Learning Techniques., Deep Learning for image recognition. She has published more than 17 papers indexed in SCOPUS/Web of Science/UGC approved Journals/Conferences, etc. Email: patil.dipali07@gmail.com

Ravikumar R. Natarajan is an experienced academician with over 12+ years of teaching and research expertise across institutions in India, the UAE, and the Maldives. He currently serves as an Assistant Professor at Marwadi University, Rajkot, specializing in Machine Learning, Recommender Systems, and AI-driven applications in healthcare and security. He holds a Ph.D. in Machine Learning and Recommender Systems from Amity University, Jaipur, a Master of Engineering in CSE from VMRF University, Salem, and a Bachelor of Engineering in CSE from Anna University. He also has a MiniMBA from the Swiss eLearning Institute. Dr. Ravikumar has authored more than 8+ multiple SCIE/Scopus-indexed journal articles, 25+ IEEE/Springer conference papers, and co-authored a book on Artificial Intelligence. He has also published 9 patents and contributed 10+ book chapters in emerging domains such as Federated Learning, Blockchain, and Digital Twins for IGI Global, Wiley and Springer publisher. He actively mentors students through initiatives like the Research Activities Club (RAC) and Student Research Internship Training (SRIT), promoting innovation and research culture.

Aarthi S is a dedicated Assistant Professor in Computer Science with over six years of teaching experience. She is currently pursuing her Ph.D. at Marwadi University, Gujarat, and holds an M.Phil. and MCA from Bharathiar University, Tamil Nadu. Throughout her academic career, Aarthi has specialized in programming languages such as C++, Java, and C, and contributed to research in recommender systems and AI-related data privacy. She has published journal articles, contributed to book chapters, and actively participated in technical workshops. Known for her student mentorship and engagement in academic events, Aarthi remains committed to fostering innovation and excellence in computer science education.

S. Tamilselvi is an Assistant Professor in the Department of Computer Technology at Kongu Engineering College, Perundurai. With 9 years of teaching experience, she brings a strong academic background, holding an M.C.A., M.Phil., and Ph.D., from Bharathiar University. Her areas of specialization include Advanced Networking and Cloud Computing, where she actively contributes through teaching and academic research.

Afsheen Sabir holds a Bachelor of Science degree in Software Engineering from the University of the Punjab, Lahore, Pakistan, where she achieved a commendable Cumulative Grade Point Average (CGPA). She is currently pursuing a Master of Science degree in Computer Science at Balochistan University of Information Technology, Engineering, and Management Sciences (BUITEMS), Quetta, Pakistan. Her research interests lie at the intersection of deep learning and remote sensing applications, with a specific focus on leveraging these technologies to improve healthcare outcomes. Her work aims to harness the potential of cutting-edge computational techniques to drive innovation in the field of healthcare.

Varadharajan Bhoopathy is a Professor at Sree Rama Engineering College. He holds a Ph.D. in Information and Communication Engineering from Anna

University and has over 16 years of academic experience. Specializing in wireless sensor networks and IoT, he has published numerous research papers and holds several international patents. Dr. Bhoopathy is a recipient of prestigious awards for his contributions to education and research. His commitment to academia continues to drive innovation and excellence in the field of computer science.

Mengyao Yang is a graduate student at Hainan University, specializing in remote sensing research.

Index

A

B

C

D

E

F

L

M

N

P

R

S

T

U

W

www.ingramcontent.com/pod-product-compliance
Lightning Source LLC
LaVergne TN
LVHW081532300926
849161LV00006B/1393
9798337335018